# WOOD CHEMISTRY

# WOOD CHEMISTRY
## Fundamentals and Applications
### SECOND EDITION

**EERO SJÖSTRÖM**

*Laboratory of Wood Chemistry*
*Forest Products Department*
*Helsinki University of Technology*
*Espoo, Finland*

**ACADEMIC PRESS**
*An Imprint of Elsevier*
San Diego   New York   Boston
London   Sydney   Tokyo   Toronto

Permissions may be sought directly from Elsevier's Science and Technology
Rights Department in Oxford, UK. Phone (44) 1865 843830, Fax: (44) 1865
853333, e-mail: permissions@elsevier.co.uk. You may also complete your
request on-line via the Elsevier homepage: http://www.elsevier.com by selecting
"Customer Support" and then "Obtaining Permissions".

Academic Press
*An Imprint of Elsevier*
525 B Street, Suite 1900, San Diego, California 92101-4495, USA
http://www.apnet.com

Academic Press Limited
24-28 Oval Road, London NW1 7DX, UK
http://www.hbuk.co.uk/ap/

Library of Congress Cataloging-in-Publication Data

Sjöström, Eero, date.
    Wood chemistry : fundamentals and applications / Eero Sjöström -
2nd ed.
            xiv. 293 p. : ill. ; 24            cm.
    Includes bibliographical references (p. 251-276) and index.
    ISBN-13: 978-0-12-647481-7  ISBN-10: 0-12-647481-8
    1. Wood--Chemistry.    I. Title.
TS932.S587    1993
674'.13 28                                                        93-23493   r97
ISBN-13: 978-0-12-647481-7                        CIP
ISBN-10: 0-12-647481-8

Transferred to Digital Printing 2007

# CONTENTS

**Preface to the Second Edition**   xi

**Preface to the First Edition**   xiii


## 1   The Structure of Wood

1.1 The Macroscopic Structure of Wood   1
1.2 The Living Tree   2
  1.2.1 Growth of the Tree   2
  1.2.2 Development of the Cell   4
  1.2.3 Annual Rings   4
  1.2.4 Cell Types   5
  1.2.5 Pits   6
  1.2.6 Softwood Cells   7
  1.2.7 Hardwood Cells   10
  1.2.8 Sapwood and Heartwood   11
1.3 Wood Ultrastructure   12
  1.3.1 Building Elements   12
  1.3.2 Cell Wall Layers   13
  1.3.3 Pits   17
1.4 Reaction Wood   18

## 2   Introduction to Carbohydrate Chemistry

2.1 Definitions and Nomenclature    21
2.2 Monosaccharides    23
    2.2.1 Configuration of Monosaccharides    23
    2.2.2 Ring Structures of Monosaccharides    26
    2.2.3 Mutarotation    29
    2.2.4 Conformation of Monosaccharides    30
2.3 Monosaccharide Derivatives    33
    2.3.1 Glycosides    33
    2.3.2 Alkylidene Derivatives    34
    2.3.3 Ethers    35
    2.3.4 Anhydro Sugars    36
    2.3.5 Esters    37
2.4 Oligo- and Polysaccharides    38
2.5 Reactions of Carbohydrates    39
    2.5.1 Oxidation    39
    2.5.2 Reduction    42
    2.5.3 Addition and Condensation Reactions
         of Carbonyl Groups    42
    2.5.4 The Influence of Acid    44
    2.5.5 The Influence of Alkali    46
2.6 Methods    50

## 3   Wood Polysaccharides

3.1 Biosynthesis    51
3.2 Cellulose    54
    3.2.1 Molecular Structure    54
    3.2.2 Polymer Properties    58
3.3 Hemicelluloses    63
    3.3.1 Softwood Hemicelluloses    63
    3.3.2 Hardwood Hemicelluloses    67
    3.3.3 Isolation of Hemicelluloses    69
    3.3.4 Distribution of Hemicelluloses    70

## 4   Lignin

4.1 Isolation    71
4.2 Biosynthesis and Structure    73

4.2.1 Phenylpropane—The Basic Structural Unit
of Lignin    73
4.2.2 Biosynthesis of Lignin Precursors    73
4.2.3 Polymerization of Lignin Precursors    77
4.2.4 Types of Linkages and Dimeric Structures    80
4.2.5 Functional Groups    82
4.2.6 Lignin Formula    83
4.2.7 Lignin–Carbohydrate Bonds    84
4.3 Classification and Distribution    86
4.4 Polymer Properties    88

## 5  Extractives

5.1 Terpenoids and Steroids    92
5.1.1 Occurrence    92
5.1.2 Chemical Composition    92
5.2 Fats and Waxes    103
5.2.1 Occurrence    103
5.2.2 Chemical Composition    104
5.3 Phenolic Constituents    104
5.4 Inorganic Components    107
5.5 Changes Caused by Wood Storage    107

## 6  Bark

6.1 Anatomy of Bark    109
6.1.1 Inner Bark    110
6.1.2 Outer Bark    111
6.2 Chemistry of Bark    111
6.2.1 Soluble Constituents (Extractives)    112
6.2.2 Insoluble Constituents    113
6.2.3 Inorganic Constituents    113

## 7  Wood Pulping

7.1 Background and Definitions    114
7.2 Sulfite Pulping    119
7.2.1 Cooking Chemicals and Equilibria    119
7.2.2 Impregnation    122

7.2.3    Morphological Factors    122
7.2.4    General Aspects of Delignification    124
7.2.5    Lignin Reactions    126
7.2.6    Carbohydrate Reactions    132
7.2.7    Reactions of Extractives    135
7.2.8    Side Reactions    136
7.2.9    Composition of Sulfite Spent Liquors    137
7.2.10  Recovery and Conversion of Sulfite Cooking
        Chemicals    138
7.3  Kraft Pulping 140
7.3.1    Cooking Chemicals and Equilibria    140
7.3.2    Impregnation    142
7.3.3    General Aspects of Delignification    142
7.3.4    Lignin Reactions    145
7.3.5    Reactions of Polysaccharides    150
7.3.6    Stabilization of Polysaccharides against Alkaline
         Degradation    155
7.3.7    Sulfur-Free Pulping    156
7.3.8    Reactions of Extractives    157
7.3.9    Composition of Black Liquor    158
7.3.10   Recovery and Conversion of Kraft Cooking
         Chemicals    160
7.4  Special Features of High-Yield Pulping    161

# 8   Pulp Bleaching

8.1  Lignin-Removing Bleaching    166
     8.1.1  Background    166
     8.1.2  Bleaching Chemicals    170
     8.1.3  General Aspects of Bleaching    177
     8.1.4  Lignin Reactions    180
     8.1.5  Carbohydrate Reactions    188
     8.1.6  Reactions of Extractives    193
     8.1.7  Spent Liquors after Bleaching    194
8.2  Lignin-Preserving Bleaching    198
8.3  Yellowing of High-Yield Pulps    201

# 9   Cellulose Derivatives

9.1    Reactivity and Accessibility of Cellulose    204
9.2    Swelling and Dissolution of Cellulose    206

9.3   Swelling Complexes—Alkali Celluloses    208
9.4   Esters of Inorganic Acids    209
      9.4.1   Cellulose Nitrate    210
      9.4.2   Cellulose Sulfate    210
      9.4.3   Other Inorganic Cellulose Esters    211
9.5   Esters of Organic Acids    211
      9.5.1   Cellulose Acetate    211
      9.5.2   Other Esters of Organic Acids    213
9.6   Ethers    214
      9.6.1   Alkyl Ethers    214
      9.6.2   Hydroxyalkyl Ethers    215
      9.6.3   Carboxymethylcellulose    216
      9.6.4   Cyanoethylcellulose    217
9.7   Cellulose Xanthate    218
9.8   Cross-Linking of Cellulose    218
9.9   Grafting on Cellulose    219
9.10  Cellulose Ion Exchangers    222

# 10   Wood-Based Chemicals and Pulping By-Products

10.1  Wood-Based Chemicals and Fuels    226
      10.1.1  Extractives    227
      10.1.2  Hydrolysis Products and Their Further
              Processing    230
      10.1.3  Thermal Treatment    234
      10.1.4  Other Treatments    236
10.2  Chemicals from Pulping Liquors    236
      10.2.1  Sulfite Spent Liquors    237
      10.2.2  Kraft Black Liquors    240


**Appendix**    249


**Bibliography**    251


**Index**    277

# PREFACE TO THE SECOND EDITION

The first edition of "Wood Chemistry: Fundamentals and Applications" was published in 1981. Because there has been considerable activity in this field since then, it seemed appropriate to prepare a second edition. As in the first edition there are ten chapters, but most of them have either been extensively modified or completely rewritten. In spite of the increased coverage of the book, I have attempted to discuss the topics as concisely as possible with consideration to the theoretical background and current ideas.

Chapter 1 (Structure of Wood) has been kept almost unchanged. In Chapter 2 (Introduction to Carbohydrate Chemistry), most of the illustrations are new and the text has been modified. Relatively few additions and changes have been made to Chapter 3 (Wood Polysaccharides). Chapter 4 (Lignin) has been partly rewritten to include more information about the biosynthesis of lignin. Chapter 5 (Extractives) has been completely revised, while Chapter 6 (Bark) has been kept practically unchanged. In Chapter 7 (Wood Pulping), the principles of pulping and alternative pulping methods are discussed more thoroughly than in the first edition. However, little attention has been paid to the pulping methods using organic solvents ("organosolv pulping") or enzymes because most of them are still theoretical and without significance for industrial applications. Chapter 8 (Pulp Bleaching) has been rewritten with particular consideration to the radical changes in modern bleaching practices caused by environmental needs. Finally, Chapter 9 (Cellulose Derivatives) was kept practically unchanged, whereas Chapter 10 (Wood-Based Chemicals and Pulping By-Products) has been rewritten.

For readers who would like more information on the topics discussed, a bibliography (books and articles are listed separately) appears in chapter order at the end of the book. Although more references have been included, these selected examples represent only a fraction of the vast literature available.

This new edition is intended for the same category of readers as the first edition, namely for teachers and students with wood chemistry in their curricula, as well as more generally for chemists, biochemists, biologists, and environmental scientists working either in industry or in research institutes. Wood chemistry plays an important role whether the purpose is to study the growth of a tree or to utilize renewable forest resources ("biomass") for production of fibers, chemicals, and energy.

I am grateful for the generous help of friends and colleagues in providing material and reading portions of the manuscript. Special thanks are due to B. Holmbom and R. Ekman, Åbo Akademi, Åbo, whose suggestions resulted in considerable improvement of the chapter on extractives. Other improvements were possible because of the help and comments of the following persons: J. Blackwell, Case Western Reserve University, Cleveland, Ohio and D. P. Delmer, The Hebrew University of Jerusalem, Jerusalem (crystalline structure and biosynthesis of cellulose); G. Brunow, Helsinki University, Helsinki, and G. Gellerstedt, Royal Institute of Technology, Stockholm (lignin and bleaching); T. Norin, Royal Institute of Technology, Stockholm (extractives); J. Janson, The Finnish Pulp and Paper Research Institute, Espoo, (pulping); and R. Grundelius and P. Haglund, STORA, Falun (bleaching, especially environmental aspects).

I am also indebted to my earlier students and co-workers, including the personnel of the Laboratory of Wood Chemistry, for their help and encouragement. Special thanks to R. Alén and T. Vuorinen who offered their help in several areas, including the content and illustrations for Chapters 10 and 2, respectively. K. Niemelä helped in the choice of literature references. I am also thankful to Arja Siirto who made the computer-drawn figures.

Finally, I would like to thank the personnel at Academic Press for their cooperation and skillful editorial work.

*Eero Sjöström*

# PREFACE TO THE FIRST EDITION

Despite the rapid development of the disciplines allied with wood chemistry, nearly two decades have elapsed since the last English-language book devoted specifically to this subject appeared (B. L. Browning, ed., "The Chemistry of Wood," Wiley-Interscience, New York, 1963). Two years later, Sven Rydholm in his book "Pulping Processes" (Wiley-Interscience, New York, 1965) gave a unique exposé of pulping, discussing this matter comprehensively also from the standpoint of wood chemistry. The present book has been written in the belief that there is now a rather wide circle of readers who need a knowledge of modern wood chemistry in the form of a textbook. In addition to pulping and papermaking there are numerous potential applications in wood chemistry particularly connected with the utilization of wood and wood wastes as well as the by-products from pulping processes for production of chemicals and energy. Indeed, as a renewable raw material, wood constitutes an enormous resource for biomass conversion in the future.

This book attempts to discuss various aspects of wood chemistry in relation to applications. It is believed that the book might be useful not only for students and teachers but generally for chemists, biochemists, and others working either in the laboratory as researchers or in production and planning. Chapter 1 describes the structure and anatomy of wood. Carbohydrate chemistry belongs to the fundamentals in wood chemistry because two thirds of the wood constituents are polysaccharides. Chapter 2 therefore deals with the general structure, properties, and pertinent reactions of carbohydrates. In Chapter 3 the chemistry and polymer properties of wood polysaccharides are specifically discussed. The challenging chemistry of lignin is presented in Chapter 4 in conjunction with morphological aspects. Chapter 5 covers the interesting group of extractives, which consists of extremely diversified constituents. They cause problems in pulping and bleaching, but are also a source of valuable by-products. The anatomy and chemistry of bark also are discussed (Chapter 6). The reactions of wood constituents during sulfite and kraft pulping and bleaching are dealt with in Chapters 7 and 8. Basic inorganic reactions of the pulping and bleaching chemicals are included in addition to some general aspects pertinent to the technology of the delignification processes. Chapter 9 covers cellulose derivatives and related products. Chapter 10 finally discusses various alternatives and possibilities for utilization of solid wood (residues) as well as by-products from pulping. It is hoped that this challenging field will attract chemists for new endeavors.

Based on current concepts an attempt has been made to present a rationalized and logical account of wood chemistry with emphasis on its applications. Although not covered comprehensively, references to the relevant literature are listed at the end of each chapter. Many of these are only examples selected from the vast collection available and serve those readers who need a further guidance or information of the topics discussed.

Although much of the content is based on my earlier book in Finnish on wood chemistry, the first edition of which appeared in 1977 (Otakustantamo, Espoo), this version has been considerably improved and enlarged. Fortunately, at the early stages of preparation of the manuscript, I was encouraged by Professors K. V. Sarkanen and T. E. Timell, who generously offered their help. I am deeply indebted to them for checking the manuscript in detail and for the numerous improvements with respect to both content and language. Other friends and colleagues, including Dr. W. Brown, Professors J. Gierer, J. Gripenberg, J. J. Lindberg, T. Norin, B. Rånby, O. Theander, and Drs. J. Janson, K. Kringstad, and B. Lindgren, read portions of the manuscript and offered many useful comments. I am also grateful for the material provided by Dr. E. Back, Professor W. A. Côté, Professor D. P. Delmer, Dr. D. A. I. Goring, Mrs. M-S. Ilvessalo-Pfäffli, Professor H. Meier, and Professor T. E. Timell. Finally, I wish to thank the staff members of our laboratory, R. Alén, Christine Hagström, E. Seppälä, and T. Vuorinen who provided help in several respects and Kristiina Holm for typing and Eija Wiik and Ritva Valta for the drawings.

*Eero Sjöström*

# Chapter 1

# THE STRUCTURE OF WOOD

Trees belong to seed-bearing plants (Spermatophytae), which are sub-divided into gymnosperms (Gymnospermae) and angiosperms (Angiospermae). Coniferous woods or softwoods belong to the first-mentioned category and hardwoods to the second group. Altogether 30,000 angiosperms and 520 coniferous tree species are known; most of the former grow in tropical forests. In North America the number of species is about 1200, while in Europe only 10 softwood and 51 hardwood species exist naturally. This limited number represents species surviving the period of glaciation, during which genera such as *Sequoia* and *Pseudotsuga* completely disappeared from Europe.

## 1.1   The Macroscopic Structure of Wood

Wood is composed of elongated cells, most of which are oriented in the longitudinal direction of the stem. They are connected with each other through openings, referred to as pits. These cells, varying in their shape according to their functions, provide the necessary mechanical strength to the tree and also perform the function of liquid transport as well as the storage of reserve food supplies.

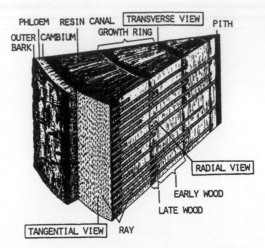

PHLOEM   RESIN CANAL   TRANSVERSE VIEW   PITH
OUTER | CAMBIUM   GROWTH RING
BARK |

RADIAL VIEW

EARLY WOOD
LATE WOOD

TANGENTIAL VIEW   RAY

**Fig. 1-1.**   Sections of a four-year-old pine stem.

Figure 1-1 shows the macroscopic structure of wood as it appears to the naked eye. The centrally located pith is discernible as a dark stripe in the middle of the stem or branches. It represents the tissues formed during the first year of growth. The xylem or wood is organized in concentric growth rings (annual increments). It also contains rays in horizontal files, extending from the outer bark either to the pith (primary rays) or to an annual ring (secondary rays). Some softwoods also contain resin canals. The inner part of a tree usually consists of dark-colored heartwood. The outer part, or sapwood, is lighter in color and conducts water from the roots to the foliage of the tree. The cambial zone is a very thin layer consisting of living cells between the wood (xylem) and the inner bark (phloem). The cell division and radial growth of the tree takes place in this region.

## 1.2   The Living Tree

### 1.2.1   Growth of the Tree

The tree grows through the division of the cells. The length of the growth period largely depends on the climate, but in many parts of North America and Scandinavia growth occurs from May to early September, and is most intensive in the spring. The majority of the cells develop into various permanent cells and only a very few are retained as growing cells capable of division.

The growth of a tree is always continuous although it becomes slower in the course of time. Giant sequoias (*Sequoiadendron giganteum*) in California can be up to 4000 years old measuring 100 meters in height and 12 meters in diameter at the base.

Longitudinal growth (primary growth), which takes place in the early season, proceeds at the end of the stem, branches and roots. The growth

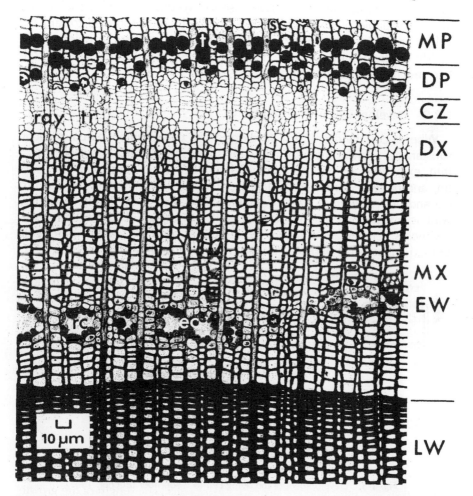

**Fig. 1-2.** Transverse section of xylem and phloem of red spruce (*Picea rubens*). CZ, cambial zone; DP, differentiating phloem; MP, mature phloem with sieve cells (sc) and tannin cells (tc); DX, differentiating xylem with ray cells and tracheids (tr); MX, mature xylem, earlywood (EW) with resin canals (rc), lined with epithelial cells (ec); LW, latewood. Note that each ray continues from the xylem, through the cambial zone, and into the phloem. Light micrograph by L. W. Rees. Courtesy of Dr. T. E. Timell.

points are located inside the buds, which have been formed during the preceding autumn.

Radial growth begins in the *cambium* which is composed of a single layer of thin-walled living cells (initials) filled with protoplasm (cf. Fig. 1-2). The *cambial zone* consists of several rows of cells, which all possess the ability to divide. On division the initial cell produces a new initial and a xylem mother cell, which in its turn gives rise to two daughter cells; each of the latter is capable of further division. More cells are produced toward the xylem on the inside than toward the phloem on the outside; phloem cells divide less frequently than xylem cells. For these reasons, trees always contain much more wood than bark.

## 1.2.2   Development of the Cell

When a cell divides, it first develops a cell plate, which is rich in pectic substances. Each of the two new cells subsequently encloses itself with a thin, extensible, primary wall, consisting of cellulose, hemicelluloses, pectin, and protein. During the following phase of differentiation, the cell first expands to its full final size, after which formation of the thick, secondary wall is initiated. At this stage, this wall consists of cellulose and hemicelluloses. Lignification begins while the secondary wall is still being formed. Figures 1-3 and 1-12 show the structures of a mature cell.

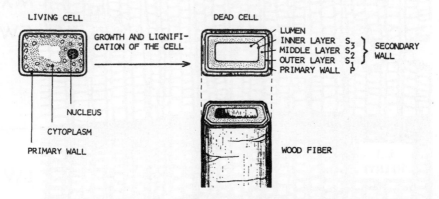

**Fig. 1-3.**   Development of the living cell to wood fiber (Bucher, 1965).

## 1.2.3   Annual Rings

At the beginning of the growth the tree requires an effective water transportation system. In softwoods thin-walled cells with large cavities are

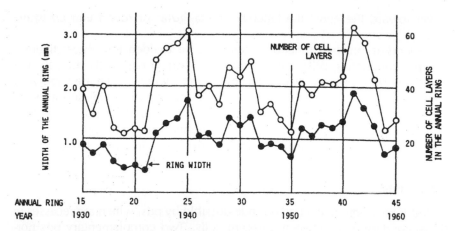

**Fig. 1-4.** Annual variations in the radial growth of a slowly grown pine stem (Ilvessalo-Pfäffli, 1967).

formed; in hardwoods special vessels take care of the liquid transportation. Comparatively light-colored and porous *earlywood* is thus formed. Later, the rate of growth decreases and *latewood* is produced. It consists of thickwalled fibers and gives mechanical strength to the stem and is darker and denser than the earlywood.

The age of a tree can be calculated from the number of growth rings at the base of the stem. With a continuous growth period (tropical woods) regular annual rings are lacking. Alternation of wet and dry periods may, however, result in the formation of growth rings. The boundary between earlywood and latewood varies. It may be very sharp as in larch or nearly nonexisting (birch, aspen, and alder). Earlywood is weaker than the thick-walled latewood. Pulp fibers from earlywood and latewood also have different papermaking properties.

The width of the annual rings varies greatly depending on tree species and growth conditions. The variation limits for Scots pine in Scandinavia may be 0.1–10 mm (Fig. 1-4). For similar reasons the proportion of latewood may vary greatly. Typical percentages for the latewood in Scandinavia are 15–50% for pine and 10–40% for spruce; the values are higher in the northern than in the southern parts of these countries.

## 1.2.4  Cell Types

On the basis of their different shape wood cells can be divided into prosenchyma and parenchyma cells. The former are thin, long cells, narrower toward the ends; the latter are rectangular or round and are short cells.

Depending on their functions, cells can be divided into three different groups: conducting cells, supporting cells, and storage cells. Conducting and supporting cells are dead cells containing cavities which are filled with water or air. In hardwoods the conducting cells consist of vessels and the supporting cells of fibers. In softwoods the tracheids perform both functions. The storage cells transport and store nutrients. They are thin-walled parenchyma cells which function as long as they remain in the sapwood.

### 1.2.5   Pits

Water conduction in a tree is made possible by pits, which are recesses in the secondary wall between adjacent cells. Two complementary pits normally occur in neighboring cells thus forming a pit pair (Fig. 1-5). Water transport between adjacent cell lumina occurs through a pit membrane which consists of a primary wall and the middle lamella. Bordered pit pairs are typical of softwood tracheids and hardwood fibers and vessels. In softwoods the pit membrane might be pressed against the pit border thus preventing water transport, since the torus is impermeable. The pits connecting tracheids, fibers, and vessels with the ray parenchyma cells are half-bordered. Simple pits without any border connect the parenchyma cells with one another.

The different shape of the pits are distinctive features in the microscopic identification of wood and fibers. Knowledge of the porous structure of

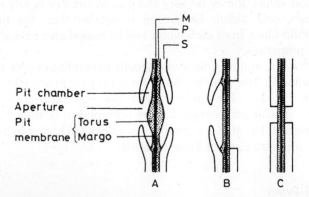

**Fig. 1-5.**   Types of pit pairs. A, bordered pit pair; B, half bordered pit pair; C, simple pit pair; M, middle lamella; P, primary wall; S, secondary wall.

wood is also of great importance for understanding the phenomena which are associated with the impregnation of wood.

### 1.2.6  Softwood Cells

The wood substance in softwoods is composed of two different cells: tracheids (90–95%) and ray cells (5–10%).

*Tracheids* give softwoods the mechanical strength required (especially the thick-walled latewood tracheids) and provide for water transport, which occurs through the thin-walled early wood tracheids with their large cavities. The liquid transport from one tracheid to another takes place through the bordered pits; their amount in earlywood tracheids is about 200 per

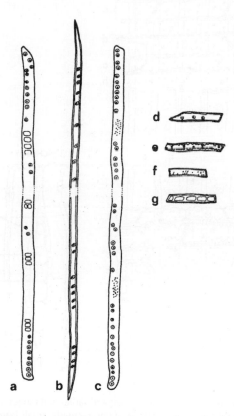

**Fig. 1-6.**  Cells of coniferous woods. An earlywood (a) and a latewood (b) pine tracheid, an earlywood spruce tracheid (c), ray tracheid of spruce (d) and of pine (e), ray parenchyma cell of spruce (f) and pine (g) (Ilvessalo-Pfäffli, 1967).

tracheid, most of them located in the radial walls in one to four lines. Latewood tracheids have only 10 to 50 rather small bordered pits.

Liquids move from the tracheids to the ray parenchyma cells through half-bordered pits. The location and nature of these pits are characteristic and used for the identification of different wood species (compare the small

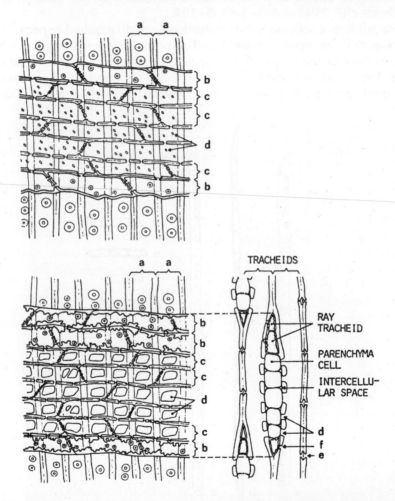

**Fig. 1-7.** Radial section of a spruce ray (above) and radial and tangential section of a pine ray (below). (a) Longitudinal tracheids. (b) Rows of ray tracheids (small bordered pits). (c) Rows of ray parenchyma. (d) Pits in the cross fields leading from ray parenchyma to longitudinal tracheids. (e) A bordered pit pair between two tracheids. (f) A bordered pit pair between a longitudial and a ray tracheid (Ilvessalo-Pfäffli, 1967).

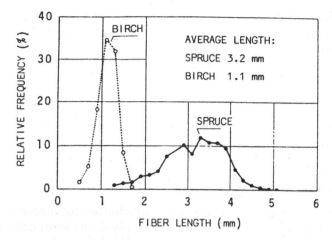

**Fig. 1-8.** Example of the distribution of fiber length in softwood (*Picea abies*) and hardwood (*Betula verrucosa*) (Ilvessalo-Pfäffli, 1977).

elliptic pits in spruce with the large window pores in Scots pine, Figs. 1-6 and 1-7).

As in other cells, the dimensions of the tracheids vary depending on genetic factors and growth conditions. Variations exist among different species and individuals as well as between different parts of the stem and within one and the same growth ring. The fiber length in the stem increases from the pith toward the cambium and reaches a maximum at the middle of the bole. Tracheids in the latewood or in narrow annual rings are usually longer and narrower than those formed more rapidly. The tangential width of the fibers varies only slightly but large differences exist in the radial direction between earlywood and latewood tracheids.

The average length of Scandinavian softwood tracheids (Norway spruce and Scots pine) is 2–4 mm and the width in the tangential direction is 0.02–0.04 mm (Fig. 1-8). The thickness of earlywood and latewood tracheids is 2–4 μm and 4–8 μm, respectively.

The width of a ray usually corresponds to one cell. Several parenchyma cell files are placed on top of one another. Ray tracheids are often located at the upper and lower edges of this tier (Fig. 1-7). Parenchyma cells are thin-walled, living cells. In Norway spruce and Scots pine their length and width vary between 0.01–0.16 mm and 2–50 μm, respectively. The ray tracheids are of the same size and also provide liquid transport in the radial direction. Rays in Scots pine, for example, contain 25–31 ray tracheids per square millimeter in a tangential section.

Resin canals are intercellular spaces building up a uniform channel network in the tree. Horizontal canals are always located inside the rays which appear together in several files (fusiform rays) (Fig. 5-2). The resin canals are lined by epithelial parenchyma cells, which secrete oleoresin into the canals. Pine wood contains more and larger resin canals than does spruce wood. In pine they are concentrated in the heartwood and root, whereas in spruce they are evenly distributed throughout the whole wood. The diameters of the resin canals in pine are on the average about 0.08 mm (vertical) and 0.03 mm (radial) (see Section 5.1.1).

### 1.2.7 Hardwood Cells

Hardwoods contain several cell types, specialized for different functions (Fig. 1-9). The supporting tissue consists mainly of libriform cells, the conducting tissue of vessels with large cavities, and the storage tissue of ray parenchyma cells. In addition, hardwood contains hybrids of the abovementioned cells which are classified as fiber tracheids. Although the term

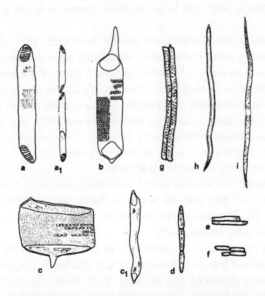

**Fig. 1-9.** Hardwood cells. Vessel elements of birch (a), of aspen (b), and of oak in earlywood (c) and in latewood ($c_1$), as well as a birch vessel ($a_1$). Longitudinal parenchyma of oak (d) and ray parenchyma of aspen (e) and of birch (f). Tracheids of oak (g) and birch (h) and a birch libriform fiber (i) (Ilvessalo-Pfäffli, 1967).

*fiber* is frequently used for any kind of wood cells, it more specifically denotes the supporting tissue, including both libriform cells and fiber tracheids. In birch these cells constitute 65 to 70% of the stem volume.

*Libriform cells* are elongated, thick-walled cells with small cavities containing some simple pits. The dimensions of birch libriform fibers are 0.8–1.6 mm or on an average 1.1–1.2 mm (length), 14–40 μm (width), and 3–4 μm (cell wall thickness). In some tropical hardwood species the average length may reach 4 mm.

*Vessels* are composed of thin-walled and rather short (0.3–0.6 mm) and wide (30–130 μm) elements, which are placed on top of one another to form a long tube. The ends have disappeared more or less completely. The channels thus formed, which might be several meters in length, are capable of a more effective water transport than the softwood tracheids. This is needed especially in the spring during the leafing. In diffuse-porous woods (aspen, birch, and maple) the vessels are evenly distributed across the annual ring. The vessels are larger and more numerous in the earlywood portion in ring-porous woods, such as ash, elm, and oak. In birch and aspen the vessels amount to about 25% of the wood volume. Several different pores are present in the walls of the vessels. These differences together with other structural features are of great help in the identification of pulp fibers. Besides the usual vessels some hardwoods contain cells resembling softwood tracheids or small vessels. Their walls are rich in bordered pits.

Hardwood rays consist exclusively of *parenchyma cells*. The ray width varies in the tangential direction. In aspen wood the rays form one row, in birch wood and oak wood 1–3 and 1–30 rows, respectively. The height varies from one up to several hundred tiers. The rays account for 5–30% of the stem volume.

### 1.2.8 Sapwood and Heartwood

At a certain age the inner wood of the stem of most trees begin to change to a completely dead heartwood and its proportion of the stem becomes successively larger as the tree grows. The dying parenchyma cells produce organic deposits such as resin, phenolic substances, pigments, etc. In softwoods the bordered pits are closed when the torus becomes pressed against either side of the border. In some hardwoods, such as oak or ash, the vessels are closed by tyloses, which enter the vessel from neighboring ray cells (Fig. 1-10). Wood with tyloses is impermeable to liquids and an excellent material for barrels. These anatomical and chemical changes often have a significant influence on the behavior of sapwood and heartwood during pulping.

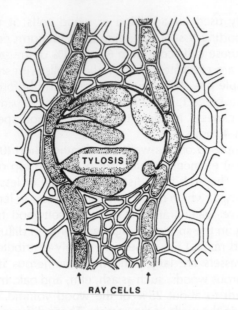

RAY CELLS

**Fig. 1-10.** Development of tyloses from ray cells. Cross-section of oak (modified from Chattaway 1949; courtesy of Mrs. M.-S. Ilvessalo-Pfäffli).

## 1.3   Wood Ultrastructure

### 1.3.1   Building Elements

The wood cell consists mainly of cellulose, hemicelluloses, and lignin (see Appendix). A simplified picture is that cellulose forms a skeleton which is surrounded by other substances functioning as matrix (hemicelluloses) and encrusting (lignin) materials.

The length of a native cellulose molecule is at least 5000 nm corresponding to a chain with about 10,000 glucose units (cf. Section 3.2.1). The smallest building element of the cellulose skeleton is considered by some to be an elementary fibril. This is a bundle of 36 parallel cellulose molecules which are held together by hydrogen bonds, but various opinions exist concerning this question. The cellulose molecules according to the "fringe micellar model" form completely ordered or crystalline regions, which without any distinctive boundary are changing into disordered or amorphous regions (Fig. 1-11). In native cellulose the length of the crystallites can

**Fig. 1-11.** Diagrammatic representation of fibrillar structure in the cell wall according to Mark (1940). Heavy lines constitute the crystalline regions. The chain molecules may pass through one or more crystalline and amorphous regions.

be 100–250 nm and the cross section, probably rectangular, is on an average $3 \times 10$ nm. According to this model the cellulose molecule continues through several crystallites.

The microfibrils, which are 10–20 nm wide, are visible in the electron microscope without pretreatment. Microfibrils are combined to greater fibrils and lamellae, which can be separated from the fibers mechanically, although their dimensions greatly depend on the method used.

Disordered cellulose molecules as well as hemicelluloses and lignin are located in the spaces between the microfibrils. The hemicelluloses are considered to be amorphous although they apparently are oriented in the same direction as the cellulose microfibrils. Lignin is both amorphous and isotropic.

## 1.3.2   Cell Wall Layers

The cell wall is built up by several layers, namely (Figs. 1-12 and 1-13), middle lamella (M), primary wall (P), outer layer of the secondary wall ($S_1$), middle layer of the secondary wall ($S_2$), inner layer of the secondary wall ($S_3$), and warty layer (W). These layers differ from one another with respect to their structure as well as their chemical composition. The microfibrils wind around the cell axis in different directions either to the right (Z helix) or to the left (S helix). Deviations in the angular directions cause physical differences and the layers can be observed in a microscope under polarized light.

The *middle lamella* is located between the cells and serves the function of binding the cells together. At an early stage of the growth it is mainly composed of pectic substances, but it eventually becomes highly lignified. Its thickness, except at the cell corners, is 0.2–1.0 μm. The *primary wall* is a thin layer, 0.1–0.2 μm thick, consisting of cellulose, hemicelluloses, pectin, and protein and completely embedded in lignin. The cellulose microfibrils form an irregular network in the outer portion of the primary wall; in the interior they are oriented nearly perpendicularly to the cell axis (Fig. 1-14). In the presence of reagents which induce strong swelling the primary

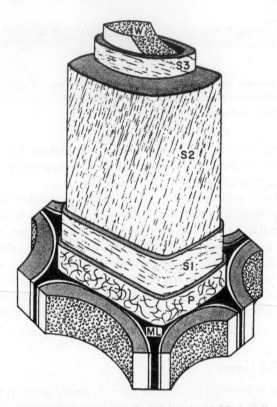

**Fig. 1-12.** Simplified structure of a woody cell, showing the middle lamella (ML), the primary wall (P), the outer ($S_1$), middle ($S_2$), and inner ($S_3$) layers of the secondary wall, and the warty layer (W) (Côté, 1967, with permission).

wall is peeled off and the belts around the fibers expand (*ballooning*) (Fig. 1-15). The middle lamella, together with the primary walls on both sides, is often referred to as the compound middle lamella. Its lignin content is high, but because the layer is thin only 20–25% of the total lignin in wood is located in this layer.

The *secondary wall* consists of three layers: thin outer and inner layers and a thick middle layer. These layers are built up by lamellae formed by almost parallel microfibrils between which lignin and hemicelluloses are located.

The *outer layer* ($S_1$) is 0.2–0.3 μm thick and contains 3–4 lamellae where the microfibrils form either a Z helix or S helix. The microfibril angle of the crossed fibrillar network varies between 50 and 70° with respect to the fiber axis.

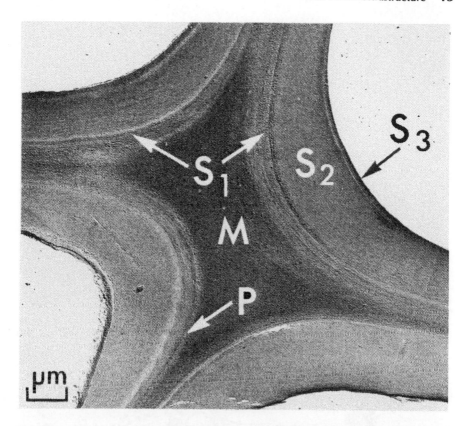

**Fig. 1-13.** Transverse section of earlywood tracheids in tamarack (*Larix laricina*), showing the middle lamella (M), the primary wall (P), and the outer (S$_1$), middle (S$_2$), and inner (S$_3$) layers of the secondary wall. Transmission electron micrograph. Scale 1 μm. Courtesy of Dr. T. E. Timell.

The *middle layer* (S$_2$) forms the main portion of the cell wall. Its thickness in softwood tracheids varies between 1 (earlywood) and 5 (latewood) μm and it may thus contain 30–40 lamellae or more than 150 lamellae. The thickness naturally varies with the cell types. The microfibrillar angle (Fig. 1-16) varies between 5–10° (latewood) and 20–30° (earlywood). It decreases in a regular fashion with increasing fiber length. The characteristics of the S$_2$ layer (thickness, microfibrillar angle, etc.) have a decisive influence on the fiber stiffness as well as on other papermaking properties.

The *inner layer* (S$_3$) is a thin layer (ca. 0.1 μm) consisting of several lamellae which contain microfibrils in both Z helices and S helices (50–90° angle). Great variations are noted among different wood species.

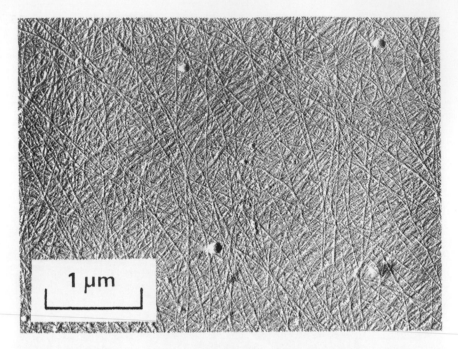

**Fig. 1-14.** Electron micrograph of a delignified primary wall (*Pinus sylvestris*) (Meier, 1958).

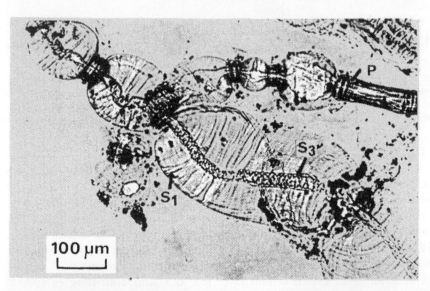

**Fig. 1-15.** Ballooning of a sulfate pulp fiber (*Pinus sylvestris*). Note the ribbonlike, unrolled primary wall (P) and the swollen secondary wall. $S_1$ is the swollen outer layer of the secondary wall, under which the microfibrils of the middle layer, nearly parallel to fiber axis, are dimly visible. $S_3$ is the inner layer of the secondary wall (Ilvessalo-Pfäffli, 1977).

**Fig. 1-16.** Electron micrograph of a delignified secondary wall ($S_2$) of *Pinus sylvestris*. Courtesy of Dr. H. Meier.

The *warty layer* (W) is a thin amorphous membrane located in the inner surface of the cell wall in all conifers and in some hardwoods, containing warty deposits of a still unknown composition. Each species has its own, characteristic warty layer.

### 1.3.3 Pits

The normal structure of the cell wall is broken by pits. Changes appear already in the growth period of the cell. For instance, early stages of pit formation in softwoods are visible in the primary wall just before the cell reaches its final dimensions (primary pit fields). The microfibril network is loosened and new microfibrils are oriented around these points. The structure in the middle of the circles is tightened and the radially oriented microfibril bundles finally form a netlike membrane, permeable to liquids (*margo*) (Fig. 1-17). The central, thickened portion of the pit membrane (*torus*) is formed after a secondary thickening of the primary wall. The torus is rich in pectic material and also contains cellulose in pine and spruce.

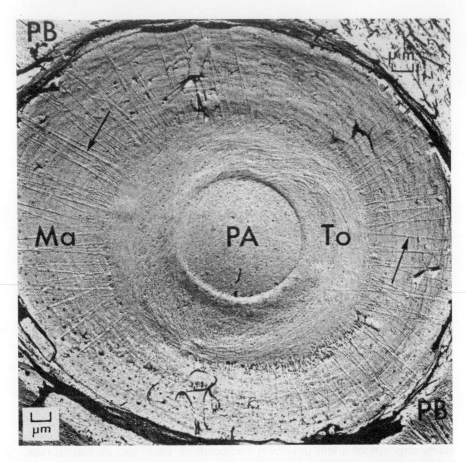

**Fig. 1-17.** Surface replica of an aspirated bordered pit in a tracheid of Douglas fir (*Pseudo-tsuga menziesii*), showing the pit aperture (PA), the torus (To), the margo (Ma), and the pit border (PB). Arrows indicate supporting cellulose strands. Transmission electron micrograph. Scale 1 μm. Courtesy of Dr. W. A. Côté, Jr.

## 1.4   Reaction Wood

As a product of living organism the structure of wood fibers is so variable and complicated that a great number of details remains to be solved for understanding the anatomy and biology even of trees grown under normal conditions. When a tree is brought out of its natural, equilibrium position in space, for example by wind or by a landslide, the tree begins to produce a special tissue, referred to as *reaction wood*. The function of this type of wood is to restore the displaced stem or branch to its original position. In a leaning stem of a conifer, *compression wood* develops on the lower side. This wood expands longitudinally as it is being formed, and the pressure exerted along the grain forces the stem to bend upward. All movements of orientation in mature conifers are effected with the aid of appropriately located compression wood. In hardwoods, *tension wood* is formed on the upper side of an inclined stem. This wood contracts as it is laid down and in this way forces the stem to end upward. Compression wood can be said to *push* a stem or a branch up; tension wood *pulls* them up.

Compression wood is heavier, harder, and denser than the normal wood. Its tracheids are short and thick-walled (even in earlywood) and in cross

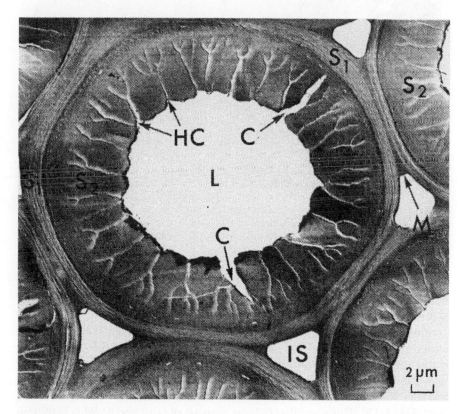

**Fig. 1-18.** Transverse section of compression wood tracheids in tamarack (*Larix laricina*), showing intercellular spaces (IS), middle lamella (M), the outer (S₁), and the inner (S₂) layer of the secondary wall, and the lumen (L). The S₂ layer contains narrow, branched helical cavities (HC) as well as two wide drying checks (C), an artifact. Transmission electron micrograph. Courtesy of Dr. T. E. Timell.

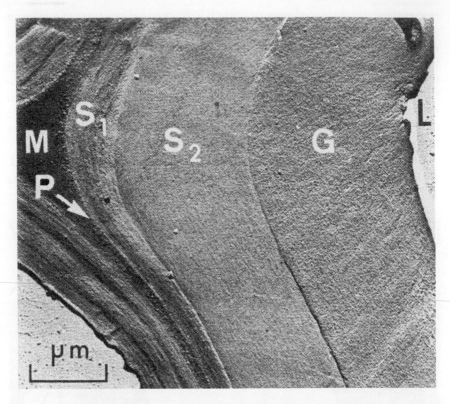

**Fig. 1-19.** Transverse section of a tension wood fiber in American beech (*Fagus grandifolia*), showing the middle lamella (M), primary wall (P), the outer (S₁), and middle (S₂) layers of the secondary wall, the thick gelatinous layer (G), and the lumen (L). Transmission electron micrograph. Scale 1 μm. Courtesy of Dr. T. E. Timell.

section rounded so that empty spaces remain between the cells. The $S_1$ layer is thicker than in a normal wood while the $S_3$ layer is absent. The $S_2$ layer contains helical cavities that parallel the microfibrils and reach from the lumen deep into the $S_2$ (Fig. 1-18). The cellulose content of compression wood is lower and the lignin content higher than for normal wood.

*Tension wood* differs less from normal wood than compression wood. It contains thick-walled fibers, terminated toward the lumen by a gelatinous layer (Fig. 1-19). This so-called G layer consists of pure and highly crystalline cellulose oriented in the same direction as the fiber axis. For this reason the cellulose content of tension wood is higher and the lignin content lower than in normal wood.

# Chapter 2

# INTRODUCTION TO CARBOHYDRATE CHEMISTRY

## 2.1 Definitions and Nomenclature

The name carbohydrate was originally derived from the general formula $C_x(H_2O)_y$, formally analogous to hydrates of carbon, but this type of simple definition does not cover the broad class of carbohydrates. Carbohydrates are polyhydroxy compounds appearing commonly in nature, either as relatively small molecules (sugars) or as large entities extending to macromolecular levels (polysaccharides). Sugars are formed in green plants as early products of photosynthesis from carbon dioxide and water and are then converted into organic plant constituents through a variety of biosynthetic pathways.

The sugars in a plant usually function as a source of energy while polysaccharides, such as starch, fulfill the need for the storage of reserve food or they (cellulose and hemicelluloses) contribute mechanical strength to the plant cell wall. In addition, a variety of carbohydrates are included as essential building elements in natural compounds performing vital functions in living organisms.

Carbohydrates may be classified into following three large groups: (1) *Monosaccharides* are simple sugars, of which D-glucose, D-mannose, D-

TABLE 2-1. Aldose Series[a]

| Carbon atom number | Trioses | Tetroses | Pentoses | Hexoses |
|---|---|---|---|---|
| 1 | CHO | CHO | CHO | CHO |
| 2 | *CHOH | *CHOH | *CHOH | *CHOH |
| 3 | $CH_2OH$ | *CHOH | *CHOH | *CHOH |
| 4 |  | $CH_2OH$ | *CHOH | *CHOH |
| 5 |  |  | $CH_2OH$ | *CHOH |
| 6 |  |  |  | $CH_2OH$ |

[a] Asterisk denotes asymmetric carbon atom.

galactose, D-xylose, and L-arabinose are the most common constituents of the cell wall polysaccharides in wood. (2) *Oligosaccharides* consist of several monosaccharide residues joined together by glycosidic linkages, named di-, tri-, tetrasaccharides, and so on. The name oligosaccharide is usually restricted to the group of carbohydrates in which the number of monosaccharide units is less than ten. (3) *Polysaccharides* are complex molecules composed of a large number of monosaccharide units joined together by glycosidic linkages. *Polyuronides* are polysaccharides containing uronic acid blocks in the main backbone and are typical components in algae (sea weeds) and pectins. *Polyhydric alcohols,* consisting of acyclic polyols (alditols, glycitols, or "sugar alcohols") and alicyclic polyalcohols (cyclitols), are classified as carbohydrates in a wider sense. Of the former, sorbitol (D-glucitol) occurs in algae as well as in higher plants and was discovered in the fresh juice of berries of mountain ash (*Sorbus aucuparia*). Of cyclitols, cyclohexanehexols or inositols, particularly *myo*-inositol, have a wide distribution and are of importance for plants, bacteria, and animals.

The monosaccharides contain either an aldehyde or a keto function and are accordingly classified as *aldoses* or *ketoses*. The aldoses and ketoses are further divided into subgroups on the basis of their number of carbon atoms. The major carbohydrates in wood consist of aldopentoses and aldohexoses (Table 2-1). A prefix deoxy is used when one of the hydroxyl groups is replaced by a hydrogen atom ($R_3COH \rightarrow R_3CH$). An ether group, usually methyl ($R_3COH \rightarrow R_3COMe$), is denoted by a prefix (*O*-methyl). Correspondingly, the prefix for an ester group, such as acetate ($R_3COH \rightarrow R_3CO$-COMe) is *O*-acetyl.

## 2.2    Monosaccharides

Most of the monosaccharides occur as glycosides and as units in oligosaccharides and polysaccharides and only comparatively few of them are present free in plants. D-Glucose is the most abundant monosaccharide in nature. It occurs in a free state in many plants, especially in fruits and can be prepared from cellulose and starch by acidic or enzymic hydrolysis. Of the other aldohexoses D-mannose and D-galactose are important components in hemicelluloses. The most common aldopentoses, abundant members of the hemicelluloses, are D-xylose and L-arabinose. D-Ribose is a constituent of nucleosides. No tetroses or trioses have been detected free in plants, but D-erythrose 4-phosphate is an important intermediate in many transformations, and D-glyceraldehyde and dihydroxyacetone are essential components in cellular metabolism. Of the heptoses, sedoheptulose 7-phosphate occurs as intermediate in photosynthesis and traces of it may be present in all plants. Of the deoxysugars, L-rhamnose (6-deoxy-L-mannose) occurs as a constituent in gum polysaccharides and traces of it are present in hemicelluloses (xylan). D-Fructose, which represents the only abundant ketose in plants, is present both free and in a combined state. Plants belonging to the Compositeae and Gramineae families store polymers of D-fructose, such as inulin, as reserve material rather than starch. D-Fructose is not a member of the cell wall polysaccharides of wood.

### 2.2.1    Configuration of Monosaccharides

An example of a simple sugar, aldotriose or glyceraldehyde is given in Fig. 2-1. It contains one "*asymmetric carbon atom*" or so-called *chiral center* since all the four substituents bound to this carbon atom are different. It follows from this that there are two stereoisomeric forms of glyceraldehyde, which are mirror images of each other and termed *enantiomers*. These enantiomers cause rotation of the plane of plane-polarized light to an equal degree but in opposite directions. The enantiomers of glyceraldehyde can be visualized by a tetrahedron in which the substituents appear at the vertices, as originally suggested by van't Hoff and later adopted in carbohydrate chemistry by Emil Fischer. A simplification of this concept is the *Fischer projection formula*, which is a two-dimensional representation of the molecule. In this the C-1 carbon is at the top of the molecule, the bond between the carbon atoms is vertical, and the bonds connecting H and OH groups to the carbon are horizontal. The symbol D is assigned to (+)-glyceraldehyde since it has the hydroxyl group on the right in the Fischer projection formula,

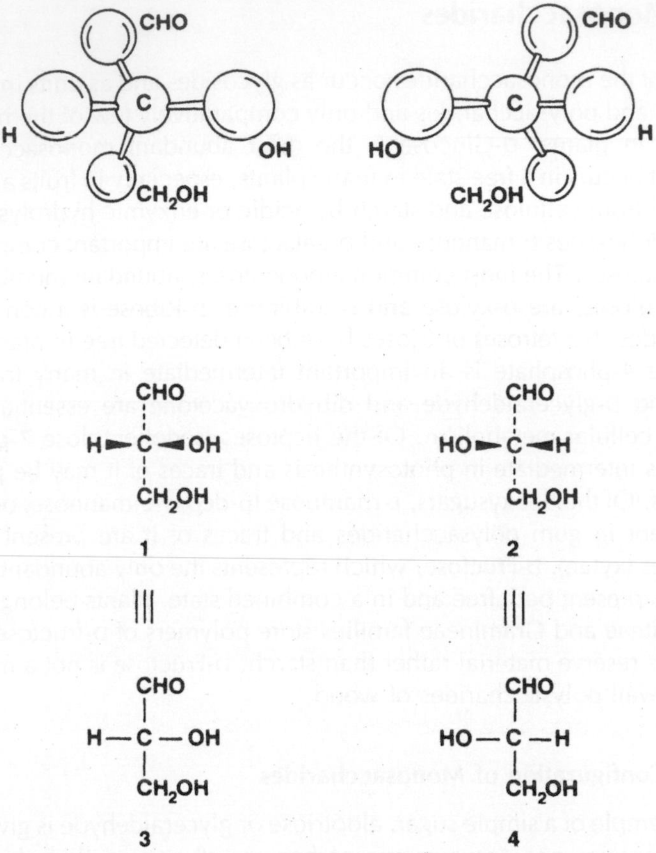

**Fig. 2-1.** D-Glyceraldehyde (1, 3) and L-glyceraldehyde (2, 4) represented by the "wedge–dash" and Fischer projection formulas. Note that the horizontal bonds from the center carbon atom to H and OH are oriented upward from the plane, whereas the vertical bonds to CHO and $CH_2OH$ groups are oriented downward.

whereas L refers to (−)-glyceraldehyde which has the hydroxyl group on the left. Figure 2-2 shows the Fischer projection formula for D-aldotetroses.

Compounds with $n$ asymmetric carbon atoms, and differing from each other only with regard to geometry, consist of $2^n$ stereoisomers or $2^{n-1}$ enantiomeric (mirror image) pairs. When the number of asymmetric carbon atoms exceeds two, so-called *diastereoisomeric* forms become possible. They possess different physical properties and are not mirror images. *Enantiomers*, however, are identical in physical properties with the exception of

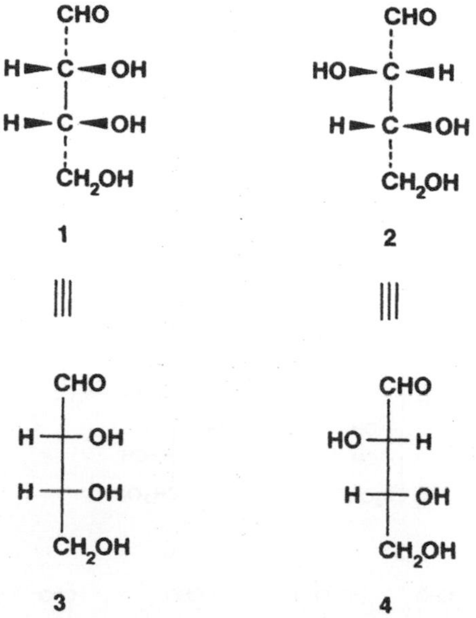

**Fig. 2-2.** D-Erythrose (1, 3) and D-threose (2, 4) represented by the corresponding wedge–dash and Fischer projection formulas.

their behavior toward polarized light. The aldohexoses comprise 16 stereoisomers (8 enantiomeric pairs) belonging to the respective series. The respective D- and L-aldohexoses are diastereomers. The aldoses in D series are shown in Fig. 2-3.

In some cases, when superimposable with its mirror image, a compound is optically inactive even though it contains chiral centers. Such an optically inactive stereoisomer may be designated by the prefix *meso* (e.g., *meso*-xylitol).

According to Rosanoff's convention the configuration is determined by the orientation of the hydroxyl group bound to the last asymmetric carbon atom in the carbon chain, which, for example in glucose is C-5. In the D form this hydroxyl group points to the right in the Fischer projection formula whereas in the L form the hydroxyl group is on the left.* The symbols D and L specify

---

*The use of DL-convention and the Fischer projection for absolute configuration has some disadvantages. According to a later and a more exact convention (Cahn-Ingold-Prelog) the configuration of a chiral center is specified using prefixes *R* and *S* (Latin: *rectus*, right; *sinister*, left). In carbohydrate chemistry, however, the DL-convention is used generally.

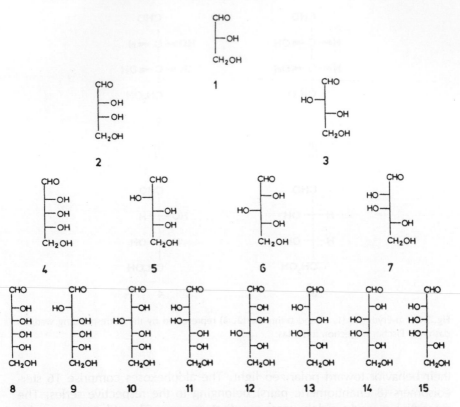

**Fig. 2-3.** Fischer projection formulas for acyclic forms of D-aldoses. Glyceraldehyde (1), erythrose (2), threose (3), ribose (4), arabinose (5), xylose (6), lyxose (7), allose (8), altrose (9), glucose (10), mannose (11), gulose (12), idose (13), galactose (14), and talose (15).

the *absolute configuration* and they bear no relationship to the direction of optical rotation, which can be separately marked by (+) or (−) after the configuration symbols.

## 2.2.2 Ring Structures of Monosaccharides

Sugars form readily cyclic hemiacetals, which in solutions are in equilibrium with the open-chain forms (cf. Fig. 2-7). Hemiacetal rings are usually composed of five or six atoms; smaller or larger rings are too strained and for thermodynamic reasons not stable with the exception of some sugars existing as seven-membered (heptanose ring) structures.

For example, as shown in Fig. 2-4, D-glucose and D-fructose can exist as

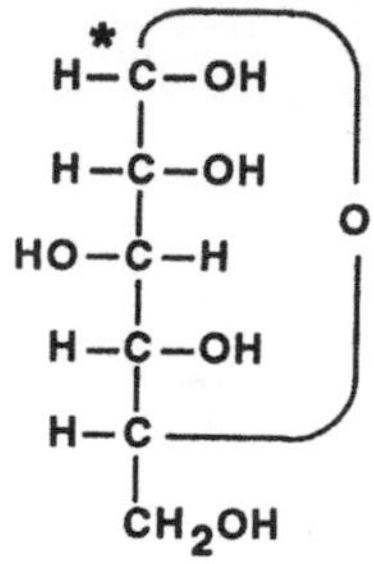

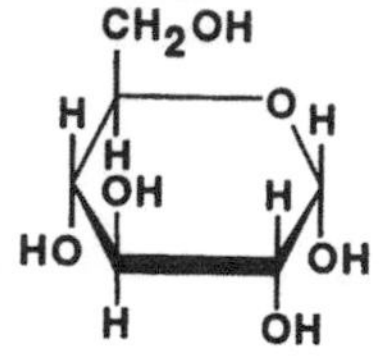

**Fig. 2-4.** The Fischer–Tollens projection and Haworth perspective formulas of α-D-glucopyranose (1) and β-D-fructofuranose (2).

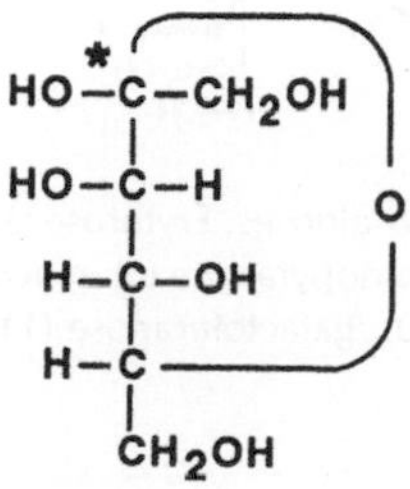

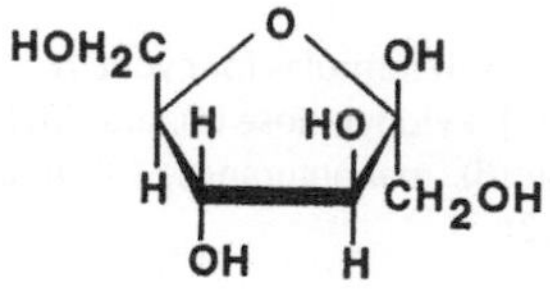

six-membered (*pyranose*) and as five-membered ring (*furanose*) structures. The formation of the hemiacetal ring generates a new chiral center and a pair of C-1 epimeric* sugars, termed *anomers*. These structures can be best seen from the Haworth perspective formula in which the plane of the pyranose or furanose ring is assumed to be perpendicular to the plane of paper. The substituents are parallel to this plane and project either above or below the plane of the ring. In the "standard formula" the bond between C-2 and C-3 (heavier line) points to the viewer out from the paper plane. When the glycosidic hydroxyl group in this formula is projecting below the plane, as in D-glucopyranose (1), the anomer is termed the α-form; in the opposite case, as in D-fructofuranose (2), the anomer is the β-form. It should be observed, however, that the reverse is true for the members in the L-series.

As examples, cyclic forms of some α-D-aldoses are visualized in Fig. 2-5 by Haworth perspective formulas.

*Diastereoisomers which differ in their configuration only at one of the carbon atoms are termed epimers. To specify the epimers the site is indicated, for example, C-2 epimers or C-3 epimers. Generally, if not indicated, the term epimer refers to diastereoisomers having reversed configuration at the asymmetric carbon atom adjacent to the anomeric center.

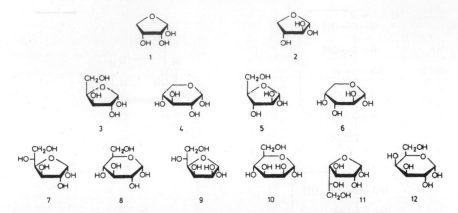

**Fig. 2-5.** Haworth formulas for cyclic α-anomers of some D-aldoses. Erythrose (1), threose (2), xylofuranose (3), xylopyranose (4), arabinofuranose (5), arabinopyranose (6), glucofuranose (7), glucopyranose (8), mannofuranose (9), mannopyranose (10), galactofuranose (11), and galactopyranose (12).

Another possibility for depicting sugar structures is to use Mills' formulas (Fig. 2-6). In this system the ring is parallel to the plane of paper. The substituents above the plane of the ring are denoted by heavy lines whereas dotted lines are used for substituents below the plane. In the case of substituents with undefined orientation, or for a mixture consisting of α and β anomers, a wiggly line is used.

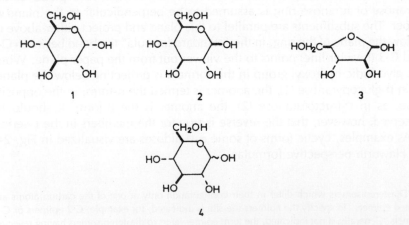

**Fig. 2-6.** Mills formulas. α-D-Glucopyranose (1), β-D-glucopyranose (2), β-D-arabinofuranose (3), and α,β-D-glucopyranose (4).

### 2.2.3  Mutarotation

On dissolution of sugars in water, the optical rotation of the solution changes continuously until an equilibrium is reached. This phenomenon, termed *mutarotation,* is accompanied by complex changes as the hemiacetal ring is opened and products with furanose and pyranose rings are formed, which in addition can either have an α or β anomeric configuration (Fig. 2-7). Some of the isomerization reactions are faster than others, and

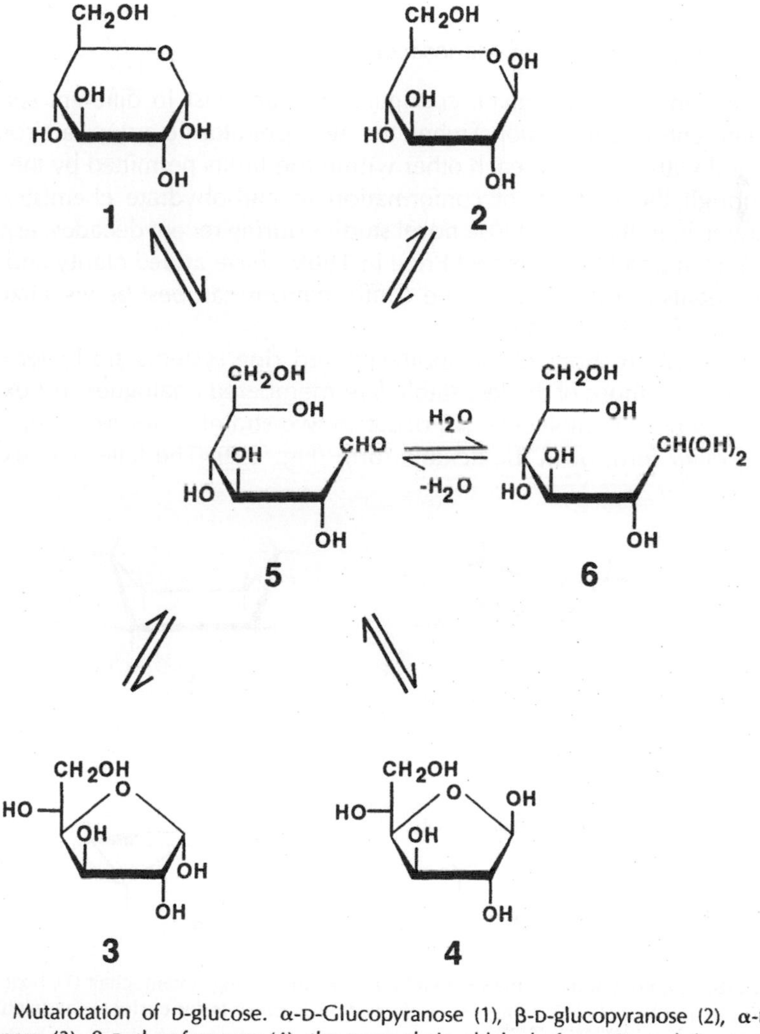

**Fig. 2-7.** Mutarotation of D-glucose. α-D-Glucopyranose (1), β-D-glucopyranose (2), α-D-glucofuranose (3), β-D-glucofuranose (4), the open-chain aldehyde form (5), and the open-chain hydrate (6).

thus a certain form may intermittently reach a high concentration, although it may be only a minor component after the equilibrium has been reached. The final proportions of the four possible ring isomers vary considerably among different sugars, depending on their thermodynamic stabilities. The equilibrium is also affected by the solvent; for instance, the proportion of the furanose form is increased in dimethyl sulfoxide since the solvation of hydroxyl groups is decreased. Traces of acids or bases accelerate this interconversion.

### 2.2.4 Conformation of Monosaccharides

Any molecule of a given configuration can exist in different spatial arrangements (*conformations*) when the atoms or atomic groups are rotated or twisted with respect to each other within the limits permitted by the bonds. Although the concept of conformation in carbohydrate chemistry is old (Haworth, in the late 1920s), novel studies during recent decades, especially by Barton and Hassel (Nobel Prize in 1969), have added clarity and important details to this concept. The conformations can best be visualized with the use of molecular models.

The conformations of the six-membered ring systems are better characterized than those of the less stable five-membered analogues. For example, the cyclohexane molecule can occur in two strainless forms, namely in the rigid *chair* form or in the flexible form (Fig. 2-8). The latter can exist in a

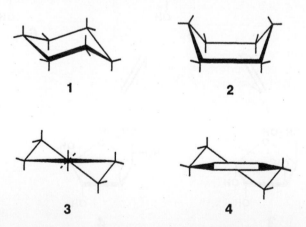

**Fig. 2-8.** Conformations of the six-membered (pyranoid) ring system: chair (1), boat (2), skew boat (3), and half-chair (4). The last conformation is possible for cyclohexene (pictured) and cyclohexane containing an oxirane ring.

variety of shapes of which only the *boat* and the *skew boat* (or *twist*) are regular and easily depictable on paper. The chair form is preferred energetically because it is usually free from steric interactions whereas the flexible forms are not. The *half-chair* conformation is possible, when a six-membered ring contains either a double bond or an oxirane ring. In the half-chair conformation four adjacent atoms are in the same plane.

The conformations of the pyranose ring are given by initial letters, that is, C (chair), B (boat), S (skew boat), and H (half-chair). The two ring atoms deviating from the reference plane are then marked by two number indexes as shown in Fig. 2-9.

In both the chair and the boat forms of the cyclohexane molecule the bonds to hydrogen atoms are oriented either *equatorially* or *axially*. The axial bonds (a) become equatorial (e), and vice versa, when the conformation of the molecule is switched between the two possible chair forms as shown in Fig. 2-10. In monosubstituted molecules the substituent favors the equatorial position as a consequence of minimum nonbonded interaction with the neighboring hydrogen atoms. This is also generally the case for derivatives with several substituents, and the molecule usually takes the conformation in which the majority of substituents are equatorial. When the above rules are applied to β-D-glucopyranose and β-D-fructopyranose, their conformations can be presented according to Fig. 2-11.

In some cases the hydroxyl groups prefer an axial orientation, for instance in molecules bearing ring oxygen atoms which participate in hydrogen bonds with the hydroxyl groups (Fig. 2-12). An exception to the tendency of the hydroxyl groups to be equatorially oriented in a pyranose ring is the substituent bound to the anomeric center where an axial position is favored (the so-called *anomeric effect*). The anomeric effect depends on the nature of the substituent and is especially strong for halogens.

The most stable conformations of cyclopentane are *envelope* and *twist* (Fig. 2-13). These two conformations are almost equally stable and the energy barrier for their interconversion is low. The relative free transforma-

**Fig. 2-9.** Nomenclature system for the chair forms. The conformations 1 and 2 are marked by the symbols $^4C_1$ and $^1C_4$ hr. The shadowed area indicates the reference plane.

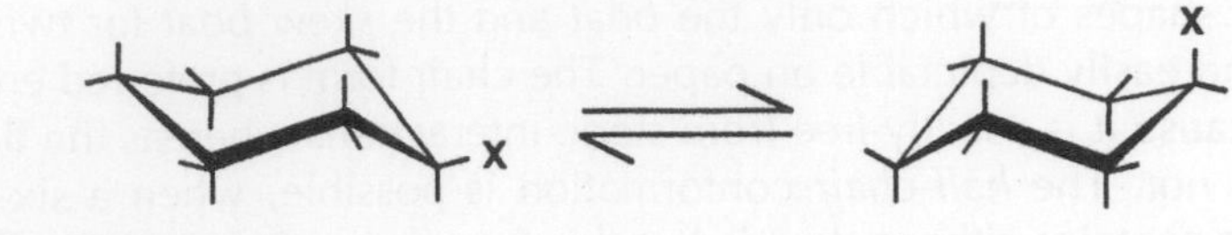

**Fig. 2-10.**   Ring inversion of a monosubstituted cyclohexane derivative: the equatorial substituent (X) (and the equatorial H atoms) becomes axial and vice versa.

**Fig. 2-11.**   The stable chair conformations of β-D-glucopyranose (1) and β-D-fructopyranose (2). The conformations are marked with the symbols $^4C_1$ and $^2C_5$, respectively.

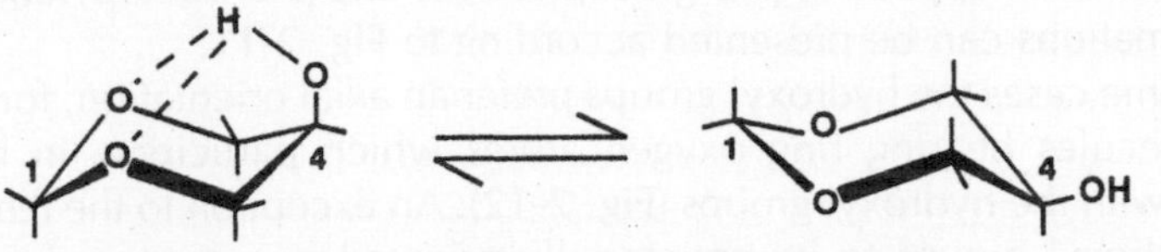

**Fig. 2-12.**   The influence of strong hydrogen-bonding interactions. The equilibrium between the two conformers of 4-hydroxy-1,3-dioxane is shifted to the left because the $^4C_1$ conformer (1) is more stable than the $^1C_4$ conformer (2).

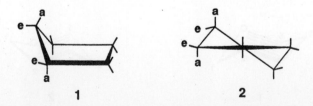

**Fig. 2-13.**   Conformations of the five-membered (furanoid) ring system: the envelope (1) and twist (2) conformations of cyclopentane. The pseudoaxial and pseudoequatorial carbon-hydrogen bonds are marked by a and e.

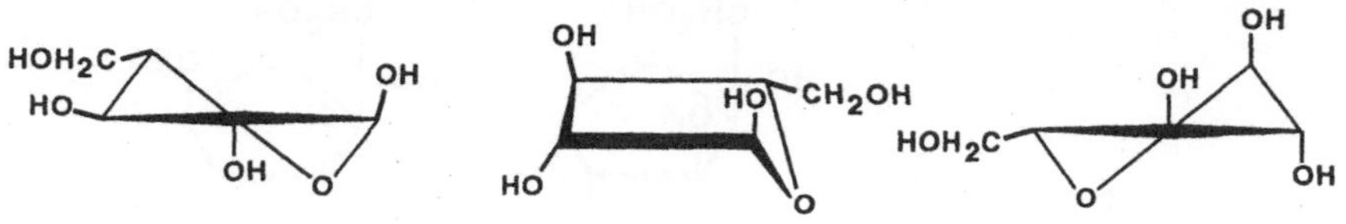

1                          2                          3

**Fig. 2-14.** Probable conformations of α-L-arabinofuranose. The conformations are marked by symbols $^4T_0$ (1), $E_0$ (2), and $^1T_0$ (3). Typical of these are a pseudoaxial anomeric hydroxyl group and a pseudoequatorial orientation of the hydroxymethyl substituent.

tion between the conformations is called *pseudorotation*. In the envelope conformation four ring atoms form a plane and the fifth ring atom is above or below this reference plane. In the twist conformation the reference plane is formed by three adjacent ring atoms and the two remaining ring atoms are above and below this plane. The substituents may have some axial or equatorial character, and this type of orientation is accordingly called pseudoaxial and pseudoequatorial.

Due to the anomeric effect, furanoses prefer conformations having pseudoaxial anomeric hydroxyl groups. On the other hand bulky substituents, such as hydroxymethyl groups, prefer a pseudoequatorial orientation. As an example, the most stable conformations of α-L-arabinofuranose are illustrated by Fig. 2-14.

The conformations of the furanose ring are named according to their initial letters, that is, E for the envelope and T for the twist, combined with number indexes for the ring atoms that are above or below the reference plane.

## 2.3    Monosaccharide Derivatives

Sugar derivatives are, in principle, formed (1) by reaction of the free carbonyl group or the anomeric hydroxyl at C-1 or (2) by reactions of the other hydroxyl groups. For reactions of free carbonyl groups, see Section 2.5.3.

### 2.3.1    Glycosides

Glycosides are cyclic sugar derivatives in which the anomeric hydroxyl group has been replaced with an alkoxyl or aroxyl group. The alkyl or aryl group attached to the anomeric oxygen atom is termed *aglycon*.

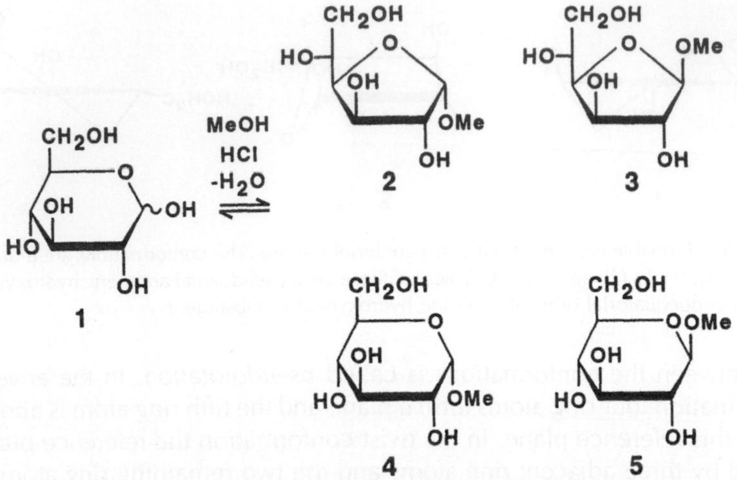

**Fig. 2-15.** Formation of methyl D-glucosides from D-glucose (1). Mainly methyl α-D-glucofuranoside (2) and methyl β-D-glucofuranoside are formed in the beginning, whereas methyl α-D-glucopyranoside (4) and methyl β-D-glucopyranoside (5) are the principal components of the equilibrium mixture.

Simple alkyl glycosides are most easily prepared by treating the free sugars with the corresponding alcohols in the presence of an acid catalyst (Fig. 2-15). Even though pure sugar anomers are often used as starting materials, a mixture of furanosides and pyranosides is obtained because mutarotation is much faster than glycosidation. Furanoses form glycosides more readily than pyranoses, whereas pyranosides are thermodynamically more stable than the furanosides. Consequently, both furanosides and pyranosides may be prepared under kinetic and thermodynamic control.

Glycosides are easily hydrolyzed by aqueous acids to free sugars but they are fairly stable toward alkali. Because of the reversibility of the glycosidation reaction, glycosides must be prepared in anhydrous solvents.

## 2.3.2 Alkylidene Derivatives

Treatment of polyhydroxy compounds, such as sugars, with acidic solutions of aldehydes or ketones results in the formation of O-alkylidene derivatives. The most stable five- and six-membered acetal rings are formed from 1,2- and 1,3-diols, respectively. Since a pair of hydroxyl groups is blocked, the O-alkylidene derivatives are extremely important intermediates in syn-

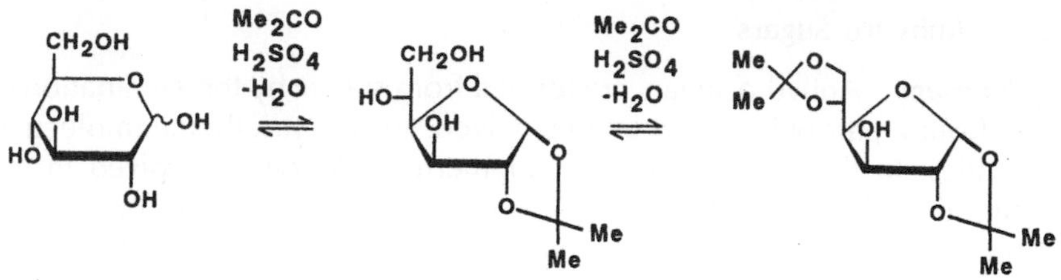

**Fig. 2-16.** Formation of 1,2-O-isopropylidene-α-D-glucofuranose (2) and 1,2:5,6-di-O-iso-propylidene-α-D-glucofuranose (3) by subsequent reactions of D-glucose (1) with acetone. Usually 3 is first recovered from the reaction mixture and then hydrolyzed by an aqueous acid to 2.

thetic carbohydrate chemistry. The most common O-alkylidene derivatives of sugars are prepared from acetone and benzaldehyde and are called O-isopropylidene and O-benzylidene sugars, respectively (Fig. 2-16). Typical of all acetals is that the O-alkylidene groups are readily hydrolyzed by acids.

## 2.3.3 Ethers

Ethers are important derivatives of both monosaccharides and polysaccharides. Etherification is often used in the determination of structures and types of linkages in oligo- and polysaccharides. Table 2-2 gives examples of the preparation of ethers. Unlike most ethers, the trityl ethers are easily hydrolyzed by acids. The trimethylsilyl ethers are also unstable under hydrolytic conditions.

**TABLE 2-2.  Some Preparation Methods of Ethers**

| Ether | | Reagent |
|---|---|---|
| Methyl | $(ROCH_3)$ | a.  $CH_3I + NaH + DMF$ |
| | | b.  $(CH_3)_2SO_4 + NaOH$ |
| | | c.  $CH_2N_2 + BF_3 \cdot Et_2O$ |
| Trimethylsilyl | $(ROSi(CH_3)_3)$ | $(CH_3)_3SiNHSi(CH_3)_3 +$ |
| | | $(CH_3)_3SiCl$ + pyridine |
| Triphenylmethyl (trityl) | $(ROC(C_6H_5)_3)$ | $(C_6H_5)_3CCl$ + pyridine |

## 2.3.4 Anhydro Sugars

*Glycosans* Anhydro sugars are formed from sugars by the elimination of water from a pair of hydroxyl groups. Glycosans are strictly intramolecular glycosides. In these derivatives the anomeric hydroxyl is involved in the formation of the anhydro linkage. These linkages are readily opened by action of acids, some of them also by bases. The 1,6-anhydroaldohexoses are the most common glycosans, and these are formed as equilibrium products of the free aldoses in aqueous acids (Fig. 2-17). The 1,2- and 1,6-aldohexoses are also important intermediates in alkaline hydrolysis of aldohexopyranosides.

*Epoxides* According to the definitions the internal ethers are derived only from alcoholic hydroxyls and the hydroxyl group in the anomeric center does not participate. They are formed when the sugar molecule contains both a good "leaving group" and a suitably located ionized hydroxyl group. Since this reaction proceeds according to the $S_N2$ mechanism, inversion takes place at the carbon atom involved (Fig. 2-18). The ring

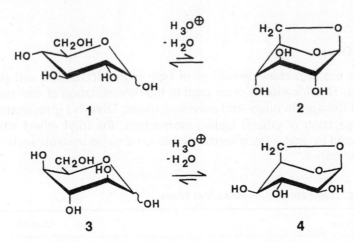

**Fig. 2-17.** Formation of 1,6-anhydro-β-D-glucopyranose (levoglucosan) (2) and 1,6-anhydro-β-D-idopyranose (4) from D-glucopyranoses (1) and D-idopyranoses (3), respectively. In both cases the reaction involves inversion of the ring conformation from $^4C_1$ to $^1C_4$. In the first reaction the axial position of the hydroxyl groups (at C-2, C-3, and C-4) destabilizes the structure 2, which is why its amount remains rather low in the equilibrium mixture. However, in the second reaction, structure 4 with equatorial hydroxyl groups is stable and the dominant component.

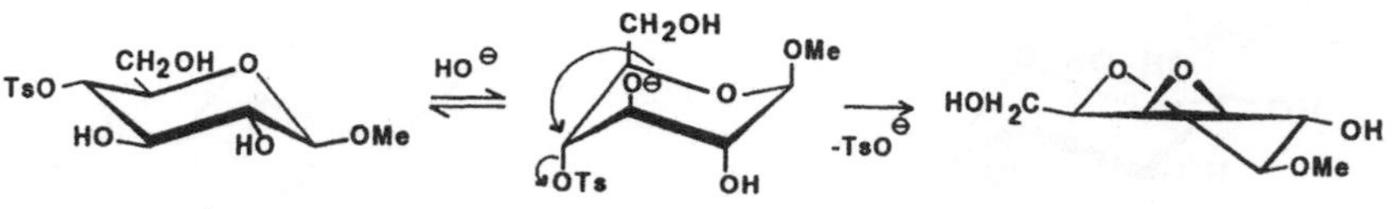

**Fig. 2-18.** Formation of methyl 3,4-anhydro-β-D-galactopyranose (2) from methyl 4-O-p-toluenesulfonyl-β-D-glucopyranoside (1). The reaction involves inversion of the ring conformation of 1 from $^4C_1$ (1a) to $^1C_4$ (1b) followed by the elimination of the O-p-toluenesulfonyl group (OTs) and formation of the oxirane with simultaneous inversion of the configuration at C-4. Observe that methyl 4-O-p-toluenesulfonyl-β-D-galactopyranoside cannot form the corresponding oxirane because of the cis orientation of 0-3 and 0-4.

size of epoxides can vary from three- to six-membered rings. The three-membered derivatives belong to the most important subclass of internal ethers and are termed *oxiranes*. A prerequisite for oxirane formation is obviously coplanarity and *trans* position of the reacting groups. Oxirane intermediates are probably formed during alkaline hydrolysis of polysaccharides such as cellulose and starch.

## 2.3.5 Esters

The hydroxyl groups of sugars can form esters both with organic and inorganic acids. The phosphate esters, such as D-glucose 6-phosphate, are important natural products and key intermediates in the biosynthesis and bioconversion of various carbohydrates. For example, O-acetylated monosaccharide units are common in plant polysaccharides and sulfated D-galactopyranosyl units are present in carrageenans of algae.

In synthetic carbohydrate chemistry esterification is mainly used for protection of hydroxyl groups (O-acetyl derivatives) or for introduction of good leaving groups (O-p-toluenesulfonyl groups). Esterification of the hydroxyl groups is brought about with acid anhydrides or halides in the presence of an acid/base catalyst.

The O-acetylation is widely applied in conjunction with GLC analysis of sugars and their derivatives. A convenient separation by HPLC is based on the reversible reaction of sugars to form borate esters. The separation is carried out under slightly alkaline conditions because the borate esters are readily decomposed by acid. In this respect they differ from most esters including acetylated sugars, which are much more readily hydrolyzed by alkali than by acid. Examples of sugar esters are given in Fig. 2-19.

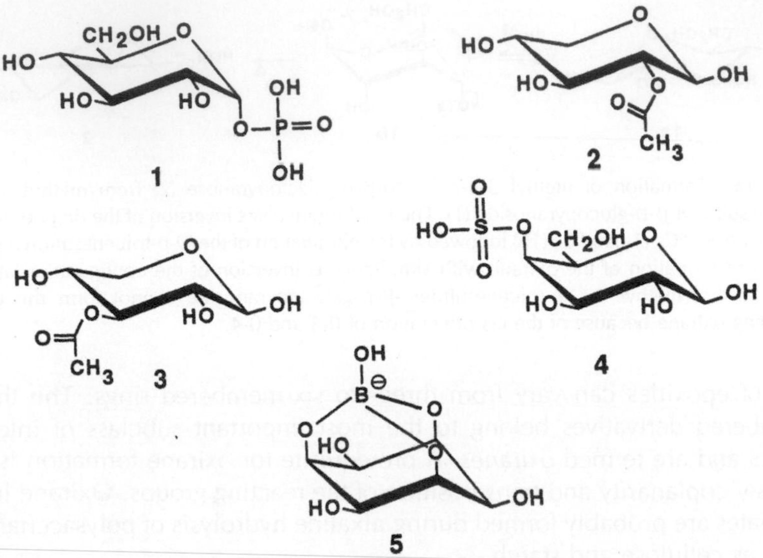

**Fig. 2-19.** Examples of sugar esters. α-D-Glucopyranose 1-phosphate (1) is an intermediate in the biosynthesis of cellulose and other wood polysaccharides. The 2- and 3-O-acetyl-β-D-xylopyranosides (2 and 3) and β-D-galactopyranose 4-sulfate (4) are components of hardwood glucuronoxylan and carrageenan, respectively. cis-Inositol forms an exceptionally stable borate ester (5).

## 2.4 Oligo- and Polysaccharides

More than 500 oligosaccharides are known today, most of them occurring as free natural substances. Oligosaccharides are also obtained by partial acidic or enzymic hydrolysis of polysaccharides. Disaccharides can be considered to be glycosides in which the aglycon part is another monosaccharide. Disaccharides are called *reducing* or *nonreducing*, depending on whether one or both reducing groups are involved in the formation of the glycosidic linkage. *Cellobiose* and *maltose* obtained by partial hydrolysis of cellulose and starch, respectively, are reducing disaccharides (Fig. 2-20). The nonreducing type is exemplified by *sucrose*, which is the most important disaccharide occurring in plants. A large number of various oligosaccharides, up to hexasaccharides, are known.

Polysaccharides are the most abundant constituents of living matter. They are in principle built up in the same manner as oligosaccharides. The chain

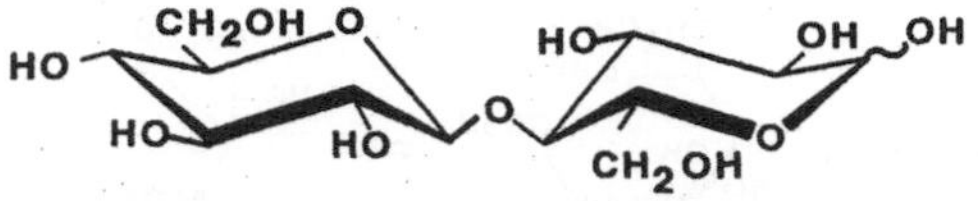

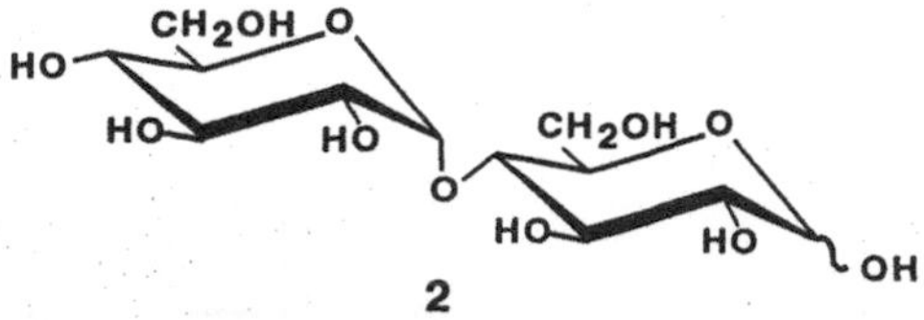

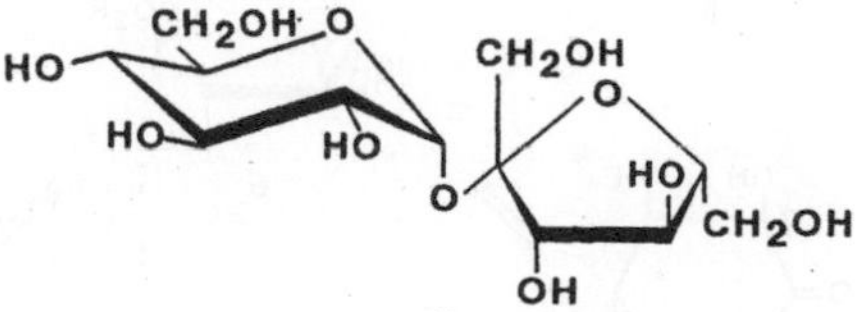

**Fig. 2-20.** 4-O-(β-D-Glucopyranosyl)-D-glucopyranose (cellobiose) (1), 4-O-(α-D-gluco-pyranosyl)-D-glucopyranose (2) (maltose), and α-D-glucopyranosyl-β-D-fructofuranoside (sucrose) (3).

molecules can be either linear or branched, a fact that markedly affects the physical properties of the polysaccharides. The carbohydrate material in plants is largely composed of cellulose and hemicelluloses. Chapter 3 deals with their structure and properties.

## 2.5 Reactions of Carbohydrates

Many reactions of wood polysaccharides (cellulose and hemicelluloses) are described in connection with pulping chemistry (Chapters 7 and 8). The following is therefore restricted to the most important and typical reactions.

### 2.5.1 Oxidation

By mild oxidants, such as aqueous bromine, aldoses (or aldehyde end groups in oligo- and polysaccharides) are oxidized to *aldonic acids* or to corresponding aldonic acid end groups (Fig. 2-21), whereas ketoses are

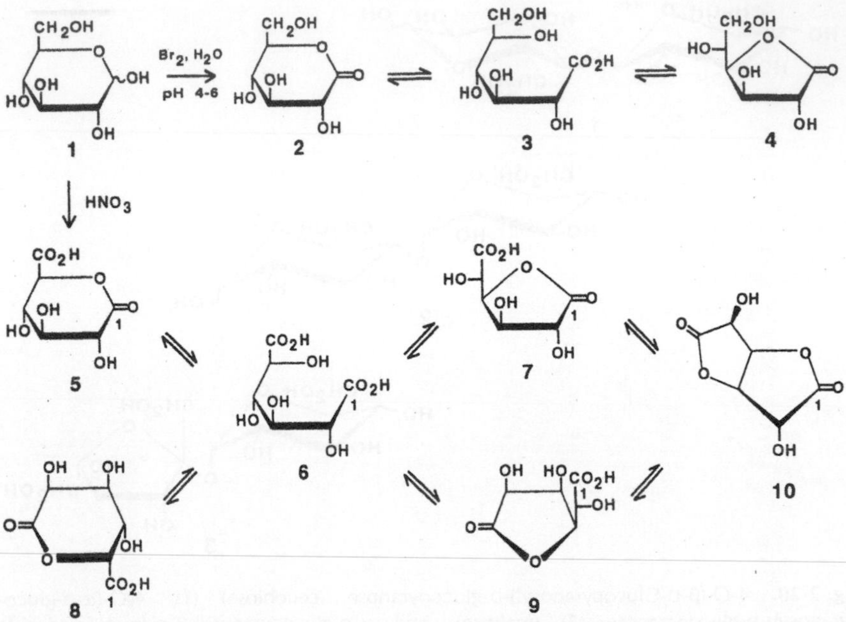

**Fig. 2-21.** Oxidation of D-glucose (1) to D-gluconic acid (3) and D-glucaric acid (6). Bromine water is a mild oxidant, and it primarily oxidizes the cyclic hemiacetal 1 to the corresponding D-glucono-1,5-lactone (2). With prolonged time 2 is hydrolyzed to 3, which is isomerized to D-glucono-1,4-lactone (4). A stronger oxidant, nitric acid, is capable of converting the primary alcohol group to a carboxyl group. This reaction probably first yields 2, which is then further (either directly or after hydrolysis and isomerization to 3 and 4, respectively) oxidized to 6. D-Glucaric acid forms a complex mixture of lactones (5, 7–10).

resistant. On the industrial scale, D-gluconic acid is produced mainly by fermentation from D-glucose. Stronger acids, such as nitric acid, convert aldoses to dicarboxylic acids, termed *aldaric acids*. Aldonic and aldaric acids occur in acidic solution mainly in the form of lactones, which are intramolecular esters. Exclusive oxidation of the primary carbon atom (C-6 in aldohexoses) to a carboxyl group, which can be accomplished using blocking groups, gives *uronic acids* (Fig. 2-22). Uronic acids are important constituents in wood polysaccharides (see Section 3.3).

The neutral oxidation products of carbohydrates include *dialdoses, aldosuloses,* and *glycodiuloses* (Fig. 2-23). They are important intermediates in the synthesis of carbohydrates and are prepared by chemical or enzymic oxidation of hydroxyl groups in the free aldoses or ketoses or their protected derivatives (cf. Section 2.3.2). Uloses can be prepared by oxidation of deriv-

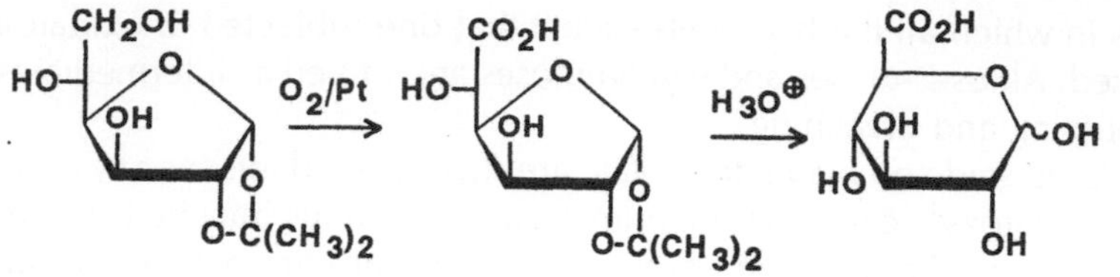

**Fig. 2-22.** Preparation of D-glucuronic acid from D-glucose. The primary hydroxyl group at C-6 is selectively oxidized after protection of the anomeric center (cf. Fig. 2-16). The most useful reagent for oxidation is oxygen in the presence of platinum metal, but potassium permanganate and dinitrogen tetroxide can also be applied.

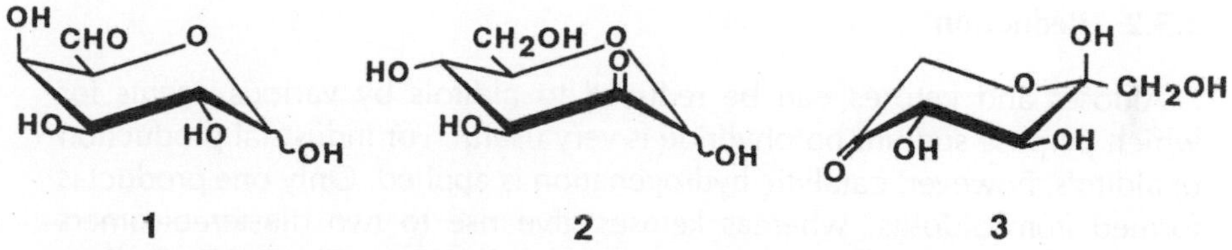

**Fig. 2-23.** Examples of neutral oxidation products of aldoses and ketoses. D-Galactose can be selectively oxidized to *meso-galacto*-hexodialdose (1) by galactose oxidase enzyme. D-Glucose and L-sorbose are oxidized by pyranose-2-oxidase to D-*arabino*-hexos-2-ulose (2) and D-*threo*-2,5-hexodiulose (3), respectively. In water the carbonyl groups are mostly hydrated.

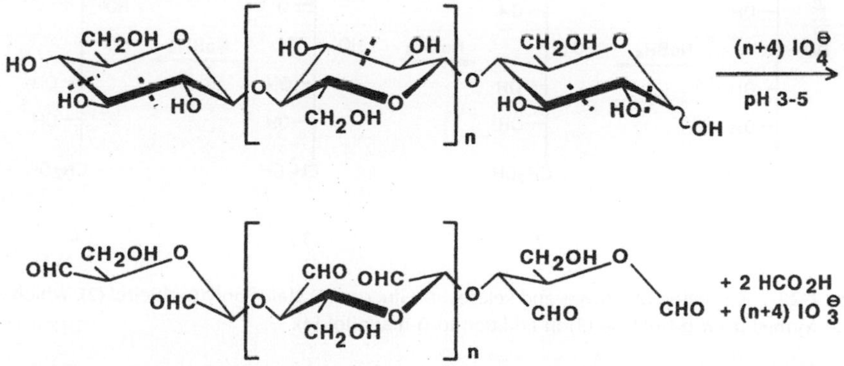

**Fig. 2-24.** Periodate oxidation of 1,4-β-D-glucan (cellulose). The carbon-to-carbon bonds of 1,2-diols are cleaved under consumption of one equivalent of periodate for each bond cleavage. The resulting oxidized polysaccharides can be reduced with sodium borohydride and then hydrolyzed in aqueous acid. Analysis of the hydrolysis products (glycolaldehyde, glycerol, and erythritol) gives additional structural information.

atives in which all the hydroxyls except that one subjected to oxidation are blocked. Aldos-2-uloses and glycodiuloses are formed as intermediates during pulping and bleaching.

*Periodic acid* and *lead tetraacetate* are specific oxidants for any combination of hydroxyl, carbonyl, or primary amine groups attached to adjacent carbon atoms (Fig. 2-24). Primary alcoholic groups are oxidized to formaldehyde, secondary to higher aldehydes, and tertiary to ketones. α-Hydroxyaldehydes are oxidized to formic acid and an aldehyde. Since this specific reaction proceeds quantitatively, it is extremely useful for structural studies (see also Section 2.6).

## 2.5.2  Reduction

Aldoses and ketoses can be reduced to alditols by various agents for which purpose sodium borohydride is very useful. For industrial production of alditols, however, catalytic hydrogenation is applied. Only one product is formed from aldoses, whereas ketoses give rise to two diastereoisomers because of the generation of a new asymmetric center (Fig. 2-25). Sodium borohydride can also be used for reduction of carbonyl groups in polysaccharides.

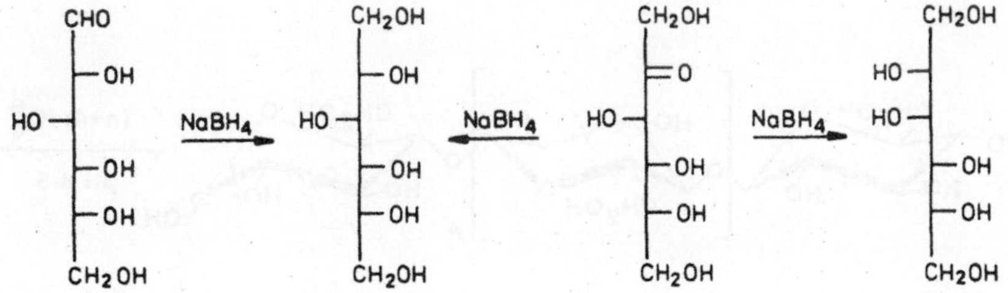

**Fig. 2-25.** Reduction of aldoses and ketoses. D-Glucose (1) yields only D-glucitol (2), which is also formed from D-fructose (3) in addition to D-mannitol (4).

## 2.5.3  Addition and Condensation Reactions of Carbonyl Groups

In the classic carbohydrate chemistry addition reactions of carbonyl groups served as valuable tools for structural studies of carbohydrates. For example, hydroxylamine, hydrazine, and phenylhydrazine react with car-

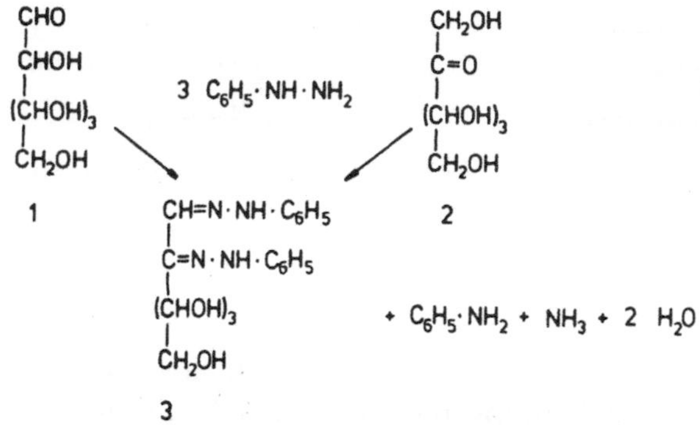

$$
\begin{array}{ccc}
\begin{array}{l}
\text{CHO} \\
|\\
\text{CHOH} \\
|\\
\text{(CHOH)}_3 \\
|\\
\text{CH}_2\text{OH}
\end{array}
&
\xrightarrow{\quad 3\ C_6H_5\cdot NH\cdot NH_2 \quad}
&
\begin{array}{l}
\text{CH}_2\text{OH} \\
|\\
\text{C=O} \\
|\\
\text{(CHOH)}_3 \\
|\\
\text{CH}_2\text{OH}
\end{array}
\end{array}
$$

1                                                     2

$$
\begin{array}{l}
\text{CH=N}\cdot NH\cdot C_6H_5 \\
|\\
\text{C=N}\cdot NH\cdot C_6H_5 \\
|\\
\text{(CHOH)}_3 \\
|\\
\text{CH}_2\text{OH}
\end{array}
\qquad + \ C_6H_5\cdot NH_2 + NH_3 + 2\ H_2O
$$

3

**Fig. 2-26.** Formation of a phenylosazone (3) from an arbitrary aldohexose (1) and ketohexose (2). Observe that the same phenylosazone is formed even if the structures at C-1 and C-2 are different.

bonyl groups to yield oximes and hydrazones. In the presence of an excess of phenylhydrazine, the C-2 position is oxidized to a carbonyl group and a *phenylosazone derivative* is formed (Fig. 2-26). Sugars, which differ only at C-1 and C-2 positions, for example, glucose, mannose, and fructose, give the same osazone (Fischer). Although mainly spectroscopic methods and chromatography are applied today for structural studies and identification

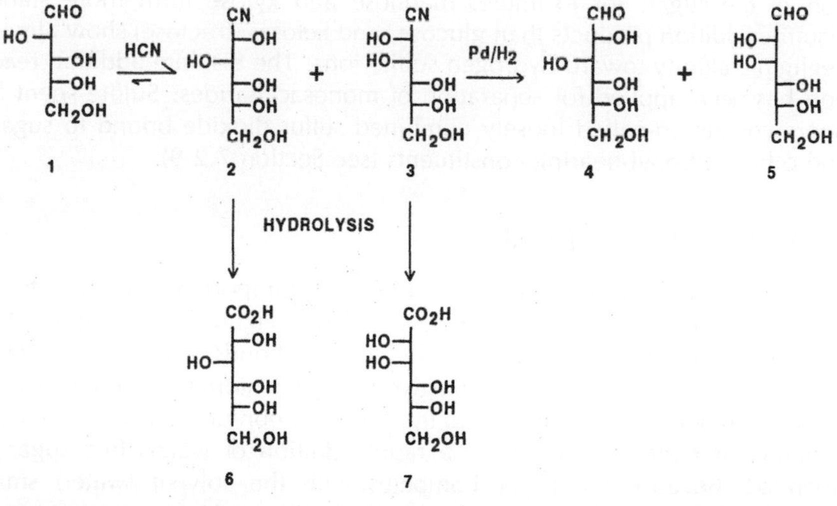

**Fig. 2-27.** Kiliani reaction. Addition of hydrogen cyanide to D-arabinose (1) yields cyanohydrins (2) and (3), which can be hydrogenated to D-glucose (4) and D-mannose (5) or hydrolyzed to D-gluconic acid (6) and D-mannonic acid (7). Because of asymmetric induction, 7 is formed preferentially.

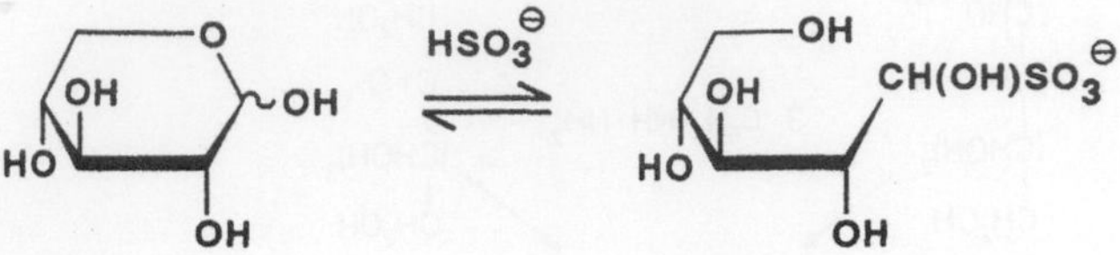

**Fig. 2-28.** Formation of epimeric α-hydroxysulfonic acids from D-xylose in the presence of hydrogen sulfite ions.

purposes, these reagents are useful for simple identification and for determination of carbonyl groups in oxidized cellulose samples.

Cyanide ions react reversibly with sugars to yield *cyanohydrins* (Kiliani) as shown in Fig. 2-27. Because of the formation of a hydroxyl group in place of the aldehyde group, a new asymmetric center is generated. Catalytic hydrogenation of the cyanohydrins gives the corresponding aldoses, and the Kiliani reaction thus opens the possibility for chain lengthening of aldoses. The cyanohydrins can also be hydrolyzed to aldonic acids. It is to be observed that the reaction is subject to so-called *asymmetric induction,* which means that the diastereoisomers are formed in unequal proportions.

Another type of addition reaction is represented by the reaction of hydrogen sulfite ions with sugars giving rise to the formation of α-hydroxysulfonic acids (Fig. 2-28). The equilibrium of this reaction depends on the configuration of the sugar; for example, mannose and xylose form more stable bisulfite addition products than glucose, and ketoses (fructose) show almost negligible affinity toward hydrogen sulfite ions. The bisulfite addition reaction has been applied for separation of monosaccharides. Sulfite spent liquors contain so-called loosely combined sulfur dioxide bound to sugars and other carbonyl-bearing constituents (see Section 7.2.9).

### 2.5.4   The Influence of Acid

The acidic hydrolysis of glycosidic bonds is of importance in many technical processes based on wood as raw material. Figure 2-29 illustrates the mechanism of the acidic cleavage of glycosidic bonds. The reaction starts with a rapid proton addition to the aglycon oxygen atom followed by a slow breakdown of the conjugate acid to the cyclic carbonium ion, which adopts a half-chair conformation. After a rapid addition of water, free sugar is liberated. Because the sugar competes with the solvent (water) small amounts of disaccharides are formed as *reversion* products in concentrated solutions.

The rate of hydrolysis of polysaccharides is affected by several factors.

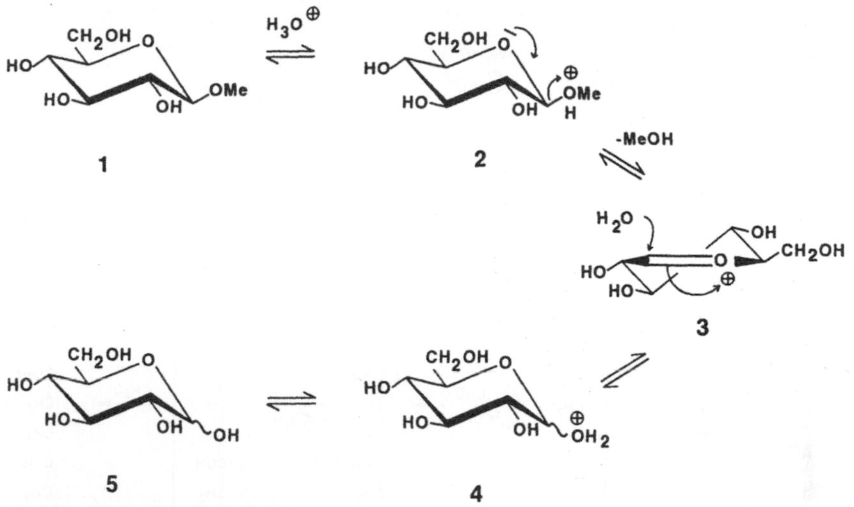

**Fig. 2-29.** A mechanism of the acid-catalyzed hydrolysis of methyl β-D-glucopyranoside (1) to D-glucose (5) involving conjugate acid (2 and 4) and cyclic carbonium ion (3) intermediates.

Because of substituent interaction effects, furanosides are hydrolyzed much more rapidly than the pyranoside analogues. Differences in the hydrolysis rates of diastereomeric glycosides are significant. For example, the relative hydrolysis rates of methyl α-D-gluco-, α-D-manno-, and α-D-galacto-pyranosides are 1.0:2.9:5.0. This can be related to the stabilities of the respective conjugate acids, which are transformed into the half-chair car-bonium ions at different rates. Also, substituents bound to the C-2 position obviously prevent the formation of the half-chair conformation.

Carboxyl groups bound to the polysaccharide chains have a considerable influence on the rate of acid hydrolysis probably mainly because of steric interaction even if inductive effects should also be considered. For example, glycuronides are hydrolyzed more slowly than glycosides. It can be as-sumed that the formation of the intermediate carbonium ion takes place more rapidly at the end than in the middle of the polysaccharide chain. In accordance with this the yield of monosaccharides after partial hydrolysis or sulfite pulping is higher than calculated on the basis of a random bond cleavage.

An opposite reaction to the acid-catalyzed hydrolysis is the above-men-tioned reversion. Acids can also catalyze the formation of anhydro sugars (see Section 2.3.4). Reversion tends to result in formation of (1 → 6)-glycosidic bonds. The degradation of pentoses and uronic acids into furfural

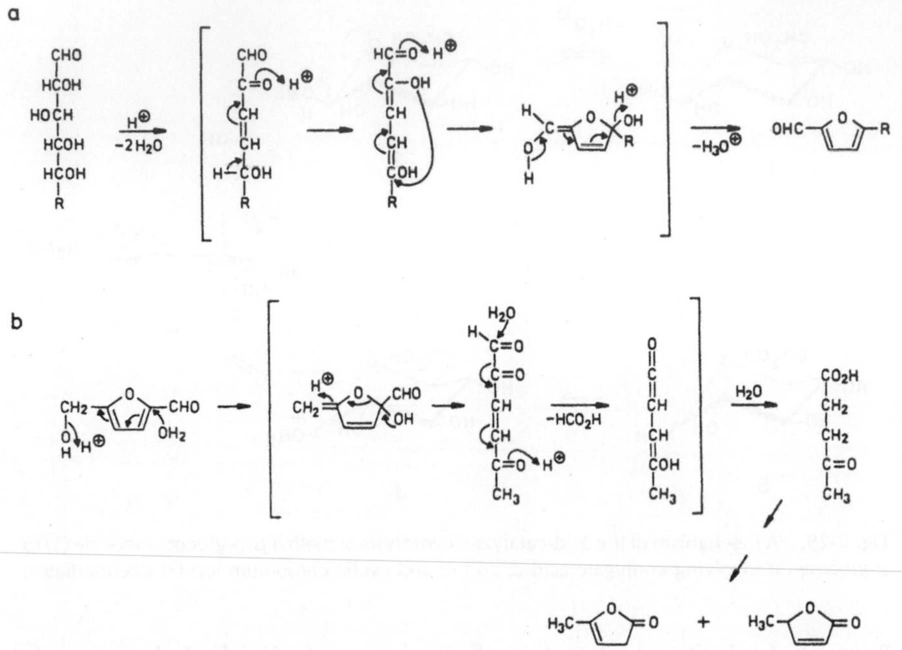

a

b

**Fig. 2-30.**   Reactions of sugars in the presence of concentrated mineral acids. (a) Pentoses (R = H) yield furfural and hexoses (R = $CH_2OH$) hydroxymethylfurfural. (b) On prolonged heating hydroxymethylfurfural is decomposed under liberation of levulinic acid. The rest of the molecule is rearranged to levulinic acid, which can be cyclized to α- and β-angelica lactones.

and of hexoses into hydroxymethylfurfural, levulinic, and formic acids are also important acid-catalyzed reactions, which, however, require concentrated acid and higher temperatures (Fig. 2-30).

## 2.5.5   The Influence of Alkali

In alkaline solutions aldoses and ketoses undergo rearrangements. An example is the *Lobry de Bruyn-Alberda van Ekenstein* transformation of aldoses (Fig. 2-31). This reaction starts by enolization of the aldose to an 1,2-enediol which can be converted to either of the two C-2 epimeric aldoses or to a ketose which also can undergo epimerization. Also aldonic acids are epimerized by alkali, especially in pyridine solutions.

Strong alkali converts monosaccharides, as well as the end groups in polysaccharides, to various carboxylic acids. (1 → 4)-Linked polysaccha-

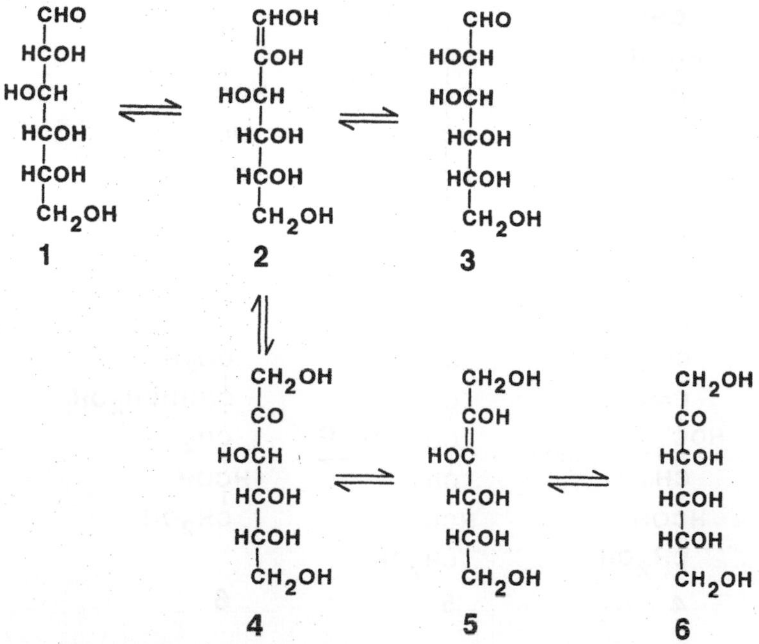

CHO
HCOH
HOCH
HCOH
HCOH
CH₂OH
**1**

⇌

CHOH
COH
HOCH
HCOH
HCOH
CH₂OH
**2**

⇌

CHO
HOCH
HOCH
HCOH
HCOH
CH₂OH
**3**

⇵

CH₂OH
CO
HOCH
HCOH
HCOH
CH₂OH
**4**

⇌

CH₂OH
COH
HOC
HCOH
HCOH
CH₂OH
**5**

⇌

CH₂OH
CO
HCOH
HCOH
HCOH
CH₂OH
**6**

**Fig. 2-31.** Lobry de Bruyn–Alberda van Ekenstein transformation of sugars. D-Glucose (1), 1,2-enediol (2), D-mannose (3), D-fructose (4), 2,3-enediol (5), and D-allulose (6).

rides, including cellulose and most hemicelluloses, are degraded by an endwise mechanism, known as the *peeling* reaction. This reaction occurs during alkaline pulping and bleaching processes, for example in kraft pulping and oxygen bleaching (see Sections 7.3.5 and 8.1.5). The reaction mechanism is outlined in Fig. 2-32. The degradation starts with the isomerization of the end group to a ketose in which the glycosidic bond is in the β position with respect to the carbonyl group. Since such a structure is labile in alkali, the glycosidic bond is cleaved with removal of the end group. This is termed "β-*alkoxy elimination*." The eliminated end group is tautomerized to a 4-deoxy-2,3-glycodiulose, which then undergoes benzilic acid rearrangement to epimeric isosaccharinic acids. In addition, a number of other acids are formed by competing mechanisms. In kraft pulping, the cellulose molecules are subjected to this endwise peeling, which results in a loss of about fifty glucose units from a single molecule. The peeling process is terminated by a so-called *stopping* reaction involving a direct β-*hydroxy elimination* from the C-3 position (Fig. 2-33). The end group undergoes a

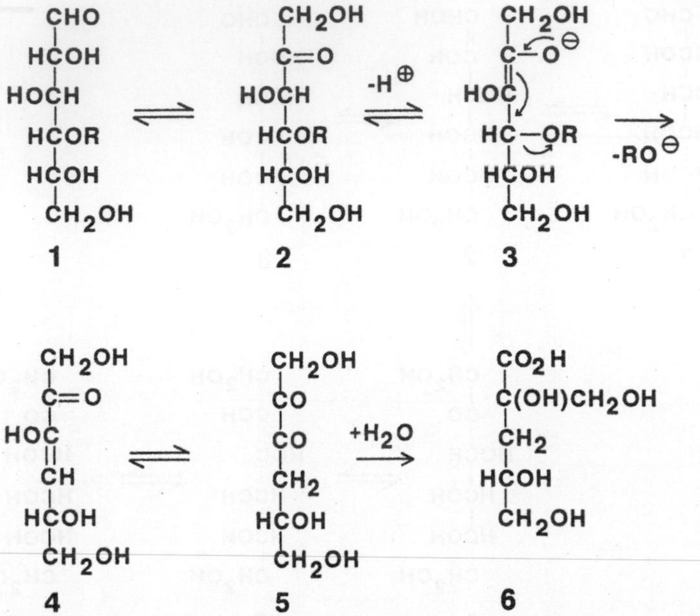

CHO
HCOH
HOCH
HCOR
HCOH
CH₂OH
**1**

⇌

CH₂OH
C=O
HOCH
HCOR
HCOH
CH₂OH
**2**

$-H^{\oplus}$ ⇌

CH₂OH
C—O$^{\ominus}$
HOC
HC—OR
HCOH
CH₂OH
**3**

$-RO^{\ominus}$ →

CH₂OH
C=O
HOC
‖
CH
HCOH
CH₂OH
**4**

⇌

CH₂OH
CO
CO
CH₂
HCOH
CH₂OH
**5**

$+H_2O$ →

CO₂H
C(OH)CH₂OH
CH₂
HCOH
CH₂OH
**6**

**Fig. 2-32.**  The endwise alkaline degradation ("peeling") of 1,4-β-D-glucan (cellulose). [R = glucan (cellulose) chain]. Reaction steps: isomerization (1 → 2), enediol formation (2 → 3), β-alkoxy elimination (3 → 4), tautomerization (4 → 5), and benzilic acid rearrangement (5 → 6) to epimeric 3-deoxy-2-C-hydroxymethylpentonic acids (glucoisosaccharinic acid) (6).

benzilic acid rearrangement to an alkali-stable metasaccharinic acid end group. Other end groups are, however, also formed. The 3-O-substituted glycosides of (1 → 4)-linked polysaccharides are rapidly stabilized in alkali through benzilic acid rearrangement because the β-alkoxy elimination takes place much easier than the β-hydroxy elimination.

The cleavage of glycosidic bonds by alkali is usually extremely slow in comparison with the acid-catalyzed hydrolysis. A suggested mechanism for this reaction is depicted in Fig. 2-34. Ionization of the C-2 hydroxyl group and conformational change results in the formation of a three-membered epoxide (oxirane) ring under simultaneous cleavage of the glycosidic bond (elimination of the alkoxy group). The opening of the oxirane ring results in the formation of a free reducing end group in the polysaccharide chain (or free sugar), or, if the steric requirements are fulfilled, a 1,6-anhydride. The mechanism explains why 1,2-trans-glycosides are more reactive than the 1,2-cis-anomers.

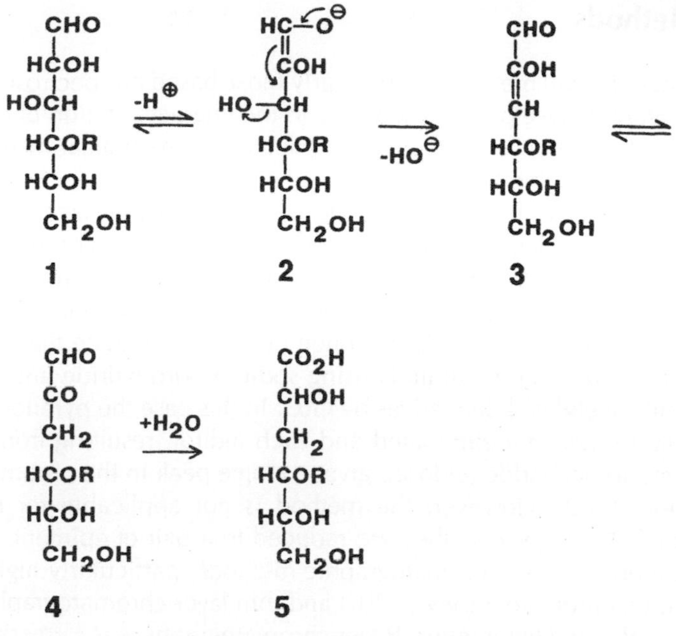

**Fig. 2-33.** Termination of the peeling reaction ("stopping reaction"). Reaction steps: 1,2-En-ediol formation (1 → 2), β-hydroxy elimination (2 → 3), tautomerization (3 → 4), and benzilic acid rearrangement (4 → 5) to epimeric 3-deoxyhexonic acid end groups (glucometasaccharinic acid) (5).

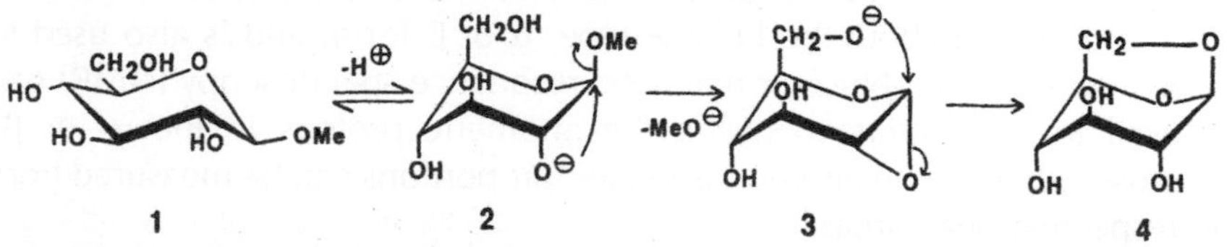

**Fig. 2-34.** Base-catalyzed hydrolysis of methyl β-D-glucopyranoside (1). The reaction starts with inversion of the ring conformation from $^4C_1$ to $^1C_4$ (2). The ionized hydroxyl group at C-2 attacks C-1, and the trans-oriented methoxyl group is eliminated. The resulting 1,2-anhydro-α-D-glucopyranose (3) is decomposed to different products directly or via 1,6-anhydro-β-D-glucopyranose (levoglucosan) (4).

## 2.6   Methods

A number of novel methods, particularly those based on spectroscopy and chromatography, have played a key role in the characterization of carbohydrates and elucidation of their structural details. Some typical methods are briefly discussed in the following.

Monosaccharides liberated on acidic hydrolysis of polysaccharides can be identified and determined by combining gas-liquid chromatography (GLC) and mass spectrometry (MS). For this purpose the monosaccharides must be transformed into volatile derivatives, such as trimethylsilyl ethers (TMS). For quantitative analysis it is often best first to reduce the monosaccharides to corresponding alditols using sodium borohydride and to separate the fully acetylated derivatives by GLC. In this case the pyranosidic and furanosidic isomers are eliminated and each alditol, resulting from the respective monosaccharide (aldose), gives a single peak in the chromatogram (cf. Section 2.5.2). However, the method is not applicable for mixtures containing ketoses because they are reduced to a pair of epimeric alditols. Also, a variety of other chromatographic methods, particularly high performance liquid chromatography (HPLC) and thin layer chromatography (TLC), are commonly used techniques. Paper chromatography was formerly widely used but has now been replaced by more sensitive methods.

A simple test with Fehling's solution is sufficient to determine whether the carbohydrate is of the reducing or nonreducing type. The etherification, usually methylation, of the free hydroxyl groups followed by hydrolysis and GLC or GLC-MS of the fragments provides information both on the position of the linkages as well as on the ring sizes. Additional evidence can be obtained by selective oxidation methods, including oxidation with periodate or lead tetraacetate.

Carbohydrates in nature are optically active and polarimetry is widely used in establishing their structure. Measurement of the specific rotation gives information about the linkage type ($\alpha$ or $\beta$ form) and is also used to follow mutarotation. Nuclear magnetic resonance spectroscopy (NMR) can be used to differentiate between the anomeric protons in the $\alpha$- or $\beta$-pyranose and furanose anomers and their proportions can be measured from the respective peak areas.

Enzymic methods are gaining wider applications. For example, maltase hydrolyzes $\alpha$-glucosidic linkages, whereas emulsin is specific for $\beta$-glucosidic bonds.

# Chapter 3

# WOOD POLYSACCHARIDES

## 3.1  Biosynthesis

As a result of photosynthesis glucose is produced in the foliage of trees from carbon dioxide and water, and then transported in the phloem to the cambial tissues. It is the basic monomer from which the wood polysaccharides are formed through a variety of biosynthetic pathways. Pioneering work by Leloir (Nobel Prize in 1970) led to the discovery of an important sugar nucleotide, UDP-D-glucose, from which cellulose is synthesized. Another important sugar nucleotide, participating in the synthesis of hemicelluloses, is GDP-D-glucose. The nucleoside moieties of these sugar nucleotides are uridine and guanosine (Fig. 3-1). The sugar nucleotides are formed from the corresponding nucleoside triphosphates and $\alpha$-D-glucose 1-phosphate by an enzymic process as illustrated for UDP-D-glucose in Fig. 3-2.

Cellulose is synthesized from UDP-D-glucose the energy content of which is used for the formation of glucosidic bonds in the growing polymer:

$$\text{UDP-D-glucose} + [(1 \rightarrow 4)\text{-}\beta\text{-D-glucosyl}]_n \rightarrow$$

$$[(1 \rightarrow 4)\text{-}\beta\text{-D-glucosyl}]_{n+1} + \text{UDP}$$

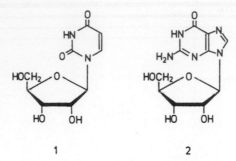

**Fig. 3-1.** Structure of two nucleosides, 1, uridine and 2, guanosine, which both contain β-D-ribofuranose residue. The aglycon part is derived from a pyrimidine and purine base, respectively.

Present concepts of the biosynthesis of cellulose are outlined in Fig. 3-3. In the synthesis of other wood polysaccharides both UDP-D-glucose and GDP-D-glucose are involved, the latter being the principal nucleotide as concerns the formation of mannose-containing hemicelluloses (galacto-glucomannans and glucomannans). The monomeric sugar components

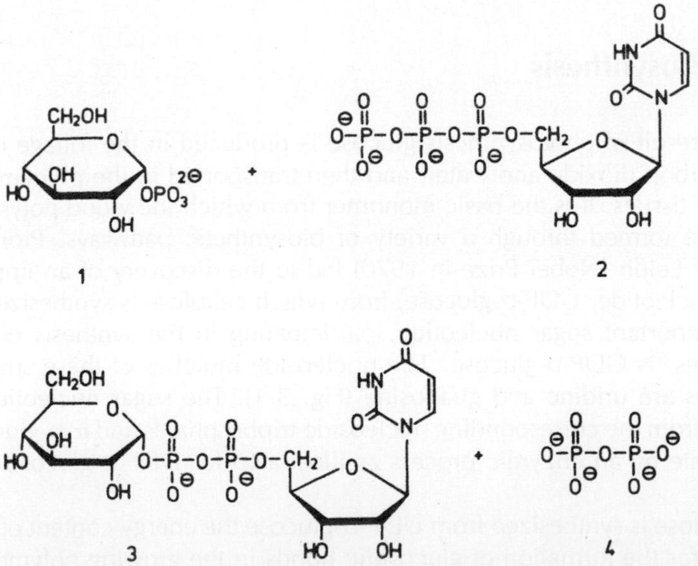

**Fig. 3-2.** Formation of uridine diphosphate glucose (UDP-D-glucose) (3) from α-D-glucopyranose 1-phosphate (1) and uridine triphosphate (2) under simultaneous release of pyrophosphate (4).

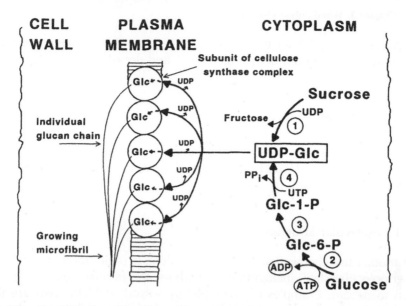

**CELL WALL**   **PLASMA MEMBRANE**   **CYTOPLASM**

Subunit of cellulose synthase complex

Glc ← UDP

Sucrose

Glc ← UDP

Fructose ← UDP (1)

Individual glucan chain →

Glc ← UDP

**UDP-Glc**

Glc ← UDP

PP$_i$ ← (4)

UTP

**Glc-1-P**

Glc ← UDP

(3)

**Glc-6-P**

Growing microfibril →

ADP ← (2)

ATP **Glucose**

**Fig. 3-3.** Hypothetical model of the mechanism of cellulose synthesis in plants. The active glucosyl precursor (UDP-glucose) is produced in cytoplasm from two sources: from sucrose by sucrose synthase (1) (reversible action) and from glucose by successive reactions which are catalyzed by hexokinase (2), phosphoglucomutase (3), and UDP-glucose pyrophosphorylase (4). After penetration into the plasma membrane, UDP-glucose transfers its glucosyl residue to the growing glucan (cellulose) chain under release of UDP. This coupling is catalyzed by active sites on subunits (symbolized by circles) of a cellulose synthase complex embedded in the plasma membrane. The glucan chains derived from one complex are assumed to associate by hydrogen bonding to form a microfibril, the size of which may vary among different cell types. As the synthesis proceeds, the orientation of the microfibrils may be determined by the movement of the complex in the fluid lipid bilayer. Such movement may be directed by microtubules found on the inner face of the plasma membrane (based upon review of Delmer, 1987).

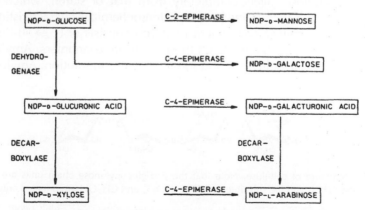

**Fig. 3-4.** Simplified representation of the formation of hemicellulose precursors from UDP-D-glucose or GDP-D-glucose. Note that NDP (nucleotide diphosphate) means either UDP or GDP.

needed are formed from the nucleotides by complex enzymic reactions involving epimerization, dehydrogenation, and decarboxylation (Fig. 3-4).

## 3.2  Cellulose

Cellulose is the main constituent of wood. Approximately 40–45% of the dry substance in most wood species is cellulose, located predominantly in the secondary cell wall (cf. Appendix).

### 3.2.1  Molecular Structure

Although the chemical structure of cellulose is understood in detail, its supermolecular state, including its crystalline and fibrillar structure, is still open to debate. Examples of incompletely solved problem areas are the exact molecular weight and polydispersity of native cellulose and the dimensions of the microfibrils.

Cellulose is a homopolysaccharide composed of β-D-glucopyranose units which are linked together by (1 → 4)-glycosidic bonds (Fig. 3-5). Cellulose molecules are completely linear and have a strong tendency to form intra- and intermolecular hydrogen bonds. Bundles of cellulose molecules are thus aggregated together in the form of microfibrils, in which highly ordered (crystalline) regions alternate with less ordered (amorphous) regions. Microfibrils build up fibrils and finally cellulose fibers. As a consequence of its fibrous structure and strong hydrogen bonds cellulose has a high tensile strength and is insoluble in most solvents. The physical and chemical behavior of cellulose differs completely from that of starch, which clearly demonstrates the unique influence of stereochemical characteristics. Like cellulose, the amylose component of starch consists of (1 → 4)-linked D-glucopyranose units, but in starch these units are α-anomers. Amylose occurs as a helix in its solid state and sometimes also in solution. Amylopectin,

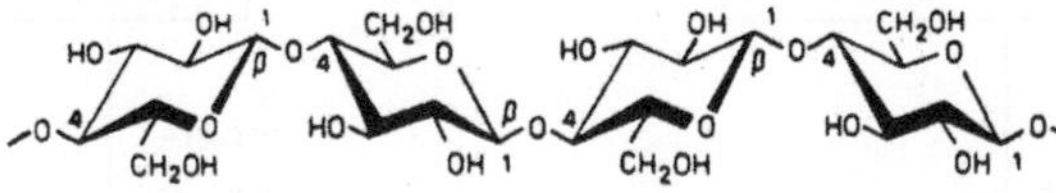

**Fig. 3-5.**  Structure of cellulose. Note that the β-D-glucopyranose chain units are in chair conformation ($^4C_1$) and the substituents HO-2, HO-3, and $CH_2OH$ are oriented equatorially.

the other starch component, is also a (1 → 4)-α-D-glucan but is highly branched. The branched structure accounts for its extensive solubility, since no aggregation can take place.

The crystalline structure of cellulose has been characterized by X-ray diffraction analysis and by methods based on the absorption of polarized infrared radiation. The unit cell of native cellulose (cellulose I) consists of four D-glucose residues (Figs. 3-6 and 3-7). In the chain direction (c), the repeating unit is a cellobiose residue (1.03 nm), and every glucose residue is accordingly displaced 180° with respect to its neighbors, giving cellulose a 2-fold screw axis. It has now been established and largely accepted that all chains in native cellulose microfibrils are oriented in the same direction, that is, they are parallel (Fig. 3-7). There are two hydrogen bonds within each cellulose chain, namely from O(6) in one glucose residue to O(2)H in the adjacent glucose and also from O(3)H to the ring oxygen, as shown in Fig. 3-8. The chains form a layer in the a-c crystallographic plane, where they are held together by hydrogen bonds from O(3) in one chain to O(6)H in the other. There are no hydrogen bonds in cellulose I between these layers, only weak van der Waals forces in the direction of the b-axis.

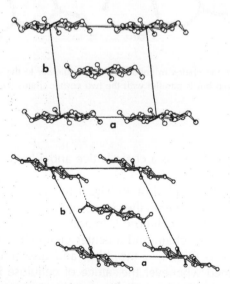

**Fig. 3-6.** Axial projections of the structures of native cellulose (cellulose I, above) and regenerated cellulose (cellulose II, below). (Reproduced from Kolpak et al., 1978, Polymer **19**, 123–131, by permission of the publishers, IPC Business Press Ltd. ©.)

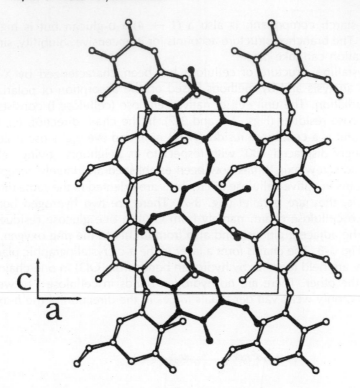

**Fig. 3-7.** Projection of the chains in cellulose I perpendicular to the a-c plane. The center chain (black) is staggered but is parallel with the two corner chains (Gardner and Blackwell, 1974).

Native cellulose therefore has a chain lattice and a layer lattice at the same time.

Regenerated cellulose (cellulose II) (Fig. 3-6) has antiparallel chains (Fig. 3-9). The hydrogen bonds within the chains and between the chains in the a-c plane are the same as in cellulose I. In addition, there are two hydrogen bonds between a corner chain and a center chain (Fig. 3-6), namely from O(2) in one chain to O(2)H in the other and also from O(3)H to 0(6). Cellulose II is formed whenever the lattice of cellulose I is destroyed, for example on swelling with strong alkali or on dissolution of cellulose. Since the strongly hydrogen bonded cellulose II is thermodynamically more stable than cellulose I, it cannot be reconverted into the latter. All naturally occur-

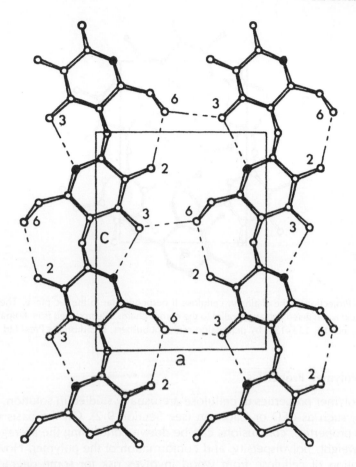

**Fig. 3-8.**   Projection of the (O2O) plane in cellulose I, showing the hydrogen bonding network and the numbering of the atoms. Each glucose residue forms two intramolecular hydrogen bonds (03-H···05' and 06···H-02') and one intermolecular bond (06-H···03). (Slightly modified from Gardner and Blackwell, 1974).

ring cellulose has the structure of cellulose I. Celluloses III and IV are produced when celluloses I and II are subjected to certain chemical treatments and heating.

The proportions of ordered and disordered regions of cellulose vary considerably depending on the origin of the sample (cf. Table 9-1). Cotton cellulose is more crystalline than cellulose in wood.

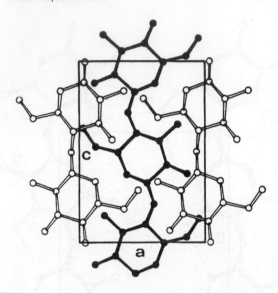

**Fig. 3-9.** Projection of the chains in cellulose II perpendicular to the a-c plane. The center chain (black) is staggered and antiparallel to the corner chains. (Reproduced from Kolpak *et al.*, 1978, *Polymer* **19**, 123–131, by permission of the publishers, IPC Business Press Ltd. ©.)

### 3.2.2   Polymer Properties

The polymer properties of cellulose are usually studied in solution, using solvents, such as CED or Cadoxen (see Section 9.2). On the basis of the solution properties, conclusions can be drawn concerning the average molecular weight, polydispersity, and conformation of the polymer. However, the isolation of cellulose from wood involves risk for some degradation resulting in a reduced molecular weight.

The distribution of molecular weights can be presented statistically as illustrated by Fig. 3-10 where the weight of polymer of a given size is plotted against the chain length. The experimental measurements give an average value of the molecular weight and some methods also give a molecular weight distribution. For any polydisperse system, these average values differ from each other depending on the method used. The *number average* molecular weight $\bar{M}_n$* can be measured using osmometry or by determining the number of reducing end groups. The *weight average* molecular weight $\bar{M}_w$

---

*The SI system (Système International d'Unités) recommends the term *relative molecular mass* instead of *molecular weight,* but because the SI term is not yet universally adopted in the polymer chemistry the latter term is used throughout this book.

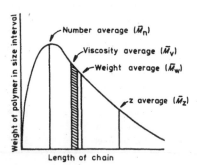

**Fig. 3-10.**  The molecular weight distribution and the average molecular weights of a typical polymer (Billmeyer, 1965).

can be deduced from light scattering data. Sedimentation equilibrium data attainable by ultracentrifugation technique give so-called $\bar{M}_z$ values. Finally, $\bar{M}_v$ refers to the molecular weight calculated on the basis of viscosity measurements. For cellulose, the relationship between molecular weight and degree of polymerization (DP) is DP = M/162, where 162 is the molecular weight of anhydroglucose unit. The ratio $\bar{M}_w/\bar{M}_n$ is a measure of polydispersity corresponding to the width of the molecular weight distribution and ranges for typical polymers from 1.5–2.0 to 20–50.

Molecular weight measurements have shown that cotton cellulose in its native state consists of about 15,000 and wood cellulose of about 10,000 glucose residues. Some polydispersity data on cellulose derivatives and polysaccharides are shown in Table 3-1. It has been suggested in the literature that the native cellulose present in the secondary cell wall of plants is monodisperse, that is, contains only molecules of one size but this does not seem probable. In such a case, number and weight average molecular weights ought to be identical. In any case, the cellulose in the primary cell wall, which has a lower average molecular weight, is evidently polydisperse, being similar in this respect to the hemicelluloses.

Based on properties in solution such as intrinsic viscosity and sedimentation and diffusion rates, conclusions can be drawn concerning the polymer conformation. Like most of the synthetic polymers, such as polystyrene, cellulose in solution belongs to a group of linear, randomly coiling polymers. This means that the molecules have no preferred structure in solution in contrast to amylose and some protein molecules which can adopt helical

**TABLE 3-1.   Polydispersity Values ($\bar{M}_w/\bar{M}_n$) of Different Polysaccharides[a]**

| Macromolecule | Source | $\bar{M}_w \times 10^{-5}$ | $\bar{M}_w/\bar{M}_n$ |
|---|---|---|---|
| Cellulose nitrate | Birch | 27[b] | 1.9[c] |
| Cellulose nitrate | Ramie | 24[b] | 1.7[c] |
| Amylose | Potato | 8.8[d] | 1.9[d] |
| Xylan | Birch | 0.8[b] | 2.3[e] |
| Xylan | Elm | 0.7[b] | 2.4[e] |
| Amylopectin | Waxy corn | 1700[b] | 116[f] |
| Hydrolyzed amylopectin | Waxy corn | 15[b] | 25[f] |
| Glycogen | Sweet corn | 190[b] | 15[f] |
| Hydrolyzed glycogen | Sweet corn | 20[b] | 6.3[f] |
| Glycogen | Rabbit liver | 390[b] | 6.6[g] |

[a] From Goring (1962).
[b] $\bar{M}_w$ by light scattering.
[c] $\bar{M}_n$ by viscometry from $[\eta] = 0.0091\ DP$.
[d] Calculated from fractionation data.
[e] $\bar{M}_n$ by osmometry.
[f] $\bar{M}_n$ from the alkali number.
[g] $\bar{M}_{ww}$ from sedimentation and diffusion used instead of $\bar{M}_n$; usually $\bar{M}_w > \bar{M}_{ww} > \bar{M}_n$.

conformations. Cellulose differs distinctly from synthetic polymers and from lignin in some of its polymer properties. Typical of its solutions are the comparatively high viscosities and low sedimentation and diffusion coefficients (Tables 3-2 and 3-3).

Any linear polymer molecule, even a reasonably stiff rod, will coil ran-

**TABLE 3-2.   Examples of Intrinsic Viscosities Corresponding to a Molecular Weight ($\bar{M}_w$) of 50,000 for Various Macromolecules[a]**

| Macromolecule | Solvent | Intrinsic viscosity $[\eta]$ ($dm^3/kg$) |
|---|---|---|
| Dioxane-HCl lignin | Pyridine | 8 |
| Lignosulfonate | 0.1 M NaCl | 5 |
| Kraft lignin | Dioxane | 6 |
| Alkali lignin | 0.1 M buffer | 4 |
| Polymethylmethacrylate | Benzene | 23 |
| Polymethylstyrene | Toluene | 24 |
| Xylan | CED | 216 |
| Cellulose | CED | 181 |

[a] From Goring (1971).

**TABLE 3-3.  Typical Polymer Parameters of Cellulose and Hydroxyethyl Cellulose Compared with Polyvinyl Acetate**[a]

| Sample | Solvent | $DP_w$ | $[\eta]$ (dm³/kg) | Sedimentation rate, $s_0$ (Svedbergs) | Diffusion rate, $D_0 \times 10^7$ |
|--------|---------|--------|-------------------|----------------------------------------|------------------------------------|
| Cellulose | Cadoxen | 2290 | 645 | 3.8 | 0.58 |
| Hydroxyethyl cellulose | Water | 2180 | 895 | 4.9 | 0.96 |
| Polyvinyl acetate | 2-Butanone | 4880 | 125 | 13.5 | 2.40 |

[a] From Brown (1966). © 1966. TAPPI. Reprinted from *Tappi* **49**(8), pp. 367–368, with permission.

domly, provided the chain is sufficiently long. In addition to the size of the monomer units, the tendency for coiling is affected by the forces between the units as well as the interaction between the polymer and the solvent. One measure of the stiffness of a polymer molecule is the end-to-end distance ($R$). For a polydisperse polymer the root-mean-square average of $R$ $(R^2)^{1/2}$ is used. $R$ is affected by the properties of the polymer itself as well as by the interaction of the solvent. The better the solvent the more the polymer swells, whereas in a poor solvent contraction occurs (Fig. 3-11). The expansion tendency of a polymer molecule is characterized by Flory's equation $R = \alpha R_0$, where $\alpha$ is the expansion coefficient. At a certain temperature in a given solvent an "ideal" state ($R = R_0$) can be reached, where the environment has no observable influence on the polymer. Such a solvent is called a theta solvent and the temperature the theta or Flory temperature.

**Fig. 3-11.**  Schematic representation of randomly coiling macromolecules in solution. In a good solvent (right) the interaction with the polymer is thermodynamically favorable, resulting in expansion whereas in a poor solvent the coil is rather compact (left) (Brown, 1966). © 1966. TAPPI. Reprinted from *Tappi* **49**(8), pp. 367–368, with permission.

The intrinsic viscosity of a polymer such as cellulose is related to the molecular weight by the Mark-Houwink equation:

$$[\eta] = K\bar{M}_v^a$$

where the coefficient $K$ and the exponent $a$ are experimental constants characteristic for the solvent and polymer type and $\bar{M}_v$ the molecular weight. The value of exponent $a$ may vary according to the conformation of the polymer and is 0.6–0.8 for both cellulose and many synthetic polymers, such as polystyrene (cf. Table 3-4). For some cellulose derivatives like nitrates, $a$ approaches the value 1.0.

At theta conditions where the molecule exists in the form of a completely closed-up (compact) random coil ($R = R_0$) $a$ is 0.5. In a good solvent $a$ is 0.8 and maximally about 1.0. For rods $a$ is 1.8.

The intrinsic viscocity $[\eta]$ is a measure of the effective hydrodynamic volume of the molecule. For a hard, nonswelling sphere, Einstein's equation is valid:

$$[\eta] = 0.025\bar{V}$$

where $\bar{V}$ is the specific volume of the material in the sphere. In the case of linear, solvent-swollen polymers such as cellulose, $V$ and thus also $[\eta]$ is much larger. A low $[\eta]$ value means that the molecule is compact and thus occupies a relatively small volume. This is typical of lignin (see Section 4.4).

**TABLE 3-4. Some Values for the Exponent $a$[a]**

| Sample | Solvent | $a$ |
|---|---|---|
| Dioxane-HCl lignin | Pyridine | 0.15 |
| Alkali lignin | Dioxane | 0.12 |
| Lignosulfonate | 0.1 M NaCl | 0.32 |
| Polymethylstyrene | Toluene | 0.71 |
| Cellulose | Cadoxen | 0.75 |
| Cellulose nitrate | Ethyl acetate | 0.99 |
| Xylan | DMSO | 0.94 |
| Xylan | CED | 1.15 |
| Einstein sphere | | 0 |
| Compact coil | | 0.5 |
| Free-draining coil | | 1 |
| Rods | | 1.8 |

[a] From Goring (1971).

# 3.3    Hemicelluloses

Hemicelluloses were originally believed to be intermediates in the biosynthesis of cellulose. Today it is known, however, that hemicelluloses belong to a group of heterogeneous polysaccharides which are formed through biosynthetic routes different from that of cellulose. In contrast to cellulose which is a homopolysaccharide, hemicelluloses are heteropolysaccharides. Like cellulose most hemicelluloses function as supporting material in the cell walls. Hemicelluloses are relatively easily hydrolyzed by acids to their monomeric components consisting of D-glucose, D-mannose, D-galactose, D-xylose, L-arabinose, and small amounts of L-rhamnose in addition to D-glucuronic acid, 4-O-methyl-D-glucuronic acid, and D-galacturonic acid. Most hemicelluloses have a degree of polymerization of only 200.

Some wood polysaccharides are extensively branched and are readily soluble in water. Typical of certain tropical trees is a spontaneous formation of exudate gums, which are exuded as viscous fluids at sites of injury and after dehydration give hard, clear nodules rich in polysaccharides. These gums, for example, gum arabic, consist of highly branched, water-soluble polysaccharides.

The amount of hemicelluloses of the dry weight of wood is usually between 20 and 30% (cf. Appendix). The composition and structure of the hemicelluloses in the softwoods differ in a characteristic way from those in the hardwoods. Considerable differences also exist in the hemicellulose content and composition between the stem, branches, roots, and bark.

Table 3-5 summarizes the main structural features of the hemicelluloses appearing in both softwoods and hardwoods.

### 3.3.1    Softwood Hemicelluloses

*Galactoglucomannans*    Galactoglucomannans are the principal hemicelluloses in softwoods (about 20%). Their backbone is a linear or possibly slightly branched chain built up of (1 → 4)-linked β-D-glucopyranose and β-D-mannopyranose units (Fig. 3-12). Galactoglucomannans can be roughly divided into two fractions having different galactose contents. In the fraction which has a low galactose content the ratio galactose:glucose:mannose is about 0.1:1:4 whereas in the galactose-rich fraction the corresponding ratio is 1:1:3. The former fraction with a low galactose content is often referred to as glucomannan. The α-D-galactopyranose residue is linked as a single-unit side chain to the framework by (1 → 6)-bonds. An important structural feature is that the hydroxyl groups at C-2 and C-3 positions in the chain units

**TABLE 3-5. The Major Hemicellulose Components**

| Hemicellulose type | Occurrence | Amount (% of wood) | Composition Units | Molar ratios | Linkage | Solubility[a] | $\overline{DP}_n$ |
|---|---|---|---|---|---|---|---|
| Galactoglucomannan | Softwood | 5-8 | β-D-Manp | 3 | 1 → 4 | Alkali, water* | 100 |
| | | | β-D-Glcp | 1 | 1 → 4 | | |
| | | | α-D-Galp | 1 | 1 → 6 | | |
| | | | Acetyl | 1 | | | |
| (Galacto)glucomannan | Softwood | 10-15 | β-D-Manp | 4 | 1 → 4 | Alkaline borate | 100 |
| | | | β-D-Glcp | 1 | 1 → 4 | | |
| | | | α-D-Galp | 0.1 | 1 → 6 | | |
| | | | Acetyl | 1 | | | |
| Arabinoglucuronoxylan | Softwood | 7-10 | β-D-Xylp | 10 | 1 → 4 | Alkali, dimethylsulfoxide*, water* | 100 |
| | | | 4-O-Me-α-D-GlcpA | 2 | 1 → 2 | | |
| | | | α-L-Araf | 1.3 | 1 → 3 | | |
| Arabinogalactan | Larch wood | 5-35 | β-D-Galp | 6 | 1 → 3, 1 → 6 | Water | 200 |
| | | | α-L-Araf | 2/3 | 1 → 6 | | |
| | | | β-L-Arap | 1/3 | 1 → 3 | | |
| | | | β-D-GlcpA | Little | 1 → 6 | | |
| Glucuronoxylan | Hardwood | 15-30 | β-D-Xylp | 10 | 1 → 4 | Alkali, dimethylsulfoxide* | 200 |
| | | | 4-O-Me-α-D-GlcpA | 1 | 1 → 2 | | |
| | | | Acetyl | 7 | | | |
| Glucomannan | Hardwood | 2-5 | β-D-Manp | 1-2 | 1 → 4 | Alkaline borate | 200 |
| | | | β-D-Glcp | 1 | 1 → 4 | | |

[a] The asterisk represents a partial solubility.

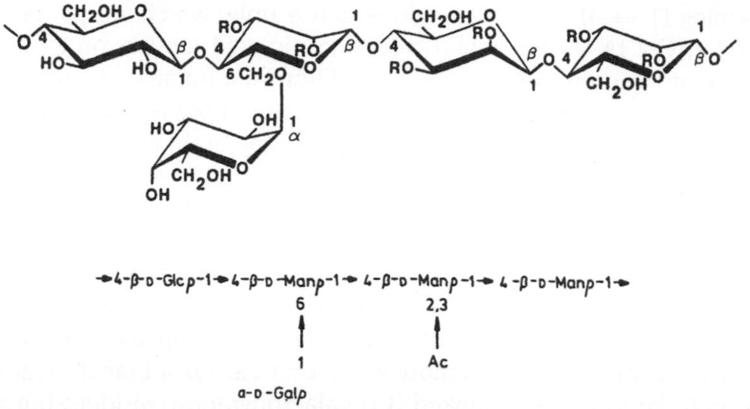

**Fig. 3-12.** Principal structure of galactoglucomannans. Sugar units: β-D-glucopyranose (Glc*p*); β-D-mannopyranose (Man*p*); α-D-galactopyranose (Gal*p*). R = CH₃CO or H. Below is the abbreviated formula showing the proportions of the units (galactose-rich fraction).

are partially substituted by O-acetyl groups, on the average one group per 3–4 hexose units. Galactoglucomannans are easily depolymerized by acids and especially so the bond between galactose and the main chain. The acetyl groups are much more easily cleaved by alkali than by acid.

*Arabinoglucuronoxylan*    In addition to galactoglucomannans, softwoods contain an arabinoglucuronoxylan (5–10%). It is composed of a framework

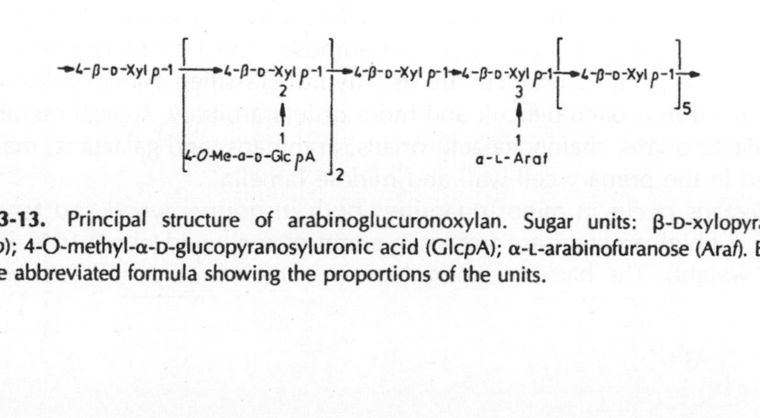

**Fig. 3-13.** Principal structure of arabinoglucuronoxylan. Sugar units: β-D-xylopyranose (Xyl*p*); 4-O-methyl-α-D-glucopyranosyluronic acid (Glc*p*A); α-L-arabinofuranose (Ara*f*). Below is the abbreviated formula showing the proportions of the units.

containing (1 → 4)-linked β-D-xylopyranose units which are partially sub-stituted at C-2 by 4-O-methyl-α-D-glucuronic acid groups, on the average two residues per ten xylose units. In addition, the framework contains α-L-arabinofuranose units, on the average 1.3 residues per ten xylose units (Fig. 3-13). Because of their furanosidic structure, the arabinose side chains are easily hydrolyzed by acids. Both the arabinose and uronic acid substituents stabilize the xylan chain against alkali-catalyzed degradation (see Sections 2.5.5 and 7.3.5).

*Arabinogalactan*   The heartwood of larches contains exceptionally large amounts of water-soluble arabinogalactan, which is only a minor consti-tuent in other wood species. Its backbone is built up by (1 → 3)-linked β-D-galactopyranose units. Almost every unit carries a branch attached to position 6, largely (1 → 6)-linked β-D-galactopyranose residues but also L-arabinose (Fig. 3-14). There are also a few glucuronic acid residues present in the molecule. The highly branched structure is responsible for the low viscosity and high solubility in water of this polysaccharide.

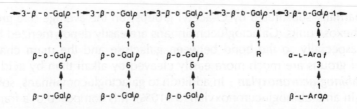

**Fig. 3-14.** Abbreviated formula of arabinogalactan. Sugar units: β-D-galactopyranose (Gal*p*), β-L-arabinopyranose (Ara*p*), α-L-arabinofuranose (Ara*f*), and R is β-D-galactopyranose or, less frequently, α-L-arabinofuranose, or a β-D-glucopyranosyluronic acid residue.

*Other Polysaccharides*   Besides galactoglucomannans, arabinoglucu-ronoxylan and arabinogalactan, softwoods contain other polysaccharides, usually present in minor quantities. Such polysaccharides include starch (composed of amylose and amylopectin) and pectic substances.

The most common units in pectic polysaccharides consist of D-galac-tosyluronic acid and D-galactose, L-arabinose, and L-rhamnose residues. Although the pectic substances are usually not classified as hemicelluloses, the distinction is often difficult and more or less arbitrary. Typical members are galacturonans, rhamnogalacturonans, arabinans, and galactans, mainly located in the primary cell wall and middle lamella.

Galactans occur in minor quantities both in normal wood and tension wood, but high amounts are present in compression wood (about 10% of the wood weight). The backbone of galactans, which is slightly branched, is built up of (1 → 4)-linked β-D-galactopyranose units substituted at C-6 with α-D-galacturonic acid residues (Fig. 3-15). The function of galactans in wood is probably more related to hemicelluloses than to pectic substances. In addition, compression wood contains about 2% laricinan, which is a (1 → 3)-β-D-glucan and occurs only in trace amounts in normal wood.

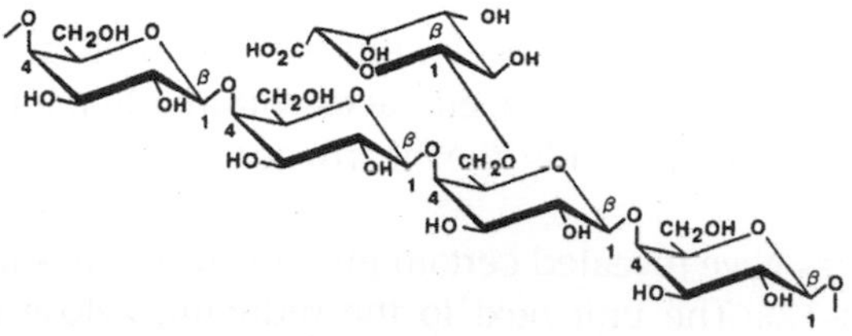

$$\rightarrow 4\text{-}\beta\text{-}D\text{-}Galp\text{-}1\rightarrow 4\text{-}\beta\text{-}D\text{-}Galp\text{-}1\rightarrow 4\text{-}\beta\text{-}D\text{-}Galp\text{-}1\overset{\big[}{\rightarrow} 4\text{-}\beta\text{-}D\text{-}Galp\text{-}1\overset{\big]}{\rightarrow}$$
$$6 \qquad\qquad 17$$
$$\uparrow$$
$$1$$
$$\beta\text{-}D\text{-}GalpA$$

**Fig. 3-15.** Principal structure of galactan in compression wood. Sugar units: β-D-galactopyranose (Galp) and β-D-galactopyranosyluronic acid (GalpA). Below is the abbreviated formula showing the proportions of the units (for further details, see Timell, 1986).

### 3.3.2   Hardwood Hemicelluloses

*Glucuronoxylan*   Even if hemicelluloses in various hardwood species differ from each other both quantitatively and qualitatively, the major component is an O-acetyl-4-O-methylglucurono-β-D-xylan, sometimes called glucuronoxylan. Often the xylose-based hemicelluloses in both softwoods and hardwoods are termed simply xylans.

Depending on the hardwood species, the xylan content varies within the limits of 15–30% of the dry wood. As can be seen from Fig. 3-16, the backbone consists of β-D-xylopyranose units, linked by (1 → 4)-bonds. Most

**Fig. 3-16.** Abbreviated formula of glucuronoxylan. Sugar units: β-D-xylopyranose (Xylp), and 4-O-methyl-α-D-glucopyranosyluronic acid (GlcpA). R is an acetyl group (CH₃CO).

of the xylose residues contain an O-acetyl group at C-2 or C-3 (about seven acetyl residues per ten xylose units). The xylose units in the xylan chain additionally carry (1 → 2)-linked 4-O-methyl-α-D-glucuronic acid residues, on the average about one uronic acid per ten xylose residues. The xylosidic bonds between the xylose units are easily hydrolyzed by acids, whereas the linkages between the uronic acid groups and xylose are very resistant. Acetyl groups are easily cleaved by alkali, and the acetate formed during kraft pulping of wood mainly originates from these groups (see Section 7.3.5). They are slowly hydrolyzed to acetic acid within a living tree as a result of the acidic nature of especially the heartwood (cf. also galactoglucomannans, p. 60).

Recent studies have revealed certain interesting features in the structure of xylans (Fig. 3-17). The unit next to the reducing xylose end group is D-galacturonic acid, linked to a L-rhamnose unit through the C-2 position. The rhamnose unit, in turn, is connected through its C-3 position to the xylan chain.

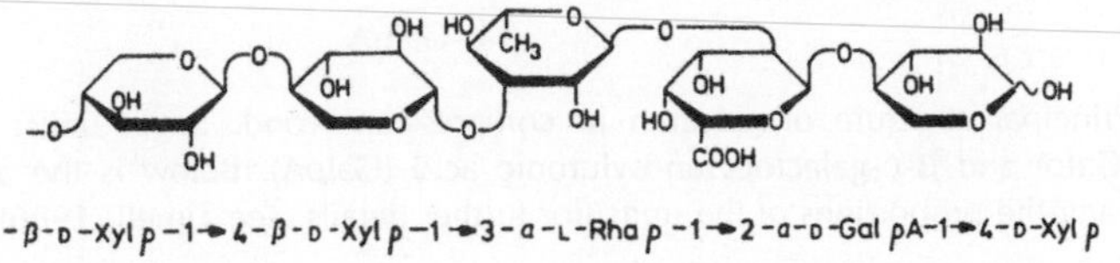

$- \beta - D - Xyl\, p - 1 \rightarrow 4 - \beta - D - Xyl\, p - 1 \rightarrow 3 - \alpha - L - Rha\, p - 1 \rightarrow 2 - \alpha - D - Gal\, pA - 1 \rightarrow 4 - D - Xyl\, p$

**Fig. 3-17.** Structure associated with the reducing end group of birch xylan (adopted from Johansson and Samuelson, 1977).

*Glucomannan*  Besides xylan, hardwoods contain 2–5% of a glucomannan, which is composed of β-D-glucopyranose and β-D-mannopyranose units linked by (1 → 4)-bonds (Fig. 3-18). The glucose:mannose ratio varies between 1:2 and 1:1, depending on the wood species. The mannosidic bonds between the mannose units are more rapidly hydrolyzed by acid than the corresponding glucosidic bonds, and glucomannan is easily depolymerized under acidic conditions.

$\rightarrow 4 - \beta - D - Glc\cdot p - 1 \rightarrow 4 - \beta - D - Man\, p - 1 \rightarrow 4 - \beta - D - Glc\, p - 1 \rightarrow 4 - \beta - D - Man\, p - 1 \rightarrow 4 - \beta - D - Man\, p - 1 \rightarrow$

**Fig. 3-18.** Abbreviated formula of glucomannan.

*Other Polysaccharides*  In addition to xylan and glucomannan minor amounts of miscellaneous polysaccharides are present in hardwoods, partly of the same type as those occurring in softwoods. They might be important components for the living tree, although of little interest when considering the technical applications.

### 3.3.3   Isolation of Hemicelluloses

Hemicelluloses can be isolated from wood, holocellulose, or pulp by extraction. Among the few neutral solvents which are effective, dimethyl sulfoxide is useful particularly for the extraction of xylan from a holocellulose. Although only a part of the xylan can be extracted, the advantage is that no chemical changes take place. More xylan can be extracted with alkali (KOH or NaOH). Addition of sodium borate to the alkali facilitates the dissolution of galactoglucomannans and glucomannans. However, alkali extractions have the disadvantage of deacetylating the hemicelluloses almost completely.

Glucomannans are more effectively extracted by sodium than by potassium hydroxide. A gradient elution at varying alkali concentrations can be used for a rough fractionation of the hemicellulose components. The solvating effect of borate ions is based on their reaction with vicinal hydroxyl groups in the *cis* position present, for example, in mannose units. The borate complex is readily decomposed on acidification. The polysaccharides can be precipitated from the alkaline extract by acidification with acetic acid. Addition of a neutral organic solvent, e.g., ethanol, to the neutralized extraction solution results in a more complete precipitation. Some more specific precipitation agents are also known, for instance, barium hydroxide for glucomannan and cetyltrimethylammonium bromide or hydroxide for glucuronoxylan. Fehling's solution can also be used for precipitation purposes.

The precipitated hemicellulose preparations can be further purified by column chromatography. Gel permeation chromatography is useful for fractionation according to the molecular weight. In most cases, however, differences in chemical properties form the basis for the separation, for example, the above-mentioned capability of certain hydroxyl groups for complex formation. Particularly useful in the separation of hemicelluloses are ion exchangers based on cellulose, dextran or agarose, e.g., diethylaminoethyl cellulose in different ionic forms. Generally, chromatography in its various forms is used for the characterization of the acidic hydrolysis products of isolated hemicelluloses, either after total hydrolysis (analysis of monosaccharides) or after partial hydrolysis (analysis of oligosaccharides). The general method for the localization of bonds is complete methylation followed by hydrolysis and identification of the methylated sugars (gas-liquid chromatography-mass spectrometry). In addition, separate determinations can be carried out for the analysis of uronic acids, pentosan, acetyl, and methoxyl groups as well as for the molecular weight and its distribution (see also Section 2.6).

### 3.3.4   Distribution of Hemicelluloses

Because of the heterogeneous structure of wood the content and composition of its constituents, including polysaccharides, vary considerably depending on the age and morphological origin of the sample. Also in this connection it is noteworthy that compression wood contains some polysaccharide constituents which can be totally absent from the normal wood (cf. Section 3.3.1). Great differences are also typical between the various cell types. For example, the proportion of xylan is much higher in softwood (spruce and pine) ray cells than in tracheids. Also the hardwood (birch) parenchyma cells are more xylan-rich than are the tracheids and fibers.

Another question of fundamental importance concerns the location of polysaccharides in the wood cells. By analyzing the polysaccharide composition of the wood cells microtomed from the cambial layer, representing different stages of growth, it has been possible to determine the distribution of polysaccharides in different layers of the cell walls. This elegant and extremely tedious micromanipulation technique was applied 30 years ago by Meier, and our knowledge about the distribution of polysaccharides in the wood cells is based on his pioneering study. However, because of experimental difficulties and due to the fact that paper chromatography was the only method available at that time, the data obtained can only roughly describe this distribution.

Both in softwood and hardwood cells the pectic polysaccharides including galacturonans, galactans, and arabinans are typically enriched in the compound middle lamella (M + P). In softwood (spruce and pine) cells the outer part of the $S_1$ layer is richest in cellulose and the galactoglucomannan content increases steadily from the outer parts of the cell walls toward lumen. The concentration profile of arabinoglucuronoxylan is more irregular but its content is highest in the $S_3$ layer.

In hardwood (birch) cells the cellulose content is highest in the inner part of the $S_2$ layer and in the $S_3$ layer. Both in the outer part of the $S_2$ layer and in the $S_1$ layer much more glucuronoxylan is present than in the other parts of the cell walls.

# Chapter 4

# LIGNIN

Anselme Payen observed in 1838 that wood, when treated with concentrated nitric acid, lost a portion of its substance, leaving a solid and fibrous residue he named cellulose. As a result of much later studies it became evident that the fibrous material isolated by Payen contained also other polysaccharides besides cellulose. The dissolved material (*"la matière incrustante"*), on the other hand, had a higher carbon content than the fibrous residue and was termed "lignin." This term, already introduced in 1819 by de Candolle, is derived from the latin word for wood (lignum).

Later, the development of technical pulping processes generated much interest in lignin and its reactions. In 1897, Peter Klason studied the composition of lignosulfonates and put forward the idea that lignin was chemically related to coniferyl alcohol. In 1907, he proposed that lignin is a macromolecular substance and, ten years later, that coniferyl alcohol units are joined together by ether linkages.

## 4.1 Isolation

Lignin can be isolated from extractive-free wood as an insoluble residue after hydrolytic removal of the polysaccharides. Alternatively, lignin can be hydrolyzed and extracted from the wood or converted to a soluble derivative. So-called Klason lignin is obtained after removing the polysaccharides from extracted (resin-free) wood by hydrolysis with 72% sulfuric acid. Other acids can be used as well for the hydrolysis, but the method has the serious drawback in that the structure of lignin is extensively changed during the hydrolysis. The polysaccharides may be removed by enzymes from finely divided wood meal. The method is tedious, but the resulting "cellulolytic enzyme lignin" (CEL) retains its original structure essentially unchanged. Lignin can also be extracted from wood using dioxane containing water and hydrochloric acid, but considerable changes in its structure occur.

Besides cellulolytic enzyme lignin, the so-called Björkman lignin, alternatively referred to as "milled wood lignin" (MWL) is the best preparation known so far, and it has been widely used for structural studies. When wood meal is ground in a ball mill either dry or in the presence of nonswelling solvents, e.g., toluene, the cell structure of the wood is destroyed and a portion of lignin (usually not more than 50%) can be obtained from the suspension by extraction with a dioxane-water mixture. MWL preparations always contain some carbohydrate material.

Dehydrogenative treatment of coniferyl alcohol with hydrogen peroxide in the presence of peroxidase enzyme yields synthetic lignin named "dehydrogenation polymer" (DHP). In addition, a new type of lignin preparation, so-called "released suspension culture lignin" (RSCL), was recently isolated from suspension cultures of spruce wood cells as a secretion products. RSCL represents a carbohydrate-free coniferous lignin.

Soluble lignin derivatives (lignosulfonates) are formed by treating wood at elevated temperatures with solutions containing sulfur dioxide and hydrogen sulfite ions (see Section 7.2). Lignin is also dissolved as alkali lignin when wood is treated at elevated temperatures (170°C) with sodium hydroxide, or better, with a mixture of sodium hydroxide and sodium sulfide (sulfate or kraft lignin) (see Section 7.3). Lignin is further converted to an alkali-soluble derivative by a solution of hydrochloric acid and thioglycolic acid at 100°C.

Softwood lignin can be determined gravimetrically by the Klason method. Normal softwood contains 26–32% lignin while the lignin content of compression wood is 35–40% (cf. Appendix). The lignin present in hardwoods is partly dissolved during the acid hydrolysis and hence the gravimetric values must be corrected for the "acid-soluble lignin" using UV spectrophotometry. Direct UV spectrophotometric methods have also been developed for the determination of lignin in wood and pulps. Normal hardwood contains 20–25% lignin, although tropical hardwoods can have a lignin content exceeding 30%. Tension wood contains only 20–25% lignin.

# 4.2   Biosynthesis and Structure

Lignins are polymers of phenylpropane units. Many aspects in the chemistry of lignin still remain unclear, for example, the specific structural features of lignins located in various morphological regions of the woody xylem. Nevertheless, the principal structural elements in lignins have been largely clarified as a result of detailed studies on isolated lignin preparations, such as milled wood lignin, using specific degradative techniques based on oxidation, reduction, or hydrolysis under acidic and alkaline conditions. Much effort has been directed toward the clarification of the biosynthesis of lignin. Detailed identification of the reaction products has been possible by novel chromatographic techniques and spectroscopic methods developed during the last two to three decades.

## 4.2.1   Phenylpropane—The Basic Structural Unit of Lignin

Methods based on classical organic chemistry led to the conclusion, already by 1940, that lignin is built up of phenylpropane units. Examples of typical reactions used in these studies are illustrated in Fig. 4-1. However, the concept of a phenylpropanoid structure failed to win unanimous acceptance, and as late as 40 years ago, some scientists were not convinced that lignin in its native state was an aromatic substance. Finally, the problem was solved by Lange in 1954, who applied UV microscopy at various wavelengths directly on thin wood sections, obtaining spectra typical of aromatic compounds.

## 4.2.2   Biosynthesis of Lignin Precursors

In addition to coniferyl alcohol, which is the main lignin precursor in all gymnosperms, sinapyl alcohol and p-coumaryl alcohol are involved in the biosynthesis of lignins in angiosperms and Gramineae (including grasses), respectively. All these precursors are derivatives of cinnamyl alcohol and they are present in the cambial tissues as glucosides. The preceding step before their polymerization to lignin is their liberation by the action of β-glucosidase enzyme.

Lignin precursors are generated from D-glucose through complex reactions catalyzed by enzymes. The first sequence is known as the *shikimate* pathway (Fig. 4-2). Phosphoenol pyruvic acid (1) and D-erythrose 4-phosphate (2), formed from D-glucose, are combined to 3-deoxy-D-*arabino*-heptulosonic acid 7-phosphate (3). Elimination of phosphate from 3 and ring closure of 4 lead to 3-dehydroquinic acid (5). After loss of water from 5 the

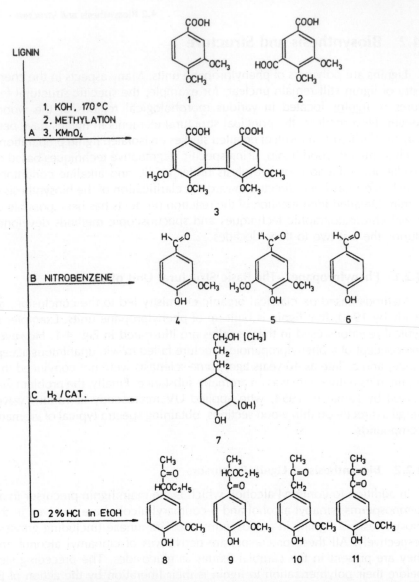

**Fig. 4-1.** Examples of classical methods indicating a phenylpropanoid structure of lignin. (A) *Permanganate oxidation* (methylated softwood lignin) affords veratric acid (3,4-dimethoxy-benzoic acid) (1) in a yield of about 10% and minor amounts of isohemipinic (4,5-dimethoxy-isophtalic acid) (2) and dehydrodiveratric (3) acid. The formation of isohemipinic acid supports the occurrence of condensed structures (e.g., β-5 or γ-5). (B) *Nitrobenzene oxidation* of softwoods in alkali results in the formation of vanillin (4-hydroxy-3-methoxybenzaldehyde) (4) (about 25% of lignin). Oxidation of hardwoods and grasses results, respectively, in syringaldehyde (3,5-dimethoxy-4-hydroxybenzaldehyde) (5) and p-hydroxybenzaldehyde (6). (C) *Hydrogenolysis* yields propylcyclohexane derivatives (7). (D) *Ethanolysis* yields so-called Hibbert ketones (8, 9, 10, and 11).

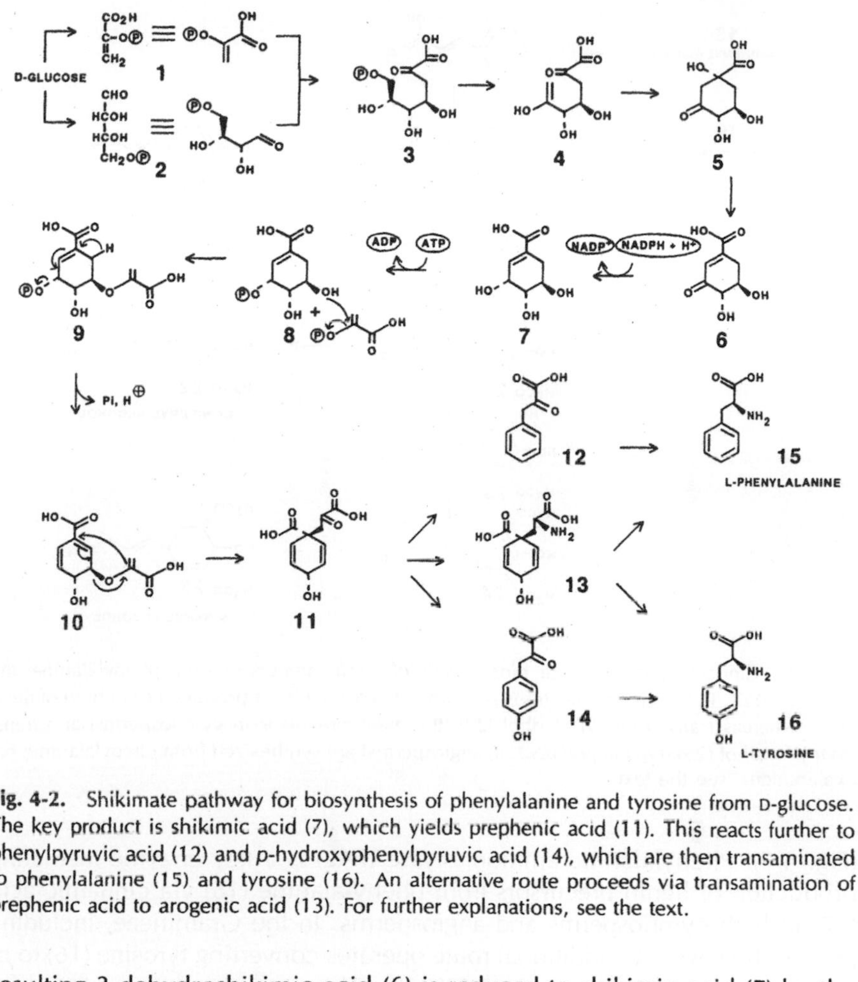

**Fig. 4-2.** Shikimate pathway for biosynthesis of phenylalanine and tyrosine from D-glucose. The key product is shikimic acid (7), which yields prephenic acid (11). This reacts further to phenylpyruvic acid (12) and p-hydroxyphenylpyruvic acid (14), which are then transaminated to phenylalanine (15) and tyrosine (16). An alternative route proceeds via transamination of prephenic acid to arogenic acid (13). For further explanations, see the text.

resulting 3-dehydroshikimic acid (6) is reduced to shikimic acid (7) by the action of NADPH-linked shikimate dehydrogenase (dehydroshikimate reductase). The phosphorylated shikimic acid (8) is combined with phosphoenol pyruvic acid to phosphochorismic acid (9). The liberated chorismic acid (10) is then rearranged to prephenic acid (11), which is converted either to phenylpyruvic acid (12) or p-hydroxyphenylpyruvic acid (14). These products are finally transaminated to L-phenylalanine (15) and L-tyrosine (16), respectively. An alternative route involves a direct transamination of 11 to arogenic acid (13), which after decarboxylation, dehydrogenation, and dehydroxylation is converted to L-phenylalanine (15) or L-tyrosine (16).

The following sequence of reactions is known as the *cinnamate* pathway

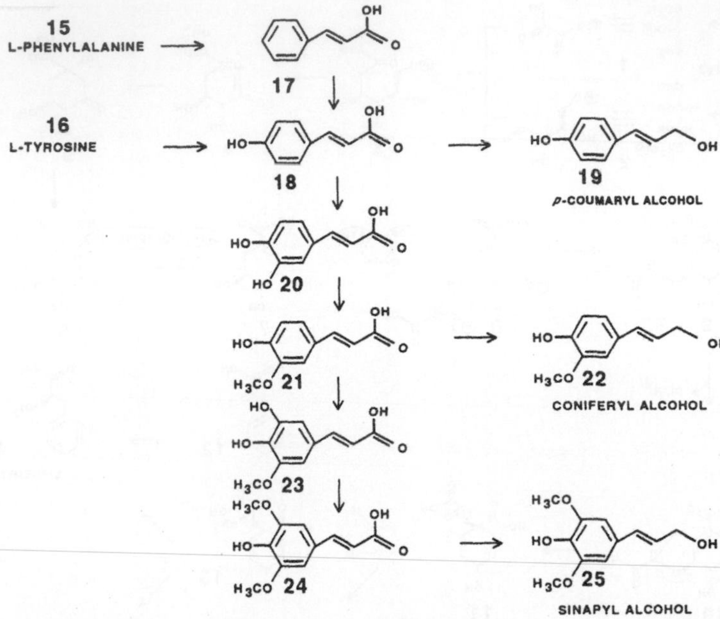

**Fig. 4-3.** Cinnamate pathway for biosynthesis of lignin precursors from phenylalanine and tyrosine pools. In the Gramineae *trans-p*-coumaryl alcohol (19) is produced from both of these pools whereas *trans*-coniferyl alcohol (22) (the main precursor in gymnosperms) and *trans*-sinapyl alcohol (25) (typical precursor in angiosperms) are synthesized from phenylalanine. For explanations, see the text.

(Fig. 4-3). Phenylalanine ammonia-lyase (PAL) is a key enzyme mediating production of lignin precursors from phenylalanine (15) via cinnamic acid (17) in both gymnosperms and angiosperms. In the Gramineae, including grasses, however, an additional route operates converting tyrosine (16) to *p*-hydroxycinnamic (*p*-coumaric acid) (18). This conversion is mediated by tyrosine ammonia-lyase (TAL). Hydroxylation and methylation of the further products in the reaction chain are mediated by hydroxylase and *O*-methyl tranferase (OMT) enzymes, respectively. In gymnosperms, OMT is essentially monofunctional, converting caffeic acid (20) almost exclusively to ferulic acid (21). In angiosperms, however, OMT is bifunctional and responsible also for the conversion of 5-hydroxyferulic acid (23) to sinapic acid (24). The pathway is completed by reduction of the respective cinnamic acid derivatives to the lignin precursors, that is, *p*-coumaryl alcohol (19), coniferyl alcohol (22), and sinapyl alcohol (25). This reduction proceeds via aldehyde intermediates.

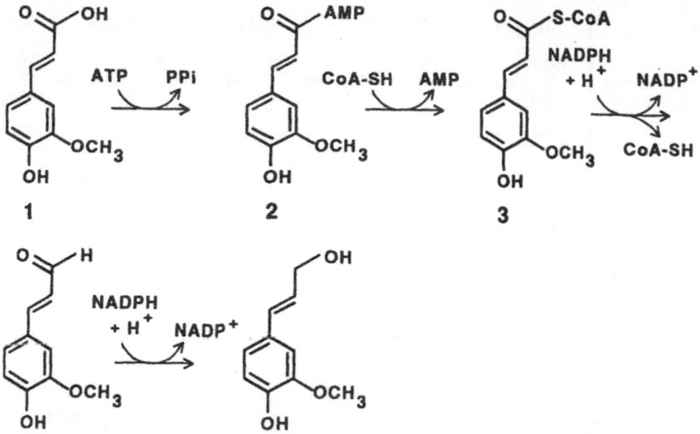

**Fig. 4-4.** Reduction of ferulic acid (1) to coniferyl alcohol (5). Enzymes involved: hydroxycin-namate:CoA ligase (step 2 to 3), hydroxycinnamoyl-CoA reductase (step 3 to 4), and cinnamyl alcohol oxidoreductase (step 4 to 5). For explanations, see the text.

As an example of the reduction of cinnamic acid derivatives to the corresponding aldehydes, the route of ferulic acid to coniferyl alcohol is illustrated in Fig. 4-4. Feruloyl CoA thiol ester (3), formed from feruloyl adenylate (2) and catalyzed by hydroxycinnamate:CoA ligase, is the activating intermediate which is then reduced to coniferyl aldehyde (4) by the action of NADPH-linked p-hydroxycinnamoyl-CoA reductase. Feruloyl-S-CoA is the best substrate, and the respective thiol esters of coumarate, sinapate, and 5-hydroxyferulate are less active in the named order. The last reduction step from coniferyl aldehyde to coniferyl alcohol (5) is catalyzed by hydroxycin-namyl alcohol oxidoreductases. These enzymes in angiosperms reduce both coniferyl and sinapyl aldehydes almost equally, but the gymnosperm enzymes are remarkably specific for coniferyl aldehyde. Therefore, like O-methyltransferase, p-hydroxycinnamyl alcohol oxidoreductase is obviously one of the key enzymes controlling the specificity of the lignin precursors, which, in turn, regulate the formation of guaiacyl and syringyl lignins in gymnosperms and angiosperms (see further, Higuchi, 1990).

## 4.2.3  Polymerization of Lignin Precursors

For an understanding of the formation and structure of lignin, investigations conducted by H. Erdtman in 1930 were of great importance. He

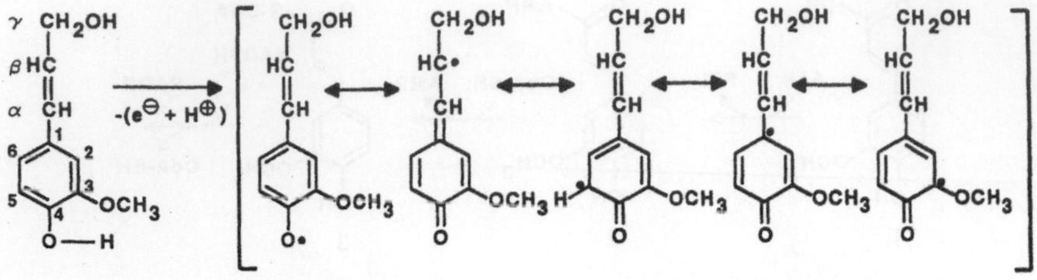

**Fig. 4-5.** One-electron transfer from coniferyl alcohol by enzymatic dehydrogenation yielding resonance-stabilized phenoxy radicals.

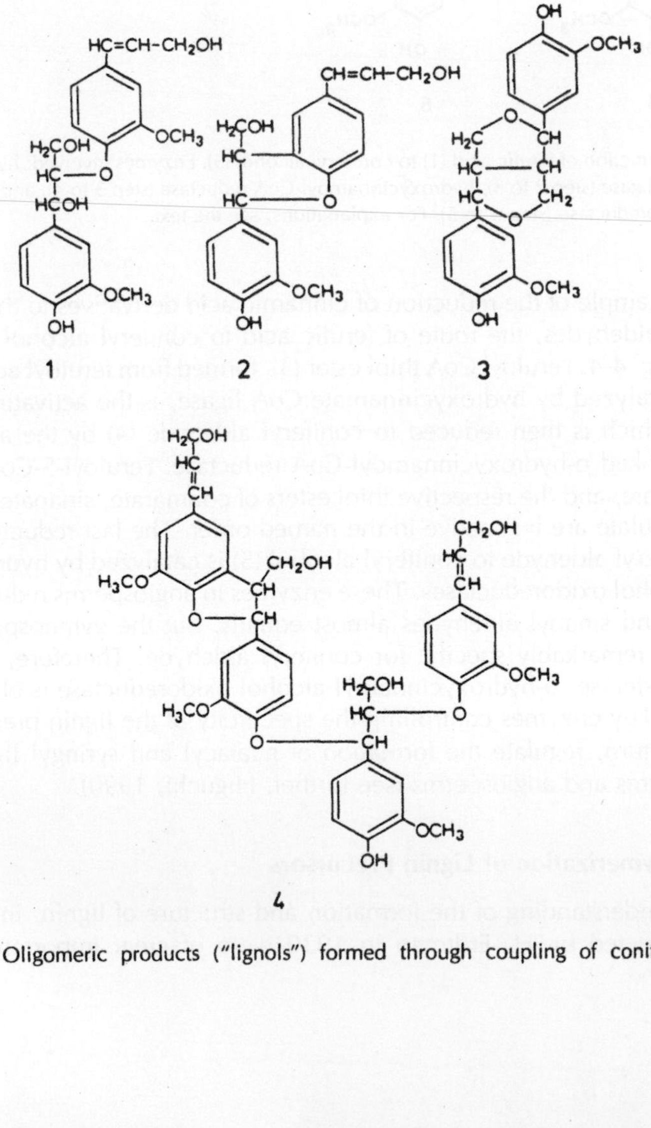

**Fig. 4-6.** Oligomeric products ("lignols") formed through coupling of coniferyl alcohol radicals.

studied the oxidative dimerization of various phenols in the biogenesis of natural products and reached the conclusion that lignin must be formed from $\alpha,\beta$-unsaturated $C_6C_3$ precursors of the coniferyl alcohol type via enzymic dehydrogenation. The polymerization of precursors to lignin in nature does indeed occur in this manner, as has been demonstrated by comprehensive studies by Freudenberg and co-workers during the period 1940 to 1970.

The role of coniferyl alcohol as the immediate precursor of softwood

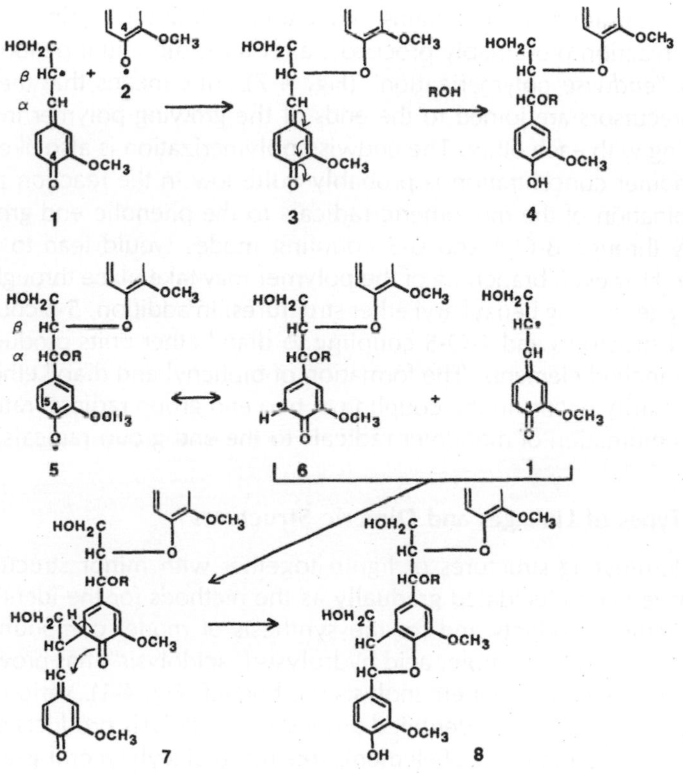

**Fig. 4-7.** Endwise $\beta$-O-4 coupling of a coniferyl alcohol radical (1) with a growing lignin end group radical (2) to an intermediate quinone methide (3), which is stabilized to a guaiacylglycerol-$\beta$-aryl ether structure (4) through addition of water (R = H). Alternatively, if a phenolic hydroxyl group (R = aryl group) is added to the quinone methide, a guiacylglycerol-$\alpha,\beta$-diaryl ether structure is obtained. A hydroxyl group of a carbohydrate can also react (R = a carbohydrate residue), resulting in a lignin–carbohydrate bond. A guiacylglycerol-$\beta$-aryl ether radical (5) formed after dehydrogenation from 4 has a resonance structure 6. This yields, through $\beta$-5 coupling with coniferyl alcohol radical 1, a quinone methide product (7), which, after intramolecular ring closure, forms a phenyl coumaran structure (8).

lignin has been demonstrated by using $^{14}$C labeling. Administration of labeled coniferyl alcohol as β-glucoside (coniferin) to seedlings of spruce results in the exclusive formation of radioactive lignin. The enzymic dehydrogenation reaction is initiated by an electron transfer which results in the formation of resonance-stabilized phenoxy radicals (Fig. 4-5). The combination of these radicals produces a variety of dimers and oligomers, termed lignols (Fig. 4-6).

It can be readily shown that further oxidative coupling of di- and oligolignols ("*bulk* polymerization") would lead to a product containing a large number of unsaturated side chains. Since their amount in lignin is relatively low, the reaction presumably proceeds, after a certain initial period, essentially as "*endwise* polymerization" (Fig. 4-7). This means that the monomeric precursors are joined to the ends of the growing polymer instead of combining with each other. The endwise polymerization is also likely since the monomer concentration is probably quite low in the reaction zone.

Combination of the monomeric radicals to the phenolic end groups exclusively through β-O-4 and β-5 coupling modes would lead to a linear polymer. However, branching of the polymer may take place through α-O-4 coupling generating benzyl aryl ether structures. In addition, 5-5 coupling to biphenyl structures and 4-O-5 coupling to diaryl ether units produce additional branched elements. The formation of biphenyl and diaryl ether structures primarily occurs in the coupling of two end group radicals rather than in the combination of monomer radicals to the end group radicals.

### 4.2.4   Types of Linkages and Dimeric Structures

The dominating structures of lignin together with minor structural elements have been elucidated gradually as the methods for the identification of degradation products and for the synthesis of model compounds have been improved. For example, acid hydrolysis ("acidolysis") has proved to be more useful than the older ethanolysis method (cf. Fig. 4-1). Various dimers and oligomers have been identified among the acidolysis products of MWL, revealing a high frequency of elements like the guaiacylglycerol-β-aryl ether and phenylcoumaran structures.

The so-called "thioacetolysis" is a method in which lignin samples are subjected to a treatment with thioacetic acid and boron trifluoride. Such treatment effectively cleaves β-O-4 bonds, resulting in an extensive fragmentation of lignin. Another method, termed "thioacidolysis," was developed quite recently. According to this method the alkyl and aryl ether bonds are cleaved in dioxane in the presence of a "hard" acid (Et$_2$O-BF$_3$) and a "soft" nucleophile (EtSH). The cleavage of the ether bonds is equally specific, but more complete than by thioacetolysis. The reaction products are separated as trimethylsilyl ethers by gas-liquid chromatography. In its latest version this method was extended for detection of condensed lignin structures. Desulfuration of the dimeric thioacidolysis products, representing condensed lignin structures, rendered their gas chromatographic separation possible.

The original permanganate oxidation method of methylated lignin (cf. Fig. 4-1) has also been considerably improved when it is performed at alkaline instead of neutral conditions. In its newer version there is also an additional oxidative treatment with hydrogen peroxide, improving the yield of products considerably. The methylated fragments are separated by gas-liquid chromatography and identified by mass spectrometry.

As a result of these studies it is clear that the phenylpropane units are joined together both with C-O-C (ether) and C-C linkages. The ether linkages dominate; approximately two thirds or more are of this type, and the rest are of the carbon-to-carbon type. Detailed knowledge about the charac-

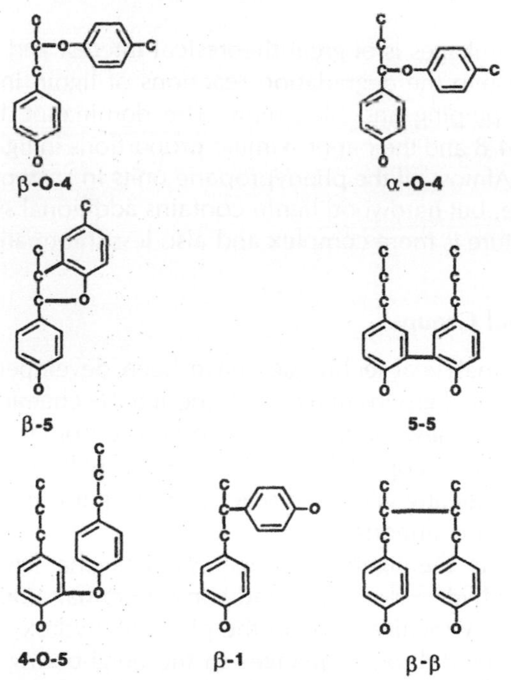

**Fig. 4-8.** Common linkages between the phenylpropane units. For proportions, see Table 4-1.

**TABLE 4-1.** Proportions of Different Types of Linkages
Connecting the Phenylpropane Units in Lignin[a]

| | | Percent of the total linkages | |
| Linkage type[b] | Dimer structure | Softwood | Hardwood |
|---|---|---|---|
| β-O-4 | Arylglycerol-β-aryl ether | 50 | 60[d] |
| α-O-4 | Noncyclic benzyl aryl ether | 2–8[c] | 7 |
| β-5 | Phenylcoumaran | 9–12 | 6 |
| 5-5 | Biphenyl | 10–11 | 5 |
| 4-O-5 | Diaryl ether | 4 | 7 |
| β-1 | 1,2-Diaryl propane | 7 | 7 |
| β-β | Linked through side chains | 2 | 3 |

[a] Approximative values based on the data of Adler (1977) obtained for MWL from spruce
(*Picea abies*) and birch (*Betula verrucosa*).
[b] For the corresponding structures, see Fig. 4-8.
[c] Low values have been reported recently.
[d] Of these structures about 40% are of guaiacyl type and 60% of syringyl type.

teristics of these linkages is of great theoretical interest and necessary for a
thorough insight into the degradation reactions of lignin in technical processes, such as pulping and bleaching. The dominating bond types are
depicted in Fig. 4-8 and their approximate proportions in lignin can be seen
from Table 4-1. Almost all the phenypropane units in softwood lignin are of
the guaiacyl type, but hardwood lignin contains additional syringyl units. In
this case the picture is more complex and also less information is available.

### 4.2.5    Functional Groups

Sophisticated analytical techniques have been developed for the determination of functional groups in lignin. Although the chemical methods still
are important, much new information has emerged from the physical methods, especially spectroscopy (UV, IR, $^1$H NMR, and $^{13}$C NMR). Because the
functional groups greatly affect the reactivity of lignin, reliable and quantitative information is important.

As its precursors the lignin polymer contains characteristic methoxyl
groups, phenolic hydroxyl groups, and some terminal aldehyde groups in
the side chain. Only relatively few of the phenolic hydroxyls are free; most
of them are occupied through linkages to the neighboring phenylpropane
units. Especially the syringyl units in hardwood lignin are extensively etherified. In addition to these groups, alcoholic hydroxyl groups and carbonyl

TABLE 4-2.  Functional Groups of Lignin (per 100 $C_6C_3$ Units)

| Group | Softwood lignin | Hardwood lignin |
|---|---|---|
| Methoxyl | 92–97 | 139–158 |
| Phenolic hydroxyl | 15–30 | 10–15 |
| Benzyl alcohol | 30–40 | 40–50 |
| Carbonyl | 10–15 | |

groups are introduced into the final lignin polymer during the dehydrogenative polymerization process. In some wood species substantial amounts of the alcoholic hydroxyl groups are esterified with p-hydroxybenzoic acid or p-hydroxycinnamic acid. Esters of p-hydroxybenzoic acid are typical in aspen lignin, whereas those of p-hydroxycinnamic acid are abundant in bamboo and grass lignins. These acids seem to form esters preferentially with the γ-hydroxyl group in the lignin side chain.

Despite intensive studies and a large number of analytical data today available on the composition of functional groups, this information cannot be summarized in a simple way because of large individual variations among the wood species. The composition also vary within the cell walls. Therefore, the literature should be consulted in each special case. Table 4-2 gives a rough idea of the frequencies of functional groups in lignin.

### 4.2.6  Lignin Formula

Based on the information obtained from studies of biosynthesis as well as analysis of various linkage types and functional groups, structural formulas for lignin have been constructed. The formula presented in its final shape in 1968 by Freudenberg for softwood lignin (spruce) has attained general acceptance, and, indeed, the formulas suggested later for softwood lignin by Adler and by others do not much deviate from it. In addition, formulas for hardwood lignins have been suggested as well.

Adler's formula is shown in Fig. 4-9. This formula consists of 16 phenylpropane units and it represents only a segment of the lignin macromolecule. Most of the linkages are of the same types already shown in Fig. 4-8. However, additional details are the glyceraldehyde-2-aryl ether group attached to unit 12 and the β-6 linkage between units 14 and 15. Of course, a model of this limited size cannot give a strictly quantitative picture. For example, incorporation of one syringyl unit (13) in the formula has only a qualitative meaning because, in reality, only low amounts (ca. 1%) of such

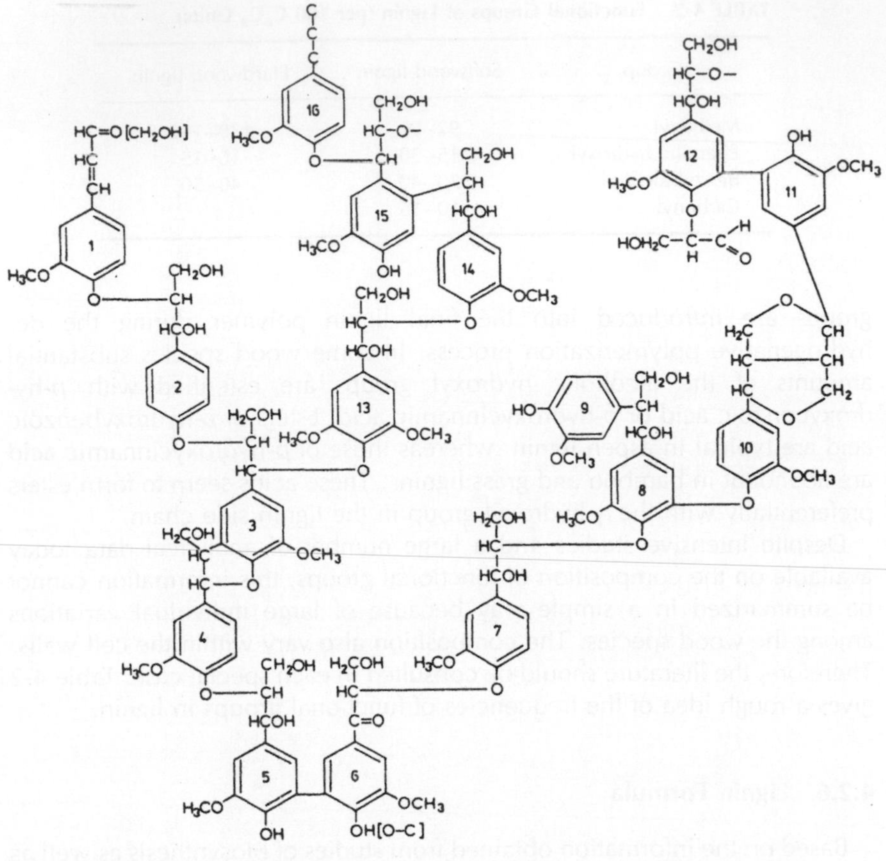

**Fig. 4-9.** A structural segment of softwood lignin proposed by Adler (1977).

units are present in softwood lignin. Likewise, one dimer entity of pinoresinol type (10–11) probably overemphasizes the presence of this substructure. In the Adler's formula no linkages between lignin and carbohydrates (see Section 4.2.7) or other wood constituents have been indicated.

### 4.2.7 Lignin–Carbohydrate Bonds

The possible existence of covalent bonds between lignin and polysaccharides has been a subject of much debate and intensive studies. This question is of great interest especially when considering the need to separate polysaccharides from lignin as selectively as possible. Because of other association forces of physical type between lignin and carbohydrates, it has been diffi-cult to verify whether there really are chemical bonds. However, it is ob-vious and now generally accepted that such chemical bonds must exist, and the term "lignin–carbohydrate complex (LCC)" is used for the covalently bonded aggregates of this type.

Chemical bonds have been reported between lignin and practically all the hemicellulose constituents. There are even indications of lignin and cel-lulose bonds. These linkages can be either of ester or ether type and even glycosidic bonds are possible. For example, instead of a free benzylic alco-hol group, it can be occupied through an ester linkage to a 4-O-meth-ylglucuronic acid group present in xylan or through an ether linkage to an arabinose or mannose unit present in arabinoxylan and glucomannan, re-spectively (Fig. 4-10). The ester linkages are easily cleaved by alkali.

More common and also much more stable than the ester bonds are the ether linkages between lignin and carbohydrates. The α-position is even in this case the most probable connection point between lignin and the hemi-cellulose blocks. In softwood xylans the bridging group can be the arabinose unit (HO-2 or HO-3). In galactoglucomannans, the galactose unit (HO-3) has been proposed to transmit this bridging. There are also indications that

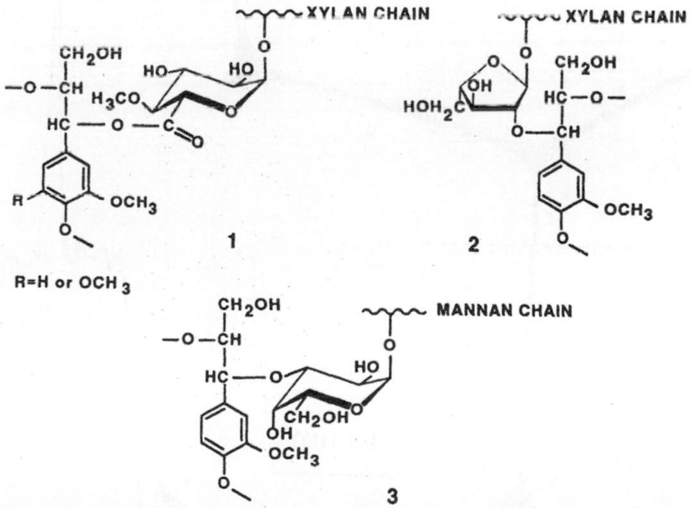

**Fig. 4-10.** Examples of suggested lignin–carbohydrate bonds: an ester linkage to xylan through 4-O-methyl glucuronic acid as a bridging group (1), an ether linkage to xylan through an arabinofuranose unit (2), and an ether linkage to galactoglucomannan through a galac-topyranose unit (3). (See also Figs. 3-12 and 3-13.)

lignin in the middle lamella and primary wall of the cell wall is associated with the pectic polysaccharides (galactan and arabinan) through ether linkages. In these cases the primary alcohol groups, that is, HO-6 in galactose units and HO-5 in arabinose units, seem to participate in this bridging.

Although less experimental evidence is available, it has been suggested that even glycosidic linkages are uniting lignin and polysaccharides. In addition to the benzylic alcohol group, which is the most probable connection point, the phenolic hydroxyl group may also be partly occupied through glycosidation. The glycosidic linkages are easily cleaved with acid.

## 4.3 Classification and Distribution

Lignins can be divided into several classes according to their structural elements. So-called "guaiacyl lignin" which occurs in almost all softwoods is largely a polymerization product of coniferyl alcohol. The "guaiacyl-syringyl lignin," typical of hardwoods, is a copolymer of coniferyl and sinapyl alcohols, the ratio varying from 4:1 to 1:2 for the two monomeric units. An additional example is compression wood, which has a high proportion

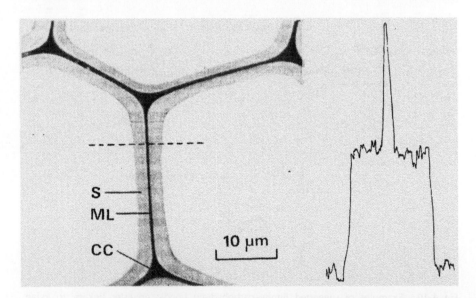

**Fig. 4-11.** Transverse section of a spruce tracheid photographed in UV light (240 nm). The densitometer tracing has been taken across the tracheid wall along the dotted line. S, secondary wall; ML, compound middle lamella; CC, cell corner (Fergus et al., 1969).

**TABLE 4-3. Distribution of Lignin in Spruce Tracheids[a]**

| Wood | Morphological region[b] | Tissue volume (%) | Lignin (% of total) | Lignin concentration (%) |
|------|-------------------------|-------------------|---------------------|--------------------------|
| Earlywood | S | 87 | 72 | 23 |
| | ML | 9 | 16 | 50 |
| | CC | 4 | 12 | 85 |
| Latewood | S | 94 | 82 | 22 |
| | ML | 4 | 10 | 60 |
| | CC | 2 | 9 | 100 |

[a] Adopted from Fergus et al. (1969). Approximate values for black spruce (Picea mariana).
[b] For explanations, see Fig. 4-11.

of phenylpropane units of the p-hydroxyphenyl type in addition to the normal guaiacyl units. The terms "syringyl lignin" and "p-hydroxyphenyl lignin" are sometimes used to denote the respective structural elements even if probably no natural lignins are exclusively composed of these units.

The lignin concentration is high in the middle lamella and low in the secondary wall. Because of its thickness, at least 70% of the lignin in softwoods is, however, located in the secondary wall as shown by quantitative UV microscopy (Fig. 4-11 and Table 4-3). The picture is very similar for the hardwoods (Table 4-4) although in this case analytical uncertainties are

**TABLE 4-4. Distribution of Lignin in Birch Xylem[a]**

| Cell | Morphological region[b] | Type of lignin[c] | Tissue volume (%) | Lignin (% of total) | Lignin concentration (%) |
|------|-------------------------|-------------------|-------------------|---------------------|--------------------------|
| Fiber | S | Sy | 73 | 60 | 19 |
| | ML | SyGu | 5 | 9 | 40 |
| | CC | SyGu | 2 | 9 | 85 |
| Vessel | S | Gu | 8 | 9 | 27 |
| | ML | Gu | 1 | 2 | 42 |
| Ray cell | S | Sy | 10 | 11 | 27 |

[a] Adopted from Fergus and Goring (1970). Approximate values for white birch (Betula papyrifera).
[b] For explanations, see Fig. 4-11.
[c] Sy, syringyl lignin; SyGu, syringyl-guaiacyl lignin; Gu, guaiacyl lignin.

involved because of the more heterogeneous nature of the wood and the presence of both guaiacyl and syringyl units in the lignin. The measurements so far indicate that the lignin located in the secondary wall of hardwood fibers has a high content of syringyl units whereas larger amounts of guaiacyl units are present in the middle lamella lignin. The vessels in birch seem to contain only guaiacyl lignin, whereas syringyl lignin predominates in parenchyma cells.

## 4.4    Polymer Properties

One experimental difficulty in studying the macromolecular properties of lignin is the fact that lignin has a very low solubility in most solvents. Instead of native lignin much of the research has been concentrated on its soluble reaction products, such as lignosulfonates and kraft lignins obtainable after pulping processes. The polymeric properties are also extremely important when evaluating the suitability of lignin by-products for technical applications.

Another problem is the isolation of lignin from wood without causing degradation and the possibility that the polymer properties of lignin may vary depending on its location in the cell wall. Very little information is available concerning this question.

The methods for characterizing the polymer properties of lignins include vapor pressure osmometry, light scattering, and ultracentrifugation. For practical purposes viscometry is mostly used, and size exclusion (gel permeation) chromatography is a relatively simple method used for the determination of the distribution of the molecular weights.

TABLE 4-5.   Molecular Weight and Polydispersity
of MWL Preparations and Dehydrogenase Polymers (DHP)
of Coniferyl Alcohol[a]

| Preparation | $\bar{M}_w$ | $\bar{M}_w/\bar{M}_n$ |
|---|---|---|
| MWL | | |
| Eastern spruce | 20,600 | 2.6 |
| Western hemlock | 22,700 | 2.4 |
| DHP | 3,700 | 1.5 |
| | 11,000 | 2.2 |
| | 31,400 | 3.7 |

[a] From Obiaga (1972).

The weight-average molecular weights ($\bar{M}_w$) of softwood lignins seem to be of the order of 20,000, whereas lower values have been reported for hardwood lignins. Compared with cellulose the polydispersity of lignin is relatively high, roughly 2.5–3.0 for softwood MWL. It is possible that the polydispersity is increased with the increased molecular weight, as has been observed for the dehydrogenase polymers (DHP) of coniferyl alcohol (synthetic lignin) (Table 4-5).

Solutions of isolated lignins, lignosulfonates, and kraft lignins typically have a low viscosity that means a compact, spherical structure of the dissolved lignin molecules. Lignin thus behaves quite differently in solution in comparison with cellulose (cf. Section 3.2.2). As can be seen from Table 3-2, the intrinsic viscosity of lignin is only about one-fortieth of that of polysaccharides and one-fourth of synthetic polymers. The exponent a for lignin according to the Mark–Houwink equation (p. 62) ranges from 0.1 to 0.5, corresponding to an intermediate form between an Einstein sphere and a compact coil (Table 3-4).

# Chapter 5

# EXTRACTIVES

A large variety of wood components, although usually representing a minor fraction, are soluble in neutral organic solvents or water. They are called *extractives*. The extractives comprise an extraordinarily large number of individual compounds of both lipophilic and hydrophilic types. The extractives can be regarded as nonstructural wood constituents, almost exclusively composed of extracellular and low-molecular-weight compounds. Similar type of constituents are present in so-called exudates, which are formed by the tree through secondary metabolism after mechanical damage or attack by insects or fungi. Although there are similarities in the occurrence of wood extractives within families, there are distinct differences in the composition even between closely related wood species.

As a rule, various parts of the same tree, that is, stem, branches, roots, bark, and needles, differ markedly with respect to both their amount and composition of extractives. In the case of pines, the heartwood typically contains much more extractives than the sapwood (Fig. 5-1).

The extractives occupy certain morphological sites in the wood structure. For example, the resin acids are located in the resin canals, whereas the fats and waxes are in the ray parenchyma cells. Phenolic extractives are present mainly in the heartwood and in bark.

Different types of extractives are necessary to maintain the diversified

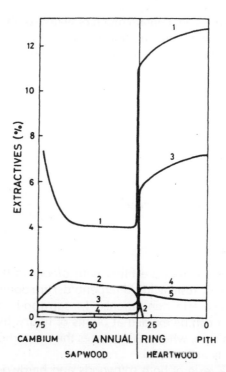

**Fig. 5-1.** Differences in the amount and composition of extractives aross the stem of a Scots pine (*Pinus sylvestris*) tree. Total extractives (1), triglycerides (2), resin acids (3), fatty acids (4), pinosylvin plus mono methyl ether (5) (Assarsson, unpublished; Lindgren and Norin, 1969).

biological functions of the tree. For example, fats constitute the energy source of the wood cells, whereas lower terpenoids, resin acids, and phenolic substances protect the wood against microbiological damage or insect attacks. Traces of certain metal ions are present usually as functional parts of the enzymes which are needed as catalysts for biosynthesis.

Although not very precise, the term "resin" is often used as a collective name for the lipophilic extractives (with the exception of phenolic substances) soluble in nonpolar organic solvents but insoluble in water. Quantitative determination of extractives in wood and pulps is carried out by standardized methods after extraction with organic solvents, such as hexane, dichloromethane, diethyl ether, acetone, or ethanol. The content of extractives is usually less than 10%, but it can vary from traces up to 40% of the dry wood weight. For analytical purposes and for identification of individual components, gas-liquid chromatographic methods in combination with mass spectrometry play a key role.

Normally, wood does not contain much of water-soluble substances, even if high amounts of tannins and arabinogalactans are present in some species. However, arabinogalactans are hemicellulose constituents and are not considered as extractives.

The extractives are important not only for understanding the taxonomy and biochemistry of the trees, but also when considering technological aspects. The extractives constitute a valuable raw material for making organic chemicals and they play an important role in the pulping and paper-making processes.

## 5.1 Terpenoids and Steroids

### 5.1.1 Occurrence

The softwood resin canals are filled with *oleoresin* (Fig. 5-2). Monoterpenoids and especially resin acids (diterpenoids) are dominant and commercially important oleoresin constituents. In sapwood, oleoresin is under pressure and thus it can be exuded at points of injury. In pines oleoresin is the dominant resin type, whereas in spruces the proportion of oleoresin and parenchyma resin is more equal.

The *parenchyma resin* of both softwoods and hardwoods contains triterpenoids and steroids, mainly occurring as fatty acid esters (cf. Section 5.2). A group of acyclic polyterpenoids, so-called betulaprenols, are constituents in birch wood. Some trees produce rubber, gutta percha, and balata, which are polyterpenes.

### 5.1.2 Chemical Composition

*Classification and Biosynthesis* Terpenoids and steroids are formally derived from isoprene units and are therefore sometimes called isoprenoids. Historically the name *terpene* was given to hydrocarbons which were detected in turpentine oil. Today terpenes are known as the large group of hydrocarbons made up of isoprene units ($C_5H_8$). Their respective derivatives with hydroxyl, carbonyl, and carboxyl functions are not hydrocarbons but strictly speaking terpenoids. For simplicity, it has been proposed to call both the terpene hydrocarbons and their derivatives collectively terpenoids. They represent a wide variety of natural products; more than 7500 structures have been clarified.

Steroids are structurally related to terpenoids, but some specific pathways

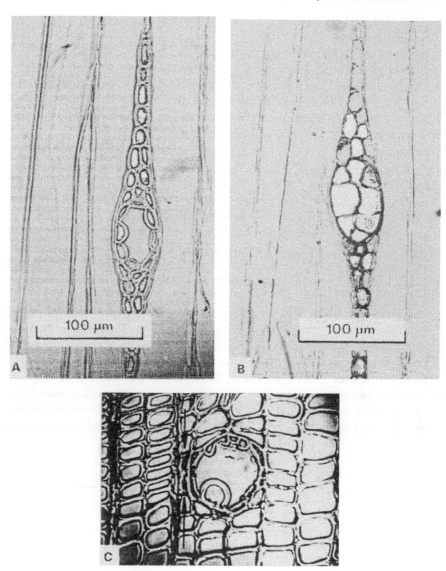

**Fig. 5-2.** Resin canals in Norway spruce (*Picea abies*) (Back, 1969). (A) Horizontal resin canal in a ray (tangential section) originating from the inner annual rings. The canal is surrounded by epithelial cells which secrete resin into the canal cavities. (B) Horizontal resin canal in a ray (tangential section) originating from the outer annual rings. The canal is filled with epithelial cells because of their swelling during sample preparation. (C) Vertical resin canal (cross section).

in their biosynthesis have resulted in certain structural characteristics and biological functions.

Terpenoids can be divided into subgroups according to the number of isoprene units (Table 5-1). Even if the biosynthetic relationships are obvious, the total number of carbon atoms can sometimes deviate from that in the parent biogenetic precursor terpenoids because of further metabolic processes involving cleavage or addition reactions. Mono-, sesqui-, di-, tri-, and polyterpenoids are the most abundant terpenoids in wood.

According to the *isoprene rule* the isoprene units are joined together in a regular head-to-tail way. This rule has been of fundamental importance for the structural studies of this class of compounds. Figures 5-3 and 5-4 show the biogenetic coupling ("prenylation") and the principle of the stepwise formation of various classes of terpenoids and steroids.

In addition to the head-to-tail coupling, other pathways are possible. For example, squalene, the precursor for triterpenoids and steroids, is synthesized through tail-to-tail dimerization of farnesyl pyrophosphate (FPP). An analogous dimerization of geranylgeranyl pyrophosphate (GGPP) results in phytoene, which is the precursor of tetraterpenoids and carotenoids.

The prenylation reaction is stereospecific, mediated by a type of prenyl transferase enzyme which generates mainly isoprenoids with *trans* configuration. In contrast, rubber has an all-*cis* configuration because its synthesis is controlled by another type of prenyl transferase enzyme.

*Monoterpenoids*   The compounds in this group are dominant in the volatile terpenoid fraction ("essential oil") recoverable as turpentine from different parts of the tree by steam distillation or from the digester relief condensates after softwood kraft pulping (cf. Section 10.2.2). Monoterpenoids

TABLE 5-1.   **Classification of Terpenes (Terpenoids)**

| Prefix | Number of carbon atoms | Number of isoprene $(C_5H_8)$ units |
|---|---|---|
| Hemi[a] | 5 | 1 |
| Mono | 10 | 2 |
| Sesqui | 15 | 3 |
| Di | 20 | 4 |
| Sester[a] | 25 | 5 |
| Tri | 30 | 6 |
| Tetra[a] | 40 | 8 |
| Poly | >40 | >8 |

[a] Rare components in woody tissues.

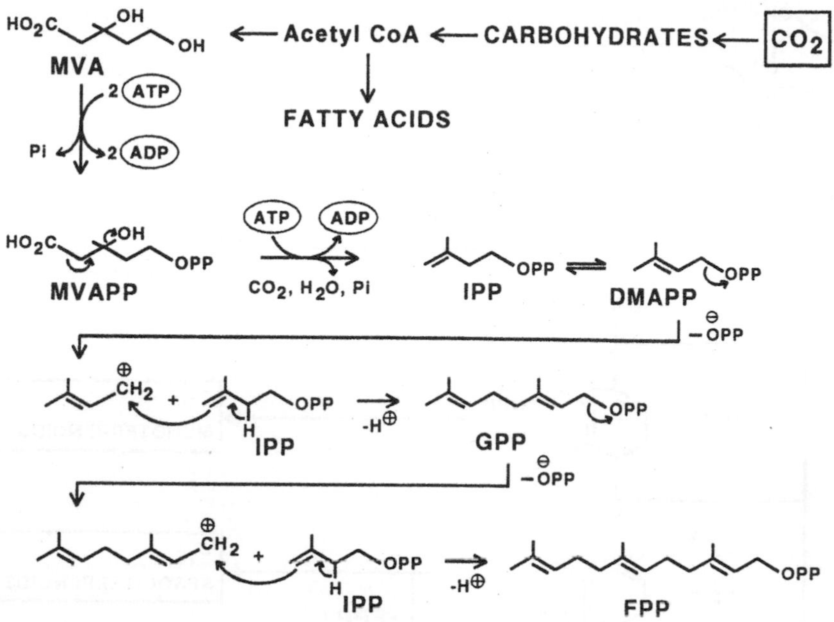

**Fig. 5-3.** Biosynthesis of extractives. Head-to-tail coupling mechanism of terpenoids and steroids (isoprenoids). Mevalonic acid (MVA), the precursor of isoprenoids, is phosphorylated and MVAPP is converted to isopentenyl pyrophosphate (IPP). This is in equilibrium with dimethylallyl pyrophosphate (DMAPP), which is the dominant component formed after enzymatic isomerization. After loss of pyrophosphate ($^-$OPP) from DMAPP, the resulting carbonium ion is coupled with IPP to yield geranyl pyrophosphate (GPP). This product is involved both in the biosynthesis of monoterpenoids and in the further sequence of reactions leading to higher terpenoids. Chain lengthening through coupling of IPP with the respective carbonium ion from GPP results in the formation of farnesyl pyrophosphate (FPP).

occur mainly in softwood oleoresin, either as hydrocarbons or their derivatives. Certain monoterpenoids, such as bornyl acetate, are typical compounds of needles, and are more seldom present in wood.

Monoterpenoids can be divided into acyclic, monocyclic, bicyclic, and tricyclic structural types. They are derived from geranyl pyrophosphate (GPP). So-called tropolones are monoterpenoids having a seven-membered ring system.

Some common monoterpenoids are shown in Fig. 5-5. Of the acyclic types β-myrcene (1) is a minor constituent of turpentine. It is useful for the preparation of flavors and fragrances and can also be synthesized by pyrolysis from β-pinene (cf. Section 10.1.1).

Limonene, a monocyclic monoterpenoid, exists in both of its optically

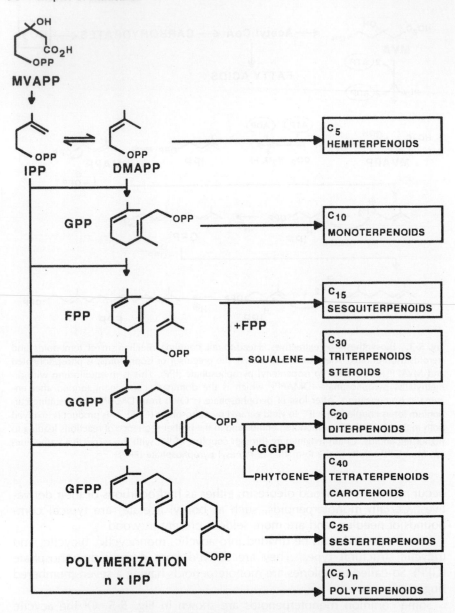

**Fig. 5-4.**  Biosynthesis of terpenoids and steroids.

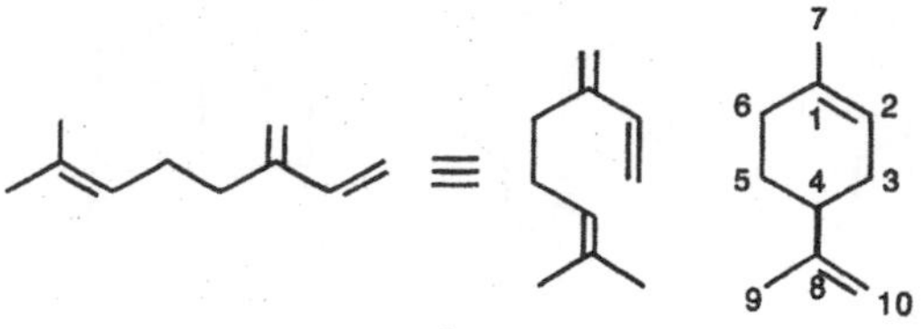

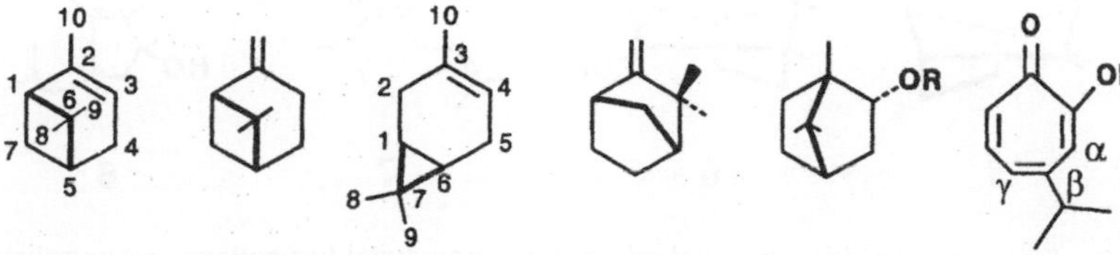

**Fig. 5-5.** Common monoterpenoids present in essential oils and in commercial turpentines. *Acyclic and monocylic types:* β-myrcene (1), (−)-limonene (2), and (−)-β-phellandrene (3). *Bicyclic types:* α-pinene (4), β-pinene (5), 3-carene (6), camphene (7), borneol (R = H) and bornyl acetate (R = acetyl group) (8), and β-thujaplicin (9).

active (+) and (−) forms. The racemized mixture is named dipentene. Both (−)-limonene (2) and (−)-β-phellandrene (3) are present in pines.

The bicyclic monoterpenoids are major turpentine constituents (cf. Table 10-4). The predominant members are α-pinene (4) and β-pinene (5), which can occur both in (+) form and in (−) form. Slash pine (*Pinus elliottii*) contains predominantly (−)-α-pinene. Other bicyclic monoterpenoids are 3-carene (6), a prominent component in *Pinus* species, camphene (7), a minor hydrocarbon, and borneol (8) (occurring as bornyl acetate) in pine wood. Finally, β-thujaplicin (9) is an example of $C_{10}$-tropolones, found in Western red cedar (*Thuja plicata*) heartwood. Tropolones are typical in many decay-resistant cedars. Owing to similarities in the structures, tropolones resemble phenols. Because tropolones form strong metal complexes, they can cause digester corrosion problems in pulping.

*Sesquiterpenoids* More than 2500 sesquiterpenoids have been identified, representing a wide variety of compounds of different skeletal types from acyclic to tetracyclic systems. However, they occur usually only in small amounts and are therefore commercially less important.

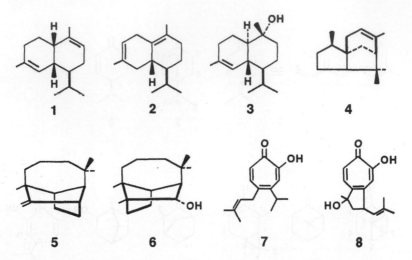

**Fig. 5-6.** Sesquiterpenoids in essential oils and in commercial turpentines. α-Muurolene (1), δ-cadinene (2), α-cadinol (3), α-cedrene (4), longifolene (5), juniperol (6), nootkatin (7), and chanootin (8).

Examples of common sesquiterpenoids are shown in Fig. 5-6. α-Muurolene (1), δ-cadinene (2), and α-cadinol (3) represent a structural type named cadalanes (1,10-cyclization). α-Cedrene (4), longifolene (5), and juniperol (6) are examples of more complex structures. Nootkatin (7) and chanootin (8), occurring in the heartwood of Alaska cedar (*Chamaecyparis nootkatensis*), belong to the group of $C_{15}$-tropolones.

**Diterpenoids**   This group can be divided into acyclic, bicyclic, tricyclic, tetracyclic, and "macrocyclic" structural types. Diterpenoids are present either as hydrocarbons or as derivatives with hydroxyl, carbonyl, or carboxyl groups. Diterpenoids constitute a major part of oleoresin.

Some typical neutral diterpenoids are shown in Fig. 5-7. Of the acyclic types (phytanes), phytol is a part of chlorophyll. Attached to the porphyrin component, it is an important part for the biological function of chlorophyll. However, the phytanes are not very common components in woody tissues. One example is geranyl-linalool (1), which is present in the oleoresin of Norway spruce (*Picea abies*).

Of the bicyclic, so-called labdane type diterpenoids, β-epimanool (2) and cis-abienol (3) occur in Pinaceae. Manoyloxide (4) is a similar type, but it has a heterocyclic ring and is, in fact, a tricyclic compound.

Pimaral (5) and pimarol (6) are the most abundant neutral tricyclic diterpenoids in pines. Cembrene (thunbergene) is a macrocyclic diterpenoid (7). It occurs in spruce and is formed through tail-to-end cyclization.

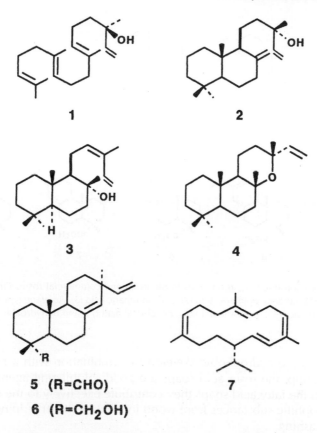

**Fig. 5-7.** Neutral diterpenoids of acyclic, bicyclic, tricyclic, and macrocyclic types. Geranyl linalool (1), β-epimanool (2), *cis*-abienol (3), manoyloxide (4), pimaral (5), pimarol (6), and cembrene (thunbergene) (7).

The *resin acids* are dominant constituents in pine wood oleoresin and in commercially important rosins (cf. Section 10.2.2). The most common resin acids in softwood are tricyclic terpenoids, and they are classified into *pimarane* and *abietane* types (Fig. 5-8). The predominant pimarane type acids are pimaric acid (1) and its isomers sandaracopimaric acid (2) and isopimaric acid (3). They have vinyl and methyl groups at the C-13 position. Abietic acid (4), levopimaric acid (5), palustric acid (6), neoabietic acid (7), and dehydroabietic acid (8) are the major abietane type acids. They have an isopropyl or an isopropenyl group at the C-13 position.

The resin acids of the abietane type with conjugated dienoic structure (4–7) are less stable against isomerization and oxidation than those of pimarane

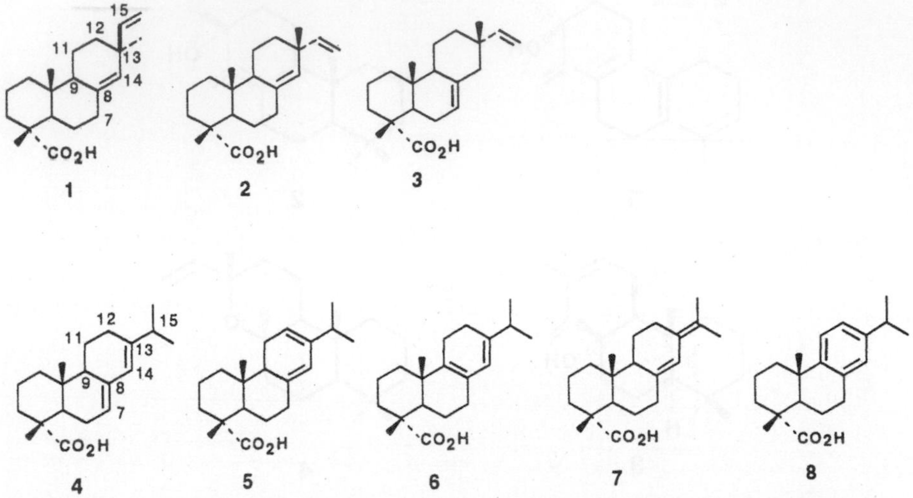

**Fig. 5-8.** Common (tricyclic) resin acids in oleoresin and commercial rosin. *Pimarane type:* Pimaric acid (1), sandaracopimaric acid (2), and isopimaric acid (3). *Abietane type:* Abietic acid (4), levopimaric acid (5), palustric acid (6), neoabietic acid (7), and dehydroabietic acid (8).

type. Due to the hydrophobic skeleton in combination with a hydrophilic carboxyl group, the resin acid soaps are good solubilizing agents, and together with the fatty acid soaps they contribute effectively to the removal of neutral lipophilic substances from wood in alkaline (kraft) pulping and subsequent washing.

Besides the common resin acids, some bicyclic resin acids are known (Fig. 5-9). So-called *labdane* type resin acids are present in some pines. Among these are lambertianic (antidaniellic) acid (*Pinus lambertiana*) (1), communic (elliotinic) acid (*P. elliottii*) (2), and mercusic (dihydroagathic) acid (3) (*P. mercusii*), having two carboxyl groups. Secodehydroabietic acid (4), also formed from levopimaric acid during kraft pulping, contains an aromatic ring.

***Triterpenoids and Steroids*** Compounds belonging to this group are widely distributed in plants. Typical members present in wood are shown in Fig. 5-10.

Triterpenoids and steroids are both structurally and biogenetically closely related, and there is no simple definition for their distinction. The biosynthesis of both triterpenoids and steroids starts from squalene, but proceeds then according to different pathways resulting in characteristic differences in their structures. Apart from the hydroxyl group, sterols are

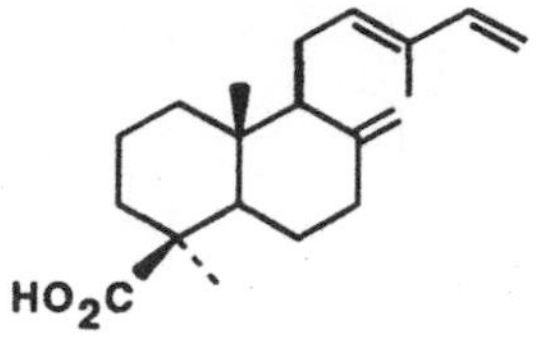

**Fig. 5-9.** Bicyclic resin acids in oleoresin and commercial rosin. Lambertianic (antidaniellic) acid (1), communic (elliotinic) acid (2), mercusic (dihydroagathic) acid (3), and secodehydroabietic acid (4). (Resin acids 1–3 are of labdane type.)

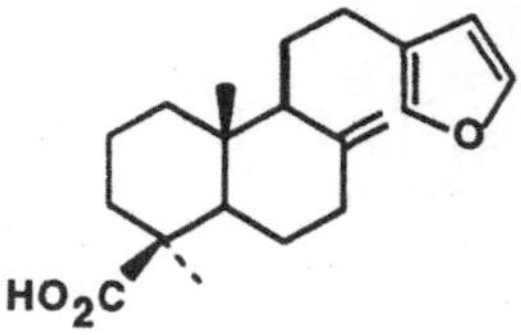

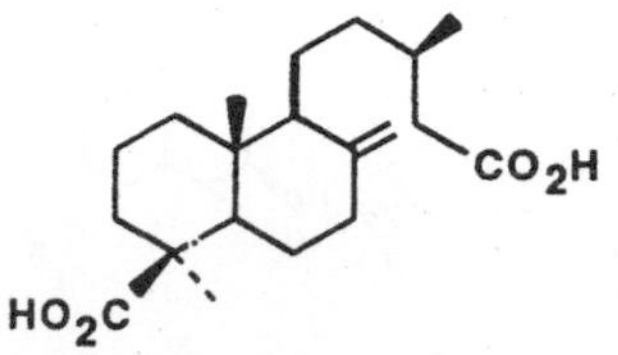

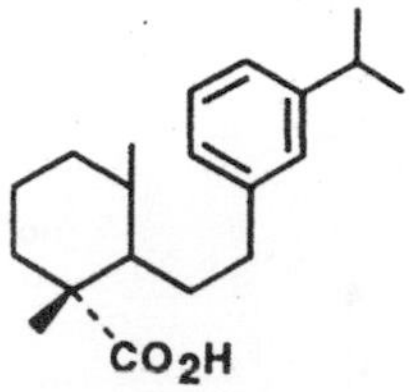

strongly hydrophobic because of the extremely lipophilic hydrocarbon skeleton. Tetracyclic triterpenoids, such as compounds 4–6 in Fig. 5-10, differ from steroids in that they have one or two methyl groups at C-4. Therefore, they are sometimes called methyl or dimethyl sterols.

All the compounds listed in Fig. 5-10 have a hydroxyl group at C-3 position, but they differ from each other in other respects. The most common steroid in wood and higher plants is sitosterol (1). Campesterol (2) is a structurally similar monoenoic steroid but is less abundant than sitosterol. Sitostanol (3) is the saturated analogue of sitosterol. Citrostadienol (4) is a dienoic sterol having a 4α-methyl group. Cycloartenol (5) and 24-methylenecycloartanol (6) are intermediates in the biosynthesis of 24-alkylsteroids.

Betulinol (betulin) (7) is a pentacyclic triterpenoid occuring in large amounts in free form in the outer bark of birch. Serratenediol (8), present in the bark of pines, is a member of a small pentacyclic triterpenoid subgroup, so-called serratanes, having a seven-membered C-ring.

Triterpenoids and steroids occur mainly as fatty acid esters (waxes) and as glycosides, but also in the free form. Being sparingly soluble hydrophobic components they can cause problems in pulping and papermaking pro-

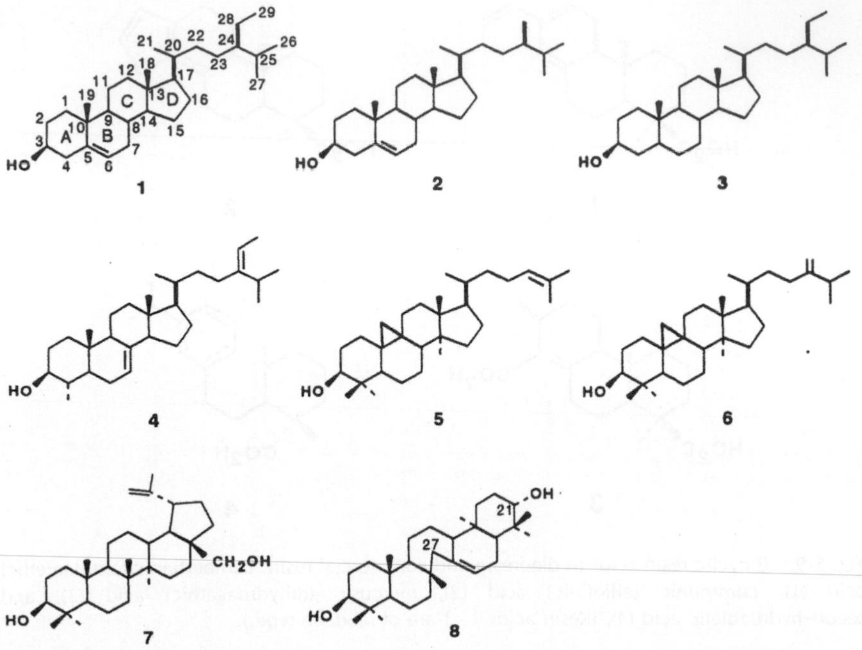

**Fig. 5-10.** Steroids and triterpenoids in wood. Sitosterol [(24R)-24-ethylcholest-5-en-3β-ol] (1), campesterol [(24R)-24-methylcholest-5-en-3β-ol] (2), sitostanol [(24R)-24-ethylcholestan-3β-ol] (3), citrostadienol [4α-methyl-24-ethylcholesta-7, 24 (28) Z-dien-3β-ol] (4), cycloartenol [4, 4-dimethyl-9β, 19-cyclocholest-24-en-3β-ol] (5), 24-methylenecycloartanol (6), betulinol (7), and serratenediol (8).

cesses. On the other hand, steroids and triterpenoids, and especially sitosterol and betulinol, are potential raw materials for making wood chemicals (cf. Chapter 10).

*Polyterpenoids* Acyclic primary alcohols of polyisoprenoids, so-called *polyprenols,* are abundant in higher plants, especially in the leaves, but not in wood. However, a special type of polyprenols, called betulaprenols (Fig. 5-11), occur as fatty acid esters in silver birch (*Betula verrucosa*). They are built up of 6 to 9 isoprene units. The double bonds have both *cis* and *trans* configurations.

Some trees produce rubber, gutta percha, or balata. The degree of polymerization (number of isoprene units) is high in these products. They differ with respect to their stereochemistry: natural rubber has the all-*cis* configuration, whereas gutta percha and balata have the all-*trans* configuration (cf. Fig. 5-11).

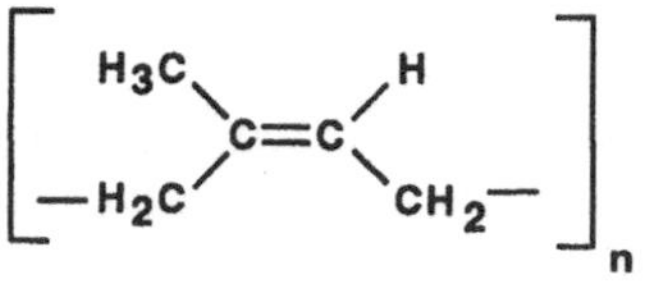

**Fig. 5-11.** Structures of polyterpenoids. Rubber (1) with all-*cis*, balata (2) with all-*trans*, and betulaprenols (3) with mixed *cis/trans* double-bond configurations (about 60% of the double bonds have *cis* form).

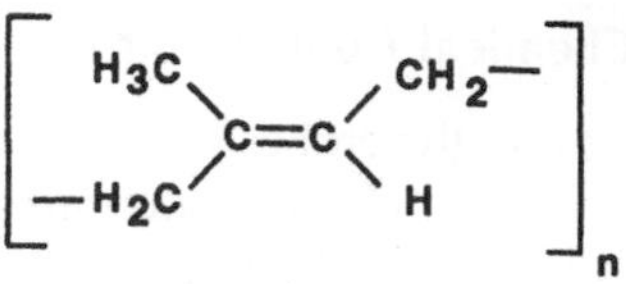

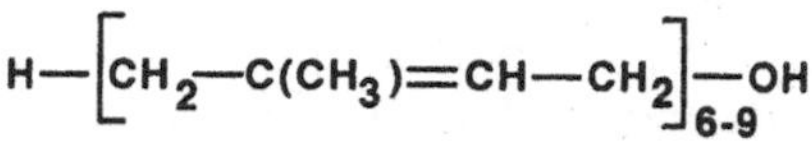

## 5.2    Fats and Waxes

### 5.2.1    Occurrence

The fats and waxes are the predominating constituents of the lipophilic material encapsulated in parenchyma cells. The chemical composition of the *parenchyma resin* is different from that of oleoresin.

In softwoods the parenchyma resin is mainly composed of fats. More than 95% of the parenchyma cells in softwoods are located in the wood rays. In spruce wood most of the ray cells are parenchyma cells, whereas in pine wood the ray tracheids dominate. When spruce wood is sulfite pulped, much of the resin remains encapsulated inside the rigid parenchyma cell walls with minute pores. Removal of ray cells through mechanical fiber fractionation is therefore an effective procedure to reduce the content of extractives of the spruce sulfite pulps to an acceptable level.

In hardwoods the parenchyma resin is virtually the only resin type. In addition to the fats, the waxes are abundant resin constituents. Removal of resin from hardwoods during pulping also depends on the pore dimensions and on the mechanical stability of the parenchyma cells. There are considerable differences among the hardwood species in these respects. For example, the pores in birch and maple parenchyma cells are narrow compared with those of aspen, beech, oak, and *Eucalyptus* species.

## 5.2.2 Chemical Composition

The fats are glycerol esters of fatty acids occuring in wood predominantly as triglycerides. In fresh wood free fatty acids are present practically only in heartwood. Fatty acids are partially liberated from triglycerides during wood storage. More than 30 fatty acids, both saturated and unsaturated, have been identified in softwoods and hardwoods. Examples of the most common fatty acids are shown in Table 5-2. Among the unsaturated $C_{18}$-fatty acids, oleic (monoenoic) acid and linoleic (dienoic) acid are the dominating components. Of the trienoic acids, linolenic acid is a common, although a minor, component in both softwoods and hardwoods. Its isomer, pinolenic acid, is a major fatty acid in pines and spruces. Eicosatrienoic acid is also a common fatty acid in conifers.

The waxes are esters of higher fatty alcohols ($C_{18}$–$C_{24}$), terpene alcohols, or sterols. Other waxy components are free fatty alcohols, among which arachinol ($C_{20}$), behenol ($C_{22}$), and lignocerol ($C_{24}$) dominate.

Fats and waxes (esters) are hydrolyzed during kraft pulping. The fatty acids, which are liberated, can be recovered together with resin acids as soap skimmings from the black liquor (cf. Section 10.2.2).

**TABLE 5-2.** Abundant Fatty Acid Components of Fats and Waxes

| Trivial name | Systematic name | Chain length |
|---|---|---|
| Saturated | | |
| Palmitic | Hexadecanoic | $C_{16}$ |
| Stearic | Octadecanoic | $C_{18}$ |
| Arachidic | Eicosanoic | $C_{20}$ |
| Behenic | Docosanoic | $C_{22}$ |
| Lignoceric | Tetracosanoic | $C_{24}$ |
| Unsaturated | | |
| Oleic | *cis*-9-Octadecenoic | $C_{18}$ |
| Linoleic | *cis,cis*-9,12-Octadecadienoic | $C_{18}$ |
| Linolenic | *cis,cis,cis*-9, 12, 15-Octadecatrienoic | $C_{18}$ |
| Pinolenic | *cis,cis,cis*-5, 9, 12-Octadecatrienoic | $C_{18}$ |
| Eicosatrienoic | *cis,cis,cis*-5, 11, 14-Eicosatrienoic | $C_{20}$ |

# 5.3 Phenolic Constituents

Especially heartwood and bark contain a large variety of complex aromatic extractives. Most of them are phenolic compounds, and many are derived from the phenylpropanoid structure. Thousands of phenolic compounds have been identified. The most important groups are (Fig. 5-12):

1. *Stilbenes,* derivatives of 1,2-diphenylethylene, possess a conjugated double bond system. A typical member is pinosylvin (1), present in pines.

2. *Lignans,* which are formed by oxidative coupling of two phenylpropane ($C_6C_3$) units. Some members are pinoresinol (2), conidendrin (3), plicatic acid (4), and hydroxymatairesinol (5). Lignans related to conidendrin are present in hemlock and spruce species, whereas western red cedar (*Thuja plicata*) contains lignans derived from plicatic acid. Hydroxymatairesinol, which exists in two stereoisomeric forms, is abundant in Norway spruce (*Picea abies*).

3. *Hydrolyzable tannins,* a group of substances which upon hydrolysis yield gallic (6) and ellagic (7) acids and sugars as main products. Tannins of this type are not very common in wood.

4. *Flavonoids,* which have a typical tricyclic, $C_6C_3C_6$, carbon skeleton. Common members are chrysin (8), present in pines, and taxifolin (dihydroquercetin) (9). Taxifolin was first found in Douglas fir heartwood, but there are other sources including *Larix* species. Catechin (10) is an important flavanol precursor of condensed tannins, and genistein (11) belongs to the subgroup of isoflavonoids.

5. *Condensed tannins,* which are polymers of flavonoids. The major source for the condensed tannins of the catechin type are quebracho and chestnut wood and wattle bark, but these polyphenols occur also in many other barks belonging to species such as *Eucalyptus* and *Betula.*

Although compounds belonging to groups 1 and 2 are common in wood, the phenolic extractives are mainly located in heartwood and bark and only traces are present in sapwood. They have fungicidal properties and thus protect the tree against microbiological attack. They also contribute to the natural color of wood. Some of the phenolic compounds, especially pinosylvin and taxifolin, are reactive components in acid sulfite pulping inhibiting the delignification (cf. Section 7.2.5). Accumulation of lignans in paper mill process waters may disturb the papermaking operations. Although the phenolic substances can negatively affect the wood processing operations and product quality, some of them, like tannins, are commercially useful products (cf. Section 10.1.1).

The biosynthesis of extractives is controlled genetically and hence each wood species tends to produce specific substances. As a result of secondary changes, heartwood contains a large variety of phenolic substances. From a chemotaxonomical point of view, chemical structures of various flavonoids, lignans, stilbenes, and tropolones are of great interest. For example, species within families of Taxodiaceae, Cupressaceae, and Pinaceae and within genera of *Pinus, Acasia,* and *Eucalyptus* can be classified according to the composition of their phenolic substances.

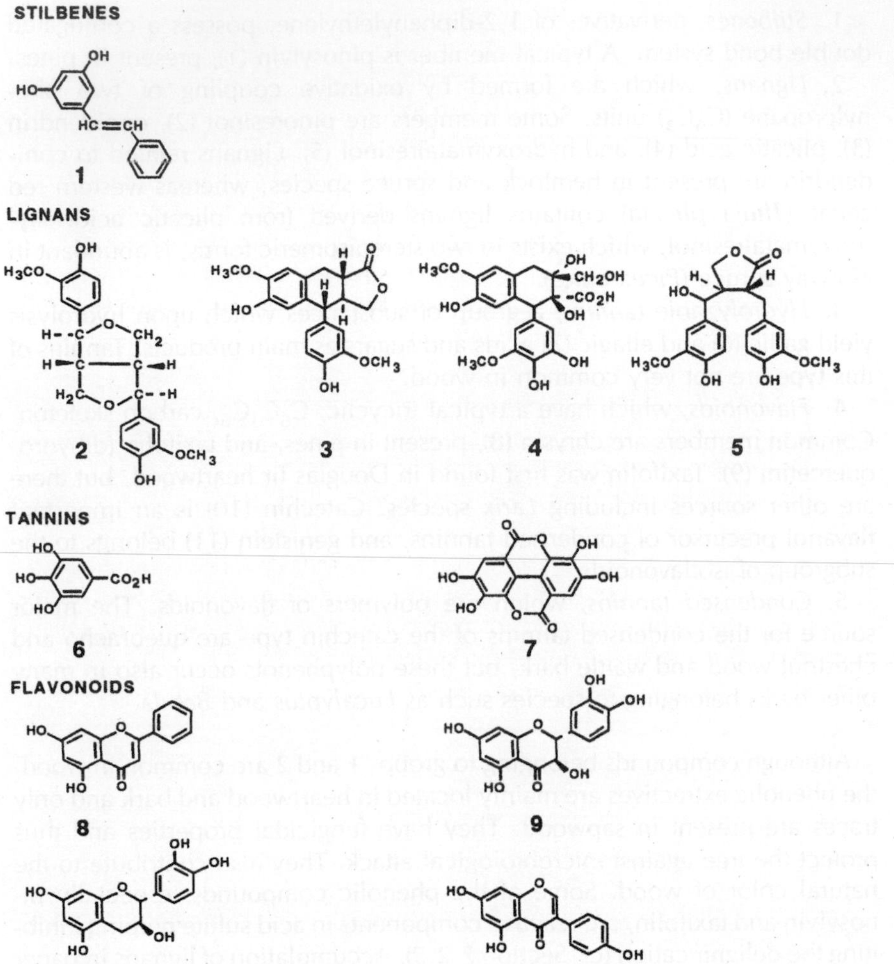

**STILBENES**

**1**

**LIGNANS**

**2**

**3**

**4**

**5**

**TANNINS**

**6**

**7**

**FLAVONOIDS**

**8**

**9**

**10**

**11**

**Fig. 5-12.** Some phenolic extractives and related constituents. *Stilbenes:* pinosylvin (1). *Lignans:* pinoresinol (2), conidendrin (3), plicatic acid (4), and hydroxymatairesinol (5). *Hydrolyzable tannins* (hydrolysis products): gallic acid (6) and ellagic acid (7). *Flavonoids:* chrysin (5,7-dihydroxyflavone) (8), taxifolin (3,5,7,3',4'-pentahydroxyflavanone) (9), catechin (10), and genistein (5,7,4'-trihydroxyisoflavone) (11).

## 5.4    Inorganic Components

Wood contains only rather low amounts of inorganic components, measured as ash seldom exceeding 1% of the dry wood weight. However, the ash content of needles, leaves, and bark can be much higher. This ash originates mainly from a variety of salts deposited in the cell walls and lumina. Typical deposits are various metal salts, such as carbonates, silicates, oxalates, and phosphates. The most abundant metal component is calcium followed by potassium and magnesium. A spectrum of other metals is also present, amounting up to 100 ppm for iron and manganese whereas most of the other metals usually occur only as traces or at least below the 10 ppm limit.

The metals are partially bound to the carboxyl groups present in xylan and pectins or, like heavy metals, such as iron and manganese, held by the wood constituents through complexing forces. Such metal ions can only be displaced and washed out from wood by aqueous acids or complexing agents. Some of the metal salts are sparingly soluble or they are present in inaccessible regions in the wood structure. Owing to these facts it is impossible, or at least extremely difficult, to wash the wood completely free from metal ions. The final pulp usually also contains more or less inorganic impurities at least partly originating from the wood raw material. Trace amounts of heavy metal ions, such as those of iron, cobalt, and manganese, detrimentally affect the bleaching process and the brightness of the final pulp.

## 5.5    Changes Caused by Wood Storage

After felling of the tree the content of lipophilic extractives starts to decrease and their composition is changed. Before sulfite pulping wood is usually stored in the form of chips for a certain time in order to reduce the "pitch problems." However, in the case of kraft pulping, fresh wood can be used. Prolonged wood storage results in negative effects, such as decreased yields of turpentine and tall oil obtained as pulping by-products. Wood polysaccharides can also be attacked by microorganisms during long storage, resulting in reduced pulp yield and low pulp quality.

Both autoxidative and enzymatic processes are taking place during wood storage. The oxygen attacks double bonds in extractives and initiates a chain reaction which generates free radicals. Hydroxyl radicals are particularly strong oxidants. In the presence of transition metal ions the intermediately formed hydroperoxides are also decomposed to hydroxyl radicals. Moreover, the autoxidation reactions are accelerated by light.

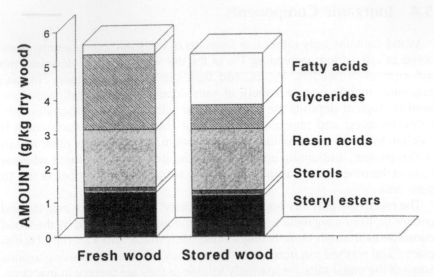

**Fig. 5-13.** Composition of the lipophilic extractives before and after storage of spruce logs in sea water (Southern Finland, four months in the summer) (Ekman, unpublished).

Resin acids with conjugated double bond systems of abietane type are oxidized readily. The unsaturated fatty acids, both free and esterified, are also oxidized. The reaction of linolenic acid proceeds more rapidly than that of linoleic acid, which in turn is oxidized faster than oleic acid.

The trienoic and dienoic fatty acids are oxidized by certain enzymes, such as lipoxidases, as a result of which the hydrophilicity and water solubility of this fraction are increased. In addition to oxidation, certain enzymes act as catalysts in the hydrolysis of fats.

The autoxidative and enzymatic reactions are largely influenced by the conditions prevailing during wood storage. These reactions are markedly faster when the wood is stored in the form of chips instead of logs. It is also known that the hydrolysis of triglycerides leading to free fatty acids proceeds faster when the conditions for wood storage are wet instead of dry. Figure 5-13 illustrates the changes in the composition of lipophilic extractives during storage of spruce logs in water.

# Chapter 6

# BARK

Bark is the layer external to the cambium which surrounds the stem, branches, and roots, amounting to about 10–15% of the total weight of the tree. Debarked wood is normally used for pulping and even traces of bark residues detrimentally affect the pulp quality. The resulting bark waste is usually burned under recovery of heat. Despite extensive studies only a small fraction of bark is used today as raw material for production of chemicals (see Section 10.1).

## 6.1 Anatomy of Bark

Bark is composed of several cell types and its structure is complicated in comparison with wood. In addition to variations occurring within the same species, depending on such factors as age and growth conditions of the tree, each species is characterized by specific features of its bark structure.

Bark can roughly be divided into living inner bark or *phloem* and dead outer bark or *rhytidome*. The tissues of the bark substance are formed either by primary or secondary growth. The primary growth means direct production of embryonal cells at the growing points of the stem apex and their further development to primary tissues. *Epidermis, cortex,* and *primary*

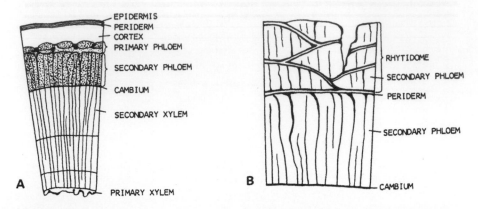

**Fig. 6-1.** Main bark tissues: young stem (A), mature bark (B) (Chang, 1954). © 1954 *TAPPI*. Reprinted from *Anatomy of Common North American Pulpwood Barks*, TAPPI Monograph 14, with permission.

*phloem* are primary tissues (Fig. 6-1). The formation of secondary tissues takes place in two special meristems, in vascular cambium, which produces the secondary phloem, and in the cork cambium (*phellogen*), which generates *periderm*. Continuous division of cells gives rise to several periderm layers. In mature bark the last-formed periderm is the boundary between the inner and outer bark.

### 6.1.1   Inner Bark

The main components of inner bark are sieve elements, parenchyma cells, and sclerenchymatous cells. *Sieve elements* perform the function for transportation of liquids and nutrients. More specifically and according to their shape the sieve elements are divided into sieve cells and sieve tubes. The former types are present in gymnosperms, the latter in angiosperms. The sieve elements are arranged in longitudinal cell rows which are connected through sieve areas. The sieve cells are comparatively narrow with tapering ends, whereas the sieve tubes are thicker and cylindrical. After 1–2 years, or after a longer time in the monocotyledons, the activity of the sieve elements ceases and they are replaced by new elements.

*Parenchyma cells* have the function of storing nutrients and are located between the sieve elements in the inner bark. Both vertical parenchyma cells and horizontal phloem rays are present. The latter are direct continuations of the xylem rays, but much shorter.

*Sclerenchymatous cells* function as the supporting tissue observable in most tree species as layers corresponding to the annual rings in xylem. According to their shape two types are distinguishable: the bast fibers, usually measuring 0.1–3 mm in length and often arranged in tangential rows, and the sclereids or stone cells, which are short and rounded and located as layers between the sieve elements.

### 6.1.2    Outer Bark

The outer bark, which consists mainly of periderm or cork layers, protects the wood tissues against mechanical damage and preserves it from temperature and humidity variations. In most woody plants a periderm replaces the epidermis within the first year of growth. The first periderm in stems usually arises from the cork cambium in the outer surface of bark, either in the subepidermal layer or in the epidermis. The following periderms are then formed in successively deeper layers of the bark or in the bast tissue. Cork tissue is predominantly formed in the outward direction, but some division also occurs inward resulting in so-called phelloderm tissue resembling parenchyma cells. Owing to this sequence the final rhytidome usually occurs as scaly bark and, in addition to the cork cells, contains the same cells as those present in the bast.

The cork cells, which consist of three thin layers and are only rarely pitted, are arranged in radial rows and die at an early stage. They are cemented together to a tight tissue resisting water and gases. Because of different growth activity in the spring and in the late summer separate layers are formed in the bark corresponding to the annual rings in the xylem.

As a dead tissue the rhytidome cannot expand and accommodate the radial growth of the stem and is therefore crushed. The resulting form of the cracked bark depends on the anatomical structure and elasticity of the rhytidome and is typical of each tree species.

## 6.2    Chemistry of Bark

The chemical composition of bark is complicated, varies among the different tree species and also depends on the morphological elements involved. Many of the constituents present in wood also occur in bark, although their proportions are different. Typical of bark is its high content of certain soluble constituents (extractives) such as pectin and phenolic compounds as well as suberins. The mineral content of bark is also much higher than that in wood.

Bark can roughly be divided into the following fractions: fibers, cork cells, and fine substance including the parenchyma cells. The fiber fraction is chemically similar to that of the wood fibers and consists of cellulose, hemicelluloses, and lignin. The other two fractions contain large amounts of extractives. The walls of the cork cells are impregnated with suberin, whereas the polyphenols are concentrated in the fine fraction.

## 6.2.1   Soluble Constituents (Extractives)

Bark extractives can roughly be divided into lipophilic and hydrophilic constituents, although these groups do not have any distinctive boundaries. The total content of both lipophilic and hydrophilic extractives is usually high in bark compared with wood and varies within wide limits among different species, corresponding to 20–40% of the dry weight of bark. These extractives constitute an extremely heterogeneous group of substances some of which are typical of bark but are rarely present in xylem (cf. Chapter 5).

The lipophilic fraction, extractable with nonpolar solvents (ethyl ether, dichloromethane, etc.) consists mainly of fats, waxes, terpenoids, and higher aliphatic alcohols (cf. Sections 5.1 and 5.2). Terpenoids, resin acids, and sterols are located in the resin canals present in the bark and also occur in the cork cells and in the pathological exudate (oleoresin) of wounded bark. Triterpenoids are abundant in bark: sitosterol occurs in waxes, as an alcohol component, and the cork cells in the outer bark (periderm) of birch contain large amounts of betulinol (cf. Fig. 5-10).

The hydrophilic fraction, extractable with water alone or with polar organic solvents (acetone, ethanol, etc.) contains large amounts of phenolic constituents (cf. Section 5.3 and Fig. 5-12). Many of them, especially the condensed tannins (often called "phenolic acids") can be extracted only as salts with dilute solutions of aqueous alkali. For example, considerable quantities of flavonoids, belonging to the group of condensed tannins, are present in the bark of hemlock, oak, and redwood. Monomeric flavonoids, including quercetin and dihydroquercetin (taxifolin), are also present in bark. Small amounts of lignans and stilbenes (e.g., piceatannol in spruce bark) occur as well. Glycosides of simple plant phenols, such as salicin and coniferin are present in barks of Salix and Picea species, respectively. Compounds belonging to the extremely heterogeneous group of hydrolyzable tannins are further phenolic constituents occurring in bark. Because the ester linkages in these tannins are partly hydrolyzed even in warm water, the resulting insoluble ellagic and gallic acids are readily precipitated (cf. Fig. 5-12).

Minor amounts of soluble carbohydrates, proteins, vitamins, etc., are present in the bark. In addition to starch and pectins, oligosaccharides, including raffinose and stachyose have been detected in phloem exudates.

## 6.2.2    Insoluble Constituents

Polysaccharides, lignin, and suberins are the principal cell wall constituents of bark.

*Polysaccharides.* The bast fibers are essentially built up by polysaccharides. Cellulose dominates (roughly 30% of the dry bark weight) in addition to the hemicelluloses, which are of the same type as in wood (see Section 3.3).

In addition, a highly branched arabinan probably occurs in many barks, and especially pines. The connecting strands of the sieve elements are surrounded by a polysaccharide called callose, which is a $(1 \rightarrow 3)$-$\beta$-D-glucan.

*Lignin.* No completely satisfactory data are available on the lignin in bark because of the difficulties to separate it from the phenolic acids. Lignin contents of about 15–30% (based on extractive-free bark weight) have been reported for coniferous bark derived from different wood species. Other studies indicate that inner bark lignin is similar to wood lignin, whereas the outerbark lignin significantly differs from it. Further work is needed, however, to confirm these differences.

*Suberins.* The cork cells in the outer bark contain polyestolides or suberins. The suberin content in the outer layer of the cork oak bark (cork) is especially high and amounts to 20–40% in the periderm of birch bark. Polyestolides are complicated polymers composed of $\omega$-hydroxy monobasic acids which are linked together by ester bonds. In addition, they contain $\alpha,\beta$-dibasic acids esterified with bifunctional alcohols (diols) as well as ferulic and sinapic acid moieties. The chain lengths vary but suberins are enriched with molecules having 16 and 18 carbon atoms. There are also double bonds and hydroxyl groups through which ester and ether cross-links are possible. The outer layer of the epidermis contains so-called cutin, which is heavily branched and has a structure similar to suberin.

## 6.2.3    Inorganic Constituents

Bark contains 2–5% inorganic solids of the dry bark weight (determined as ash). The metals are present as various salts including oxalates, phosphates, silicates, etc. Some of them are bound to the carboxylic acid groups of the bark substance. Calcium and potassium are the predominating metals. Most of the calcium occurs as calcium oxalate crystals deposited in the axial parenchyma cells. Bark also contains trace elements, such as boron, copper, and manganese.

# Chapter 7

# WOOD PULPING

## 7.1  Background and Definitions

Pulp is the basic product of wood, predominantly used for papermaking, but it is also processed to various cellulose derivatives, such as rayon silk and cellophane.

The main purpose of wood pulping is to liberate the fibers, which can be accomplished either chemically or mechanically or by combining these two types of treatments. The common commercial pulps can be grouped into *chemical, semichemical, chemimechanical,* and *mechanical* types (Table 7-1). The term *"high-yield pulp"* is often collectively used for different types of lignin-rich pulps needing mechanical defibration.

Chemical pulping is a process in which lignin is removed so completely that the wood fibers are easily liberated on discharge from the digester or at most after a mild mechanical treatment. Practically all of the production of chemical pulps in the world today is still based on the sulfite and sulfate (kraft) processes, of which the latter predominates. This chapter deals mainly with these pulping processes.

**Sulfite Pulping**  The first patent dealing with pulping of wood with aqueous solutions of calcium hydrogen sulfite and sulfur dioxide in pressurized systems was granted in 1866. This pioneering invention, made in the United

TABLE 7-1.  Commercial Pulp Types

| Pulp type[a] | Yield (% of wood) |
|---|---|
| A. Chemical | 35–65 |
|     Acid sulfite | |
|     Bisulfite | |
|     Multistage sulfite | |
|     Anthraquinone alkali sulfite | |
|     Kraft | |
|     Polysulfide-kraft | |
|     Prehydrolysis-kraft | |
|     Soda | |
| B. Semichemical | 70–85 |
|     NSSC | |
|     Green liquor | |
|     Soda | |
| C. Chemimechanical | 85–95 |
|     Chemithermomechanical (CTMP) | |
|     Chemigroundwood (CGW) | |
| D. Mechanical | 93–97 |
|     Stone groundwood (SGW) | |
|     Pressure groundwood (PGW) | |
|     Refiner mechanical (RMP) | |
|     Thermomechanical (TMP) | |

[a] Main uses: Type A: various paper qualities, board, liner, cellulose derivatives. Type B: board, liner, corrugating medium. Type C: tissue, fluff, etc. Type D: newsprint, supercalendered (SC) paper, and lightweight coated (LWC) paper.

States by B. Tilghman, can be considered to be the origin of the *sulfite pulping process*. It required almost one decade before the world's first sulfite pulp mill started its production in Sweden in 1874. This was accomplished by C. D. Ekman, who is the principal initiator of the sulfite pulp industry.

Essentially, sulfite pulping is still based on these old inventions, although several innovative modifications and technical improvements have been introduced. The later achievements during the 1950s and the 1960s concerned the introduction of so-called soluble bases, that is, replacement of calcium by magnesium, sodium, or ammonium, giving much more flexibility in adjusting the cooking conditions, extending both the raw material basis and production of different pulp types. Also, methods for the recovery of these bases as well as sulfur were developed. Although until the 1950s most of the pulp in the world was based on the sulfite process, the kraft process has gradually taken over its dominating position. However, the sulfite process is still important at least in certain countries and for some pulp qualities.

*Kraft Pulping*   Pressurized alkaline cooking systems at high temperatures were introduced in the 1850s. According to the method proposed by C. Watt and H. Burgess, sodium hydroxide solution was used as a cooking liquor and the resulting spent liquor was concentrated by evaporation and burned. The smelt, consisting of sodium carbonate, was reconverted to sodium hydroxide by calcium hydroxide (caustisizing). Since sodium carbonate was used for makeup, the cooking process was named the *soda process*.

In 1870, A. K. Eaton in the United States patented the use of sodium sulfate instead of sodium carbonate. Similar ideas were pursued by C. F. Dahl, who about 15 years later presented a technically feasible pulping process in Danzig. These inventions initiated the *sulfate (kraft) process*. However, the breakthrough of the kraft process came first in the 1930s after introduction of multistage bleaching systems. Most important was the pioneering work by G. H. Tomlinson in Canada, who developed a recovery furnace suitable for combustion of the kraft "black" liquors. In the kraft process sodium sulfate is added for makeup. It is reduced in the recovery furnace to sodium sulfide, which is the key chemical needed for delignification.

The kraft process has almost completely replaced the old soda process because of its superior delignification selectivity resulting also in a higher pulp quality. Since the 1960s the production of kraft pulps has also increased much more rapidly than that of sulfite pulps because of several factors, such as a simpler and more economic recovery of chemicals and better pulp properties in relation to market needs. The introduction of effective bleaching agents, especially chlorine dioxide, has eliminated the earlier difficulties involved in the bleaching of kraft pulps to a high brightness, and prehydrolysis of wood has made it possible to produce high-grade dissolving pulps by the kraft process.

Although today more than 80% of the chemical pulp produced in the world is kraft pulp, the kraft process still suffers from several weak sides which are difficult to master. These are the malodorous gases and the high consumption of bleaching chemicals of softwood kraft pulps. However, according to recent progress it is to be expected that new modifications will lead to improvements with regard to the environmental needs.

*High-Yield Pulping*   Cooking of wood chips at less drastic conditions or shorter time, leading only to a partial dissolution of lignin, results in *semichemical* pulps. The chemical reactions are similar to those for chemical pulping at corresponding conditions. *Neutral sulfite semichemical* (NSSC) pulps represent a usual type and they are produced by cooking the chips with sodium sulfite–bisulfite solutions prior to the mechanical defibration in disk refiners. The fiber properties of hardwood NSSC pulps make them suitable especially for corrugating medium.

*Chemimechanical* pulps (CMP) are produced by pretreating the wood chips before defibration at rather mild conditions, usually with alkaline solutions of sodium sulfite at elevated temperatures. By this treatment sulfonic acid groups are introduced into the lignin, making it more hydrophilic and increasing the degree of swelling of the pulp. In this category so-called *chemithermomechanical* pulps (CTMP) are rather recent newcomers, and the first mill producing this type of softwood pulp started in 1979 in Sweden. Because of the chemical pretreatment the physical fiber damage in the final mechanical defibration stage is less severe and the strength properties of the pulps obtained are better than those of the *thermomechanical* pulps (TMP) (see later). Although the energy demand is also lower, the process leads to an increased dissolution of the wood substance, which means special procedures for handling and purification of the dilute process solutions in order to avoid pollution of recipients.

**Mechanical Pulping**  The process of wood grinding in which barked wood logs are treated in a rotating grindstone in the presence of water spray forms the basis for *mechanical* pulping. In addition to whole fibers, the wood substance is torn off in the form of more or less damaged fiber fragments. This physical fiber damage cannot be avoided and the strength of the paper made from mechanical pulps is thus rather low. Additional drawbacks of mechanical pulping are the high energy demand and that practically only softwoods, mainly spruce, are useful raw materials.

The method of producing *stone groundwood* pulp (GW or SGW) was developed around 1840 by F. G. Keller in Germany. Newer development during the 1970s resulted in a modified groundwood process in which grinding is done at higher pressures. Since the temperature in the grinding stone is higher, lignin is softened, which favors defibration. Consequently, the pressure groundwood pulp (PGW) has somewhat better strength properties than the conventional GW pulp.

Another technique of mechanical defibration of wood is to use disk refiners, which, of course, requires a preceding chipping. An improved defibration technology was developed in the 1960s resulting in so-called *thermomechanical* pulps (TMP). This type of mechanical pulping means refining after pressurized presteaming and it results in improved strength properties of the pulp. However, a disadvantage is a high energy demand.

**Unconventional Pulping**  During recent years increasing attention in the pulp industry has been focused on minimizing pollution and saving energy.

In the sulfite pulp industry a complete recovery of sulfur dioxide from the flue gases is difficult and costly, but such emissions cannot be accepted because they cause serious environmental effects. In the kraft process this type of emissions does not cause problems, but instead an inherent disadvantage is the generation of volatile organic sulfur compounds, which, even when present as traces, emit an unpleasant odor to the surroundings. This is the reason why kraft pulp mills cannot be established near more crowded residential areas and have not been accepted in countries like West Germany and Switzerland. It is therefore understandable why alternative, "unconventional" processes and possibilities for sulfur-free pulping have attained much interest.

In the middle 1970s it was found that anthraquinone is an effective delignification agent, improving delignification markedly in alkaline conditions like hydrogen sulfide ions. This was a remarkable discovery particularly because only small amounts of anthraquinone are needed to accelerate the delignification markedly. Both anthraquinone and tetrahydroanthraquinone are important delignification agents, but they are today mostly used only as additives for kraft and alkaline sulfite pulping.

Pulping in the presence of organic solvents represents a more radical approach. Although the idea came up as early as the beginning of 1930s, it did not attain serious attention until rather recently. The laboratory and pilot plant experiments so far performed include a number of solvents and catalysts in both acidic and alkaline conditions, but so far at least "organosolv" pulping of softwood seems not very promising and the solvent recovery is problematic. It can therefore be doubted whether organosolv pulping processes can be competitive enough to be introduced more universally into large-scale operations.

Oxygen is an attractive delignification agent. It is today successfully applied for pulp bleaching, but practically no useful oxygen pulping processes have emerged from the intensive research.

Application of lignin-degradative microorganisms as delignification agents has also attained much interest. The idea is not new, but the "biopulping" processes are still unrealistic despite of the rapid development of biotechnology.

So-called steam explosion processes finally represents an unusual type of pulping in which the wood is subjected to a short treatment at high temperatures (200°–250°C) and pressures, followed by a release to atmospheric pressure. After this type of treatment wood is disintegrated, but the fiber structure is extensively damaged. The potential uses for products obtained after steam explosion may include ruminant feeding, fermentation substrates, and chemical feedstocks.

## 7.2 Sulfite Pulping

### 7.2.1 Cooking Chemicals and Equilibria

Sulfur dioxide is a two-basic acid and the following equilibria prevail in its aqueous solution:

$$SO_2 + H_2O \rightleftarrows H_2SO_3 (SO_2 \cdot H_2O) \tag{7-1}$$

$$H_2SO_3 \rightleftarrows H^+ + HSO_3^- \tag{7-2}$$

$$HSO_3^- \rightleftarrows H^+ + SO_3^{2-} \tag{7-3}$$

Since the concentrations of sulfur dioxide in its free ($SO_2$) and hydrated ($SO_2 \cdot H_2O$ or $H_2SO_3$) forms cannot be determined separately, equations (7-1) and (7-2) are combined to give expression (7-4) in which the total sulfur dioxide concentration is in the denominator. This defines the first equilibrium constant $K_1$:

$$K_1 = [H^+] [HSO_3^-] / ([SO_2] + [H_2SO_3]) \tag{7-4}$$

The second equilibrium constant derived from equation (7-3) is

$$K_2 = [H^+] [SO_3^{2-}] / [HSO_3^-] \tag{7-5}$$

By taking the logarithms of both sides of equations 7-4 and 7-5, the following expressions are obtained:

$$pK_1 = pH - \log\{[HSO_3^-] / ([SO_2] + [H_2SO_3])\} \cong 2 \tag{7-6}$$

$$pK_2 = pH + \log([HSO_3^-] / [SO_3^{2-}]) \cong 7 \tag{7-7}$$

It should be noted that $K_1$ and $K_2$ are not thermodynamic constants since the activity coefficients have been neglected, and hence they are strictly valid only at a given concentration.

It follows from these equilibria that the relative concentrations of sulfur dioxide, hydrogen sulfite, and sulfite are governed by the pH of the solution (Fig. 7-1). As can be seen, sulfur dioxide is present almost exclusively in the form of hydrogen sulfite ions at pH around 4. Below and above this value the concentrations of sulfur dioxide and sulfite ions, respectively, are successively increased. These equilibria also vary with the temperature. At temperatures used for pulping (130°–170°C), the actual pH value is higher than that measured at room temperature and this deviation is larger in the acidic region.

Because of the low solubility of calcium sulfite, a large excess of sulfur dioxide is required to avoid its formation from calcium hydrogen sulfite.

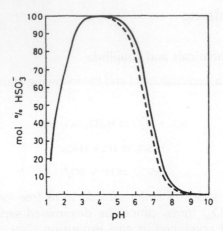

**Fig. 7-1.** The molar ratio of hydrogen sulfite ion concentration to total sulfur dioxide as a function of pH at 25°C (Sjöström et al., 1962). —, 10 g $Na_2O$/liter; ---, 50 g $Na_2O$/liter. $pK_1 \sim$ 1.7; $pK_2 \sim$ 6.6–6.8.

Calcium base is thus usable only for acid sulfite pulping. Magnesium sulfite is much more soluble. When using magnesium as base, the pH can be increased to about 4–5, but above this range magnesium sulfite starts to precipitate and in alkaline region magnesium precipitates as hydroxide. Sodium and ammonium sulfites and hydroxides are easily soluble, and the use of these bases have no limitations in the pH of the cooking liquor.

There are several modifications of the sulfite method which are designated according to the pH of the cooking liquor (Table 7-2). For the production of chemical pulps, delignification is allowed to proceed until most of the lignin in the middle lamella is removed after which the fibers can be readily separated from each other. Semichemical pulps are often produced by the neutral sodium sulfite method followed by mechanical fiberization of the partially delignified wood.

According to the usual but rather misleading convention, the total amount of sulfur dioxide is divided into "free" and "combined" sulfur dioxide. For example, sodium hydrogen sulfite solution contains equal amounts of combined and free sulfur dioxide (2 $NaHSO_3 \rightarrow Na_2SO_3 + SO_2 + H_2O$), although essentially no free sulfur dioxide exists in such a solution. The term "active base" refers to the sum of the hydrogen sulfite and sulfite ions and is usually expressed as oxide, e.g., CaO or $Na_2O$. A typical acid sulfite cooking liquor contains about 10 g and 60 g combined and free sulfur dioxide per liter, respectively.

**TABLE 7-2. Sulfite Pulping Methods and Conditions**

| Method | pH range | "Base" alternatives | Active reagents | Max. temp. (°C) | Time at max. temp. (hr) | Softwood pulp yield (%) |
|---|---|---|---|---|---|---|
| Acid (bi)sulfite | 1-2 | $Ca^{2+}$, $Mg^{2+}$, $Na^+$, $NH_4^+$ | $HSO_3^-$, $H^+$ | 125-145 | 3-7 | 45-55 |
| Bisulfite | 3-5 | $Mg^{6+}$, $Na^+$, $NH_4^+$ | $HSO_3^-$, $H^+$ | 150-170 | 1-3 | 50-65 |
| Two-stage sulfite (Stora type) | | | | | | |
| Stage 1 | 6-8 | $Na^+$ | $HSO_3^-$, $SO_3^{2-}$ | 135-145 | 2-6 | 50-60 |
| Stage 2 | 1-2 | $Na^+$ | $HSO_3^-$, $H^+$ | 125-140 | 2-4 | |
| Three-stage sulfite (Sivola type) | | | | | | |
| Stage 1 | 6-8 | $Na^+$ | $HSO_3^-$, $SO_3^{2-}$ | 120-140 | 2-3 | 35-45 |
| Stage 2 | 1-2 | $Na^+$ | $HSO_3^-$, $H^+$ | 135-145 | 3-5 | |
| Stage 3 | 6-10 | $Na^+$ | $HO^-$ | 160-180 | 2-3 | |
| Neutral sulfite (NSSC) | 6-9 | $Na^+$, $NH_4^+$ | $HSO_3^-$, $SO_3^{2-}$ | 160-180 | 0.25-3 | 75-90[a] |
| Alkaline sulfite[b] | 9-13 | $Na^+$ | $SO_3^{2-}$, $HO^-$ | 160-180 | 3-5 | 45-60 |

[a] Hardwood.
[b] Including this method in the presence of anthraquinone (AQ).

### 7.2.2 Impregnation

The cooking process begins with an impregnation stage after the chips have been immersed in the cooking liquor. This stage involves both the liquid *penetration* into wood cavities and the *diffusion* of dissolved cooking chemicals. The rate of penetration depends on the pressure gradient and proceeds comparatively rapidly, whereas diffusion is controlled by the concentration of dissolved chemicals and takes place more slowly. Penetration is influenced both by the pore size distribution and capillary forces while diffusion is regulated only by the total cross-sectional area of accessible pores.

A good impregnation is a prerequisite for a satisfactory cook. If the transport of chemicals into the chips is still incomplete after the cooking temperature has been reached, undesirable reactions catalyzed by hydrogen ions will occur. For instance, if the base concentration in an acid sulfite cook is insufficient the sulfonic acids formed are not neutralized, and the pH value of the cooking liquor drops sharply. Because of the low pH value, reactions leading to lignin condensation as well as to decomposition of the cooking acid are accelerated in the interior of the chips resulting in dark, hard cores. In the choice of the length, pressure and temperature of impregnation, consideration must be given to the wood used; for example, heartwood is much more difficult to impregnate than sapwood.

### 7.2.3 Morphological Factors

Ideally, the purpose of pulping is to remove the lignin as completely as possible and preferably from the middle lamella. In reality, however, the

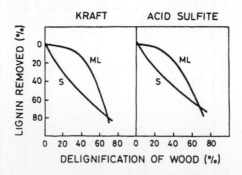

**Fig. 7-2.** Delignification of the secondary wall (S) and compound middle lamella (ML) during kraft and acid sulfite pulping (Wood and Goring, 1973). Note that the S wall is delignified faster than the ML layer at the earlier stages of the cook.

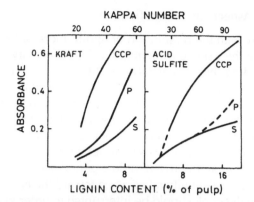

**Fig. 7-3.**   UV absorbance (222 nm, 0.5 μm section thickness) by various morphological regions of spruce fibers delignified to various lignin contents by the kraft and acid sulfite method (Wood and Goring, 1973). S, secondary wall; P, primary wall; CCP, primary wall at the cell corner.

polysaccharides located mainly in the secondary wall region are attacked by the cooking chemicals and their losses cannot be avoided.

Despite of conflicting opinions it is to be expected that the chemicals diffuse gradually from lumen through the cell layers reaching the middle lamella at the end. Indeed, this view has been supported by experimental data according to which the delignification during both kraft and sulfite pulping proceeds faster in the secondary wall than in the middle lamella (Fig. 7-2). Toward the end of the cook the differences in lignin concentration across the cell walls are reduced to a more uniform level (Fig. 7-3). Most of the lignin in the final pulp fibers is located in the secondary wall and in the cell corners (Table 7-3).

**TABLE 7-3.   Distribution of Lignin in Kraft and Acid Sulfite Fibers of Spruce Earlywood[a]**

| | Proportion of total lignin in fiber (%) | | | | |
|---|---|---|---|---|---|
| Pulping method: | Kraft | | Acid sulfite | | |
| Kappa number: | 50 | 25 | 50 | 25 | 15 |
| Secondary wall | 73 | 87 | 88 | 90 | 92 |
| Primary wall | 14 | 10 | 8 | 8 | 8 |
| Cell corner | 13 | 3 | 4 | 2 | 0 |

[a] From Wood and Goring (1973).

### 7.2.4 General Aspects of Delignification

Simultaneously with the dissolution of lignin, more or less carbohydrates are removed from the wood during pulping. The *selectivity* of delignification can be expressed as the weight ratio of the lignin and carbohydrates removed from the wood after a certain cooking time or at a given degree of delignification. A high selectivity thus means low carbohydrate losses. Figure 7-4 illustrates this situation for various pulping methods. The losses of carbohydrates are high in the beginning of the cook, which means that they are attacked even at a relatively low temperature when delignification still proceeds slowly. After an improved middle period of delignification, a rather abrupt change in the selectivity takes place toward the end of the cook. This is the point when the cook should be interrupted in order to avoid high yield losses and impairment of pulp properties.

Basically, two types of reactions, *sulfonation* and *hydrolysis,* are responsible for delignification in sulfite pulping. Sulfonation generates hydrophilic sulfonic acid groups in the hydrophobic lignin polymer, while hydrolysis breaks ether bonds between the phenylpropane units, lowering the molecular weight and creating free phenolic hydroxyl groups. Both of these

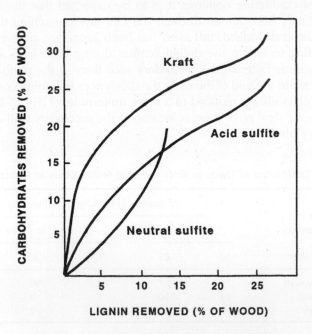

**Fig. 7-4.** Comparison of delignification selectivities.

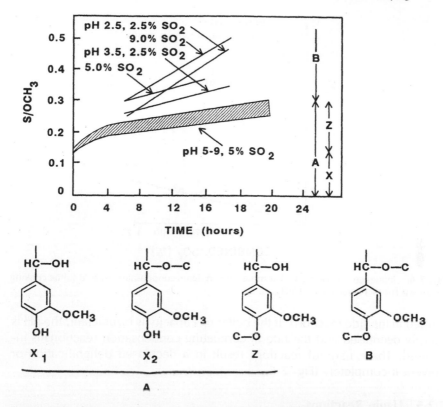

**Fig. 7-5.** The degree of sulfonation of lignin (measured as $S/OCH_3$) after treatments of softwood meal (spruce) at various pH conditions (135°C). The structures within group A are sulfonated already at a relatively neutral pH region and the rate of reaction of the X types is high. Those within group B are sulfonated only at acidic conditions. This figure roughly illustrates the sulfonation of the benzyl position in free and etherified lignin structures. (Adopted from Lindgren, 1952.)

reactions increase the hydrophilicity of the lignin, rendering it more soluble.

At the pH conditions used for neutral and alkaline sulfite pulping, the hydrolysis reactions are very slow compared with sulfonation, and the degree of sulfonation of lignin remains low (Fig. 7-5). Delignification therefore proceeds slowly. At the conditions of acid sulfite pulping hydrolysis is fast compared with sulfonation, which assumes the role of rate-determining step. Because lignin is also sulfonated to a fairly high degree, the conditions are favorable, promoting extensive dissolution of lignin.

A certain amount of base is required for neutralization of both the lignosulfonic acids and the acidic degradation products of the wood substance

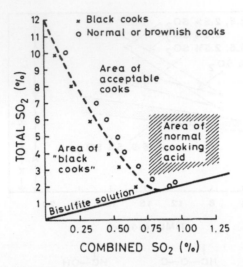

**Fig. 7-6.** Influence of cooking acid composition on lignin condensation in acid sulfite cooking of spruce (130°C) (Kaufmann, 1951).

formed in the side reactions. If the buffering capacity is insufficient, the pH is sharply decreased and the rate of competing condensation reactions is increased. These harmful reactions result in a decreased delignification or prevent it completely (Fig. 7-6).

### 7.2.5 Lignin Reactions

At a given temperature, the extent of delignification depends largely on the acidity of the cooking liquor. Conditions typical of acid sulfite pulping (140°C, pH 1–2) result in effective delignification, whereas after a corresponding treatment in neutral sulfite solution most of the lignin remains insoluble. The average degree of sulfonation of the undissolved softwood lignin, expressed as the molar ratio of sulfonic acid groups to lignin methoxyl, also remains low ($SO_3H/OCH_3 \sim 0.3$). In contrast, the degree of sulfonation of softwood lignosulfonates dissolved during acid sulfite pulping is much higher or about 0.5 (cf. Fig. 7-5). This type of difference is also typical for hardwood lignin, but the degree of sulfonation is throughout much lower than that of softwood lignin.

Most of the sulfonic acid groups introduced into the lignin replace hydroxyl or ether functions at the $\alpha$-carbon atom of the propane side chain. The sulfonation proceeds rapidly at all pH values when the phenolic hydroxyl group located at the *para* position is free. Experiments especially with dimeric model compounds representing various structures and bond types

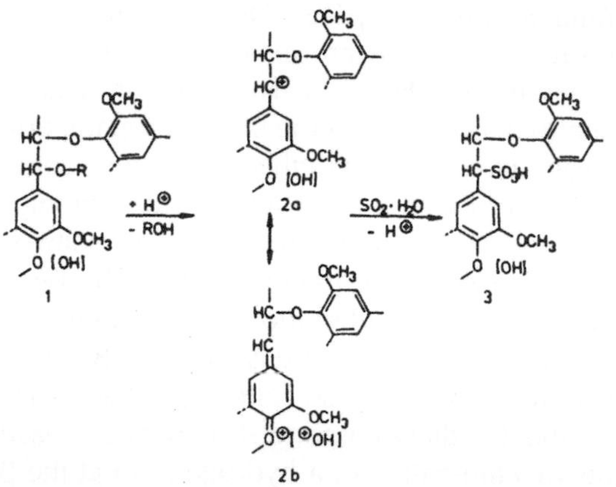

**Fig. 7-7.** Behavior of β-aryl ether and open α-ether structures during acid sulfite pulping (Gellerstedt and Gierer, 1971). R = H, alkyl, or aryl group. The first reaction step involves cleavage of the α-ether bond with formation of a resonance-stabilized carbonium ion which is then sulfonated. Note that both the phenolic and nonphenolic structures are sulfonated, while the β-aryl ether bonds are stable.

in lignin have provided insight into these reactions. Under acidic conditions the most important structural units in lignin are sulfonated irrespective of whether they are free or etherified. Under neutral conditions, however, sulfonation as well as cleavage of the ether bonds, leading to lignin fragmentation, is essentially restricted to the phenolic units.

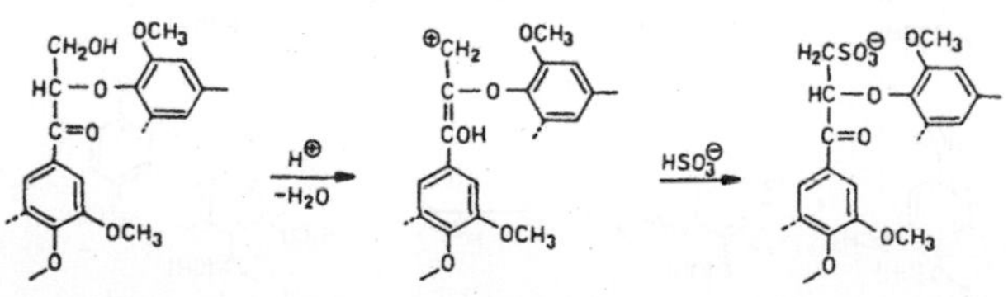

**Fig. 7-8.** Sulfonation of coniferaldehyde end groups and substituted structures containing α-carbonyl groups (see Gellerstedt, 1976). At lower temperatures aldehyde end groups can bind sulfur dioxide because of the formation of α-hydroxysulfonic acid.

*Acid Sulfite (and Bisulfite) Pulping* During acid sulfite pulping the $\alpha$-hydroxyl and the $\alpha$-ether groups are cleaved readily under simultaneous formation of benzylium ions (Fig. 7-7). This reaction takes place regardless of whether the phenolic hydroxyls of the phenylpropane units are etherified or free. The cleavage of open $\alpha$-aryl ether bonds represents the only noteworthy fragmentation of lignin during acid sulfite pulping. Although relatively few open $\alpha$-aryl ether bonds are present in softwood lignin, their cleavage results in a considerable fragmentation. The benzylium ions are sulfonated by attack of hydrated sulfur dioxide or bisulfite ions present in the cooking liquor. The coniferaldehyde end groups and $\beta$-substituted structures containing $\alpha$-carbonyl groups are also sulfonated (Fig. 7-8). The benzylium ions formed from the 1,2-diarylpropane structures are easily converted to stilbene structures by elimination of a hydrogen ion at the $\beta$-position, after which the electrophilic $\gamma$-carbon atoms can be sulfonated (Fig. 7-9).

Condensation reactions of carbonium ions compete with sulfonation and their frequency is increased with increasing acidity. Carbon-carbon bonds are formed most commonly when the benzylium ions react with the weakly nucleophilic 1- and 6-(or 5-)positions of other phenylpropane units (Fig. 7-10). Subsequently, propane side chains and hydrogen ions are eliminated,

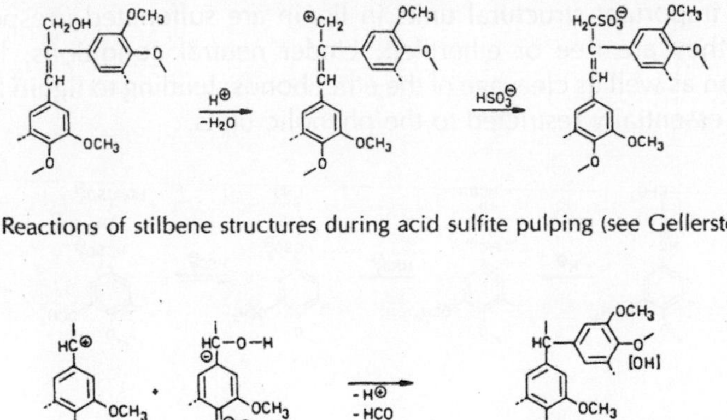

**Fig. 7-9.** Reactions of stilbene structures during acid sulfite pulping (see Gellerstedt, 1976).

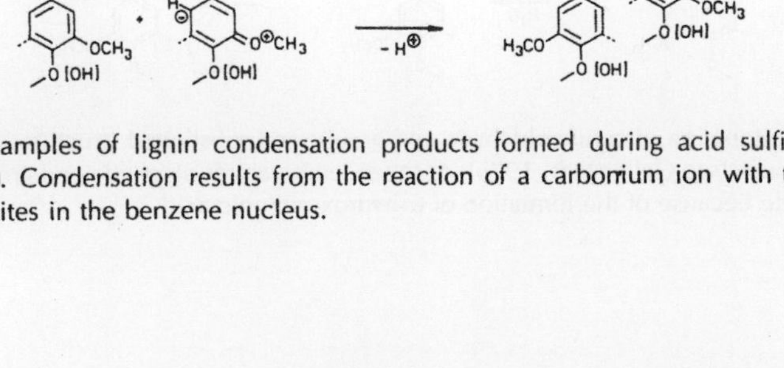

**Fig. 7-10.** Examples of lignin condensation products formed during acid sulfite pulping (Gierer, 1970). Condensation results from the reaction of a carbonium ion with the weakly nucleophilic sites in the benzene nucleus.

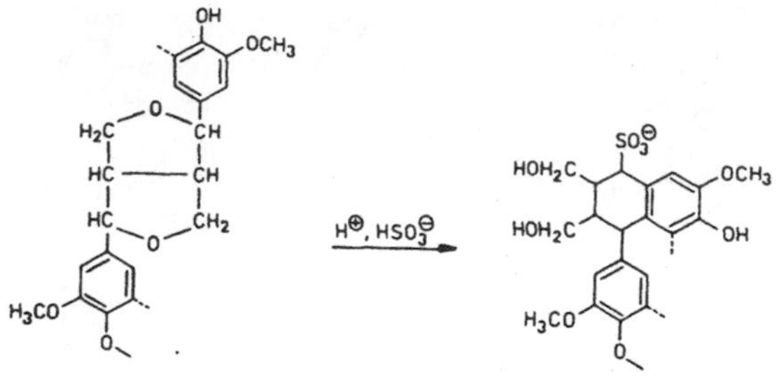

**Fig. 7-11.** Intramolecular condensation of pinoresinol structures during acid sulfite pulping (see Gellerstedt, 1976).

respectively. In general, the condensation reactions result in increased molecular weight of the lignosulfonates and the solubilization of lignin is retarded or inhibited. However, the benzylium ions formed from phenyl coumaran and pinoresinol structures can be condensed intramolecularly without increasing the molecular weight (cf. Fig. 7-11).

During acid sulfite pulping, lignin may also condense with reactive phenolic extractives. Pinosylvin and its monomethyl ether, present in pine heartwood, are examples of phenolic extractives of this type. Dual condensation of pinosylvin with lignin generates harmful cross-links. Consequently, pine heartwood cannot be delignified by the conventional acid sulfite method.

Cross-links between lignin entities may also be generated by thiosulfate present in the cooking liquor (Fig. 7-12). This results in retarded delignification and, under certain circumstances, in complete inhibition ("black cook"). For the formation of thiosulfate, see Section 7.2.8.

**Neutral and Alkaline Sulfite Pulping**   In neutral sulfite pulping the most important reactions of lignin are restricted to phenolic lignin units only. The first stage always proceeds via the formation of a quinone methide with

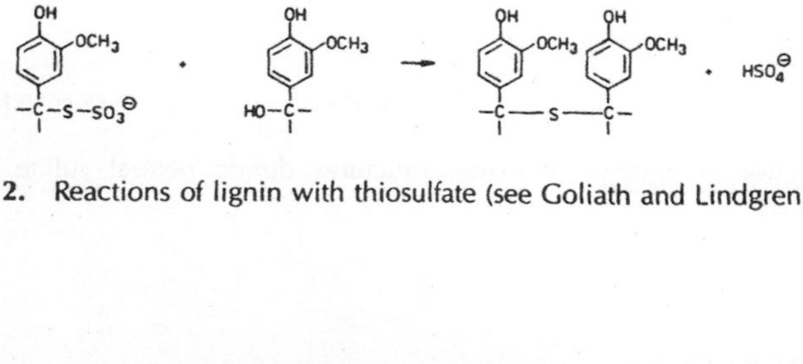

**Fig. 7-12.** Reactions of lignin with thiosulfate (see Goliath and Lindgren, 1961).

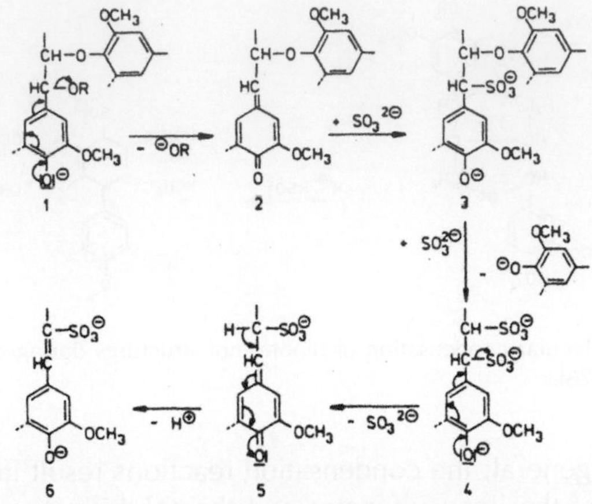

**Fig. 7-13.** Reactions of phenolic β-aryl ether and α-ether structures (1) during neutral sulfite pulping (Gierer, 1970). R = H, alkyl, or aryl group. The quinone methide intermediate (2) is sulfonated to structure (3). The negative charge of the α-sulfonic acid group facilitates the nucleophilic attack of the sulfite ion, resulting in β-aryl ether bond cleavage and sulfonation. Structure (4) reacts further with elimination of the sulfonic acid group from α-position to form intermediate (5) which finally after abstraction of proton from β-position is stabilized to a styrene-β-sulfonic acid structure (6). Note that only the free phenolic structures are cleaved, whereas the nonphenolic units remain essentially unaffected.

simultaneous cleavage of an α-hydroxyl or an α-ether group (Fig. 7-13). At least in noncyclic structures, the quinone methide is readily attacked by a sulfite or a bisulfite ion. The α-sulfonic acid group formed facilitates the nucleophilic displacement of the β-substituent in β-aryl ether structures by a sulfite or bisulfite ion. Subsequent loss of the α-sulfonate group leads to a styrene-β-sulfonic acid structure, especially at higher pH values (>7). The

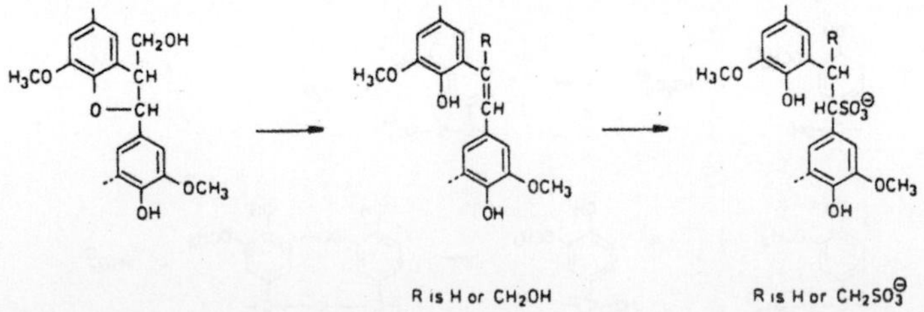

R is H or CH₂OH          R is H or CH₂SO₃⁻

**Fig. 7-14.** Reaction of phenyl coumaran structures during neutral sulfite pulping (see Gellerstedt, 1976).

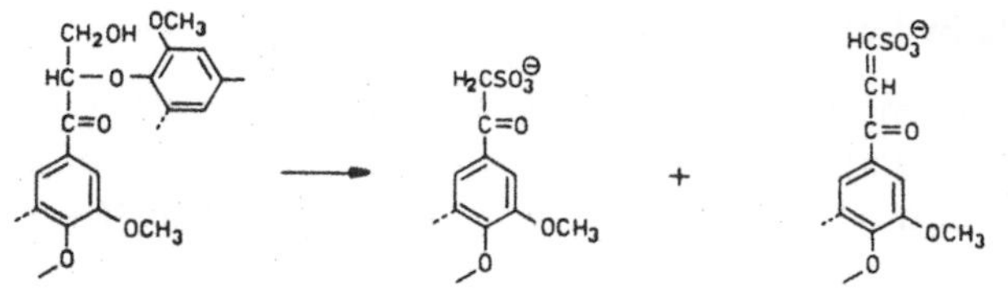

**Fig. 7-15.** Cleavage of β-aryl ether bonds in structures containing α-carbonyl groups (see Gellerstedt, 1976). Note that this reaction can take place even in nonphenolic units.

cleavage of the α- as well as the β-aryl ether bonds naturally generates new reactive phenolic units.

The quinone methides can also react by elimination of formaldehyde or hydrogen ion at the β-carbon atom, especially when the formation of conjugated diaryl structures is possible. Examples of this type of reactions are the formation of stilbenes from phenyl coumarans or 1,2-diarylpropane structures and that of 1,4-diarylbutadienes from pinoresinol structures (cf. Fig. 7-14). The conjugated structures can be further sulfonated at their α-carbon atoms. Moreover, the quinone methides can condense with nucleophilic sites of other phenylpropane units or with thiosulfate. The condensation products are similar to those formed during acid sulfite pulping.

Carbonyl groups present in lignin may have a great influence on its reactions with neutral sulfite. For example, α-carbonyl groups can activate the β-aryl ether bonds in nonphenolic units and induce their cleavage (Fig. 7-15). Coniferaldehyde end groups are also extensively sulfonated. Finally, methoxyl groups which are completely stable toward acid sulfite may, in part, be cleaved during neutral sulfite pulping with the formation of methane sulfonic acid (Fig. 7-16).

Although adequate information is lacking, the reactions of lignin with alkaline sulfite are largely related to those occurring in neutral sulfite and alkali pulping. During alkaline sulfite pulping the β-aryl ether bonds are obviously cleaved also in nonphenolic units, and the condensation reactions have been proposed to be less important compared with those during kraft pulping.

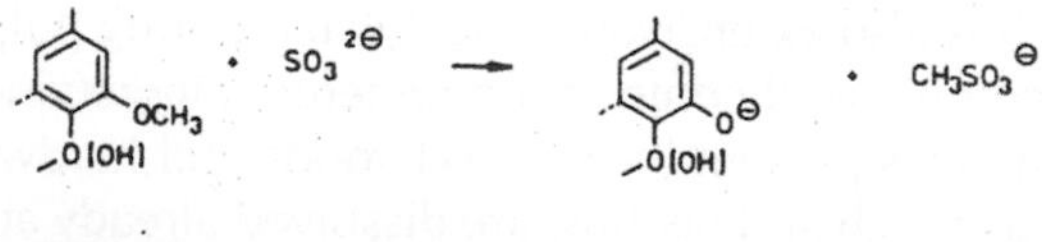

**Fig. 7-16.** Cleavage of the methyl aryl ether bond with formation of methanesulfonic acid during neutral sulfite pulping (Gierer, 1970).

## 7.2.6    Carbohydrate Reactions

The mechanisms of the reactions of carbohydrates have been dealt with in Sections 2.5.4 and 2.5.5. In this section consideration is given to the changes of wood polysaccharides during pulping as well as to the reaction products and pulp yield.

*Acid Sulfite Pulping*    Because of the sensitivity of glycosidic linkages toward acidic hydrolysis, depolymerization of wood polysaccharides cannot be avoided during acid sulfite pulping. Hemicelluloses are attacked more readily than cellulose due to their amorphous state and a relatively low degree of polymerization. Moreover, most of the glycosidic linkages of hemicelluloses are more labile toward acid hydrolysis than those of cellulose. When the hydrolysis has proceeded far enough, the depolymerized hemicellulose fragments are dissolved in the cooking liquor and are gradually hydrolyzed to monosaccharides.

The ordered structure protects cellulose. Only a moderate depolymerization of cellulose and practically no yield losses take place unless the delignification is extended to very low lignin contents and the conditions are rather drastic as is the case when producing dissolving pulp. After a certain degree of depolymerization the fiber strength weakens drastically, which, of course, is not acceptable in the case of paper-grade pulps.

The main hemicellulose component of softwoods constitutes of acetylated galactoglucomannans. The galactosidic linkages are completely hydrolyzed at normal sulfite pulping conditions. These conditions are also strong enough to remove the acetyl groups. Glucomannan is thus the component remaining in the pulp. Arabinoglucuronoxylan, the other hemicellulose constituent of softwoods, is converted to glucuronoxylan. Because the furanosidic linkages of the arabinose units are extremely labile toward acid, they are cleaved already at early stages of the cook.

The glycuronide bonds are exceptionally stable toward acid, contrary to the glycosidic bonds. However, the glucuronic acid content of the xylan fraction remaining in pulp is lower than that of the native xylan. The xylan fractions heavily substituted with high uronic acid groups are obviously more readily dissolved, whereas those having fewer side chains are preferentially retained in the pulp.

Hardwood hemicelluloses are mainly composed of acetylated glucuronoxylan, which is also extensively deacetylated during pulping. The fractions of low glucuronic acid contents are preferentially retained in the pulp.

The polysaccharides present in both softwoods and hardwoods in minor quantities, such as starch and pectins, are dissolved already at early stages of the cook. The chemical pulp is practically free from these polysaccharide components.

The hydrolyzed and soluble sugar fragments are not completely stable at the cooking conditions. In addition to various minor degradation products, about 10–20% of the monosaccharides are oxidized to aldonic acids by the hydrogen sulfite ions. This is an important reaction particularly because it gives rise to the formation of thiosulfate:

$$2R\text{-}CHO + 2HSO_3^- \rightarrow 2R\text{-}CO_2H + S_2O_3^{2-} + H_2O \qquad (7\text{-}8)$$

In addition to the formation of aldonic acids, a minor fraction of the monosaccharides is converted to sugar sulfonic acids.

Generally no cellulose is lost in the acid sulfite process. The hemicellulose yield losses are higher for hardwood than for softwood. Typical material balances resulting from acid sulfite (paper-grade) pulping of both softwood (spruce) and hardwood (birch) are given in Table 7-4. For comparison, the corresponding values for kraft pulping of pine and birch are included in this table.

*Multistage Sulfite Pulping* The carbohydrate yield can be improved considerably by applying a two-stage pulping method. According to the "Stora method" the wood chips are first precooked in a sodium bisulfite-sulfite solution, usually at pH of 6–7, and then subjected to a second cooking stage at acid sulfite pulping conditions. In the first stage lignin is sulfonated to a

**TABLE 7-4. Yields of Various Pulp Constituents after Sulfite and Kraft Pulping of Norway Spruce (*Picea abies*), Scots Pine (*Pinus sylvestris*), and Birch (*Betula verrucosa*)[a]**

| Constituents | Spruce sulfite | | Birch sulfite | | Pine kraft | | Birch kraft | |
|---|---|---|---|---|---|---|---|---|
| Cellulose | 41 | (41)[b] | 40 | (40) | 35 | (39) | 34 | (40) |
| Glucomannan | 5 | (18) | 1 | (3) | 4 | (17) | 1 | (3) |
| Xylan | 4 | (8) | 5 | (30) | 5 | (8) | 16 | (30) |
| Other carbohydrates and various components | — | (4) | — | (4) | — | (5) | — | (4) |
| Sum of carbohydrates | 50 | (69) | 46 | (74) | 44 | (67) | 51 | (74) |
| Lignin | 2 | (27) | 2 | (20) | 3 | (27) | 2 | (20) |
| Extractives | 0.5 | (2) | 1 | (3) | 0.5 | (4) | 0.5 | (3) |
| Sum of components (total yield) | 52 | (100) | 49 | (100) | 47 | (100) | 53 | (100) |

[a] The figures are calculated as percent of the dry wood.
[b] Figures in parentheses refer to the original wood composition.

certain degree but is mainly retained in the solid wood phase. Delignification is accomplished in the second, acidic cooking stage. Compared to conventional acid sulfite pulping the two-stage process results in an appreciably higher pulp (carbohydrate) yield. This yield increase strongly depends on the pH in the first stage and is maximally about 8%, calculated on the wood basis. Because the pulp properties are rather strongly dependent on the hemicellulose content of the pulp, full advantage from the higher yield can be taken only when the pulp is applied for certain purposes, such as for making transparent (glassine) papers. However, by adjusting the pH in the first stage, the yield can be varied so that optimum pulp properties are obtained for different application purposes.

An additional and perhaps the main advantage of the two-stage pulping process is the fact that pine heartwood can be processed, which is not possible when applying the conventional acid sulfite process. In the first stage the reactive groups of lignin are protected by sulfonation, which blocks their condensation reactions with phenolic extractives (pinosylvin and taxifolin).

The yield increase in two-stage pulping is mainly restricted to softwoods and associated with an increased retention of glucomannans in the pulp (Fig. 7-17). This, in turn, is a consequence of their deacetylation taking place in the first cooking stage. There is a close relationship between the acetyl content of the precooked chips (which can be regulated by adjusting the pH) and the final pulp (or glucomannan) yield. The corresponding ester linkages are extremely sensitive toward alkaline hydrolysis and they are cleaved completely when applying neutral cooking liquors at a normal cooking temperature, provided that the cooking time is sufficiently long. It should be noted that at higher temperatures the equilibrium of reaction (7-1) (p. 119) is shifted to the right and the ion product of water is increased. As a result of these two combined effects the pH of the cooking liquor is higher at the cooking temperature than at room temperature.

A probable explanation for the behavior of deacetylated glucomannans is that they are more closely hydrogen bonded to the cellulose microfibrils, or partly crystallized, in which state their resistance toward acid hydrolysis in the second pulping stage is improved. By contrast, two-stage pulping of hardwoods only moderately improves the carbohydrate (xylan) yield in comparison with conventional pulping. The glucuronic acid substituents possibly counteract association of xylan with cellulose. Two-stage sulfite pulping of hardwood is not applied in the pulp industry.

Other modifications of two-stage (or multistage) sulfite pulping are also used commercially. The "Sivola method" is suitable for producing dissolving pulps with a high cellulose content. After delignification with acid sulfite the

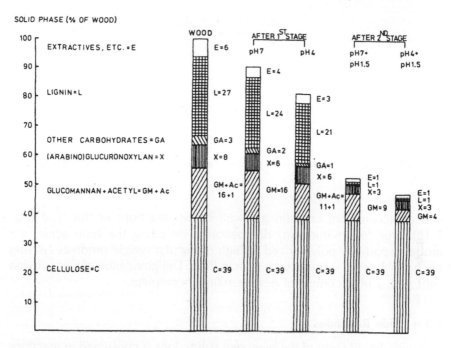

**Fig. 7-17.** Material balance for a typical two-stage cook of Scots pine (Sjöström *et al.*, 1962).

pulp is treated in the digester at a higher pH with sodium carbonate solution in order to remove hemicelluloses. In the case of pine wood, which cannot be delignified directly with acid sulfite, a neutral or alkaline pretreatment is needed as the first cooking stage.

### 7.2.7 Reactions of Extractives

During sulfite pulping the fatty acid esters are saponified to an extent determined by the conditions. Some of the resin components can also become sulfonated, resulting in increased hydrophilicity and better solubility. However, the partial removal of resin that always occurs during sulfite cooking and subsequent mechanical treatment is mainly associated with the formation of finely dispersed resin particles in stable emulsions. The dissolved lignosulfonic acids act as detergents with respect to the lipophilic resin components.

Acid sulfite pulping causes terpenes, terpenoids, and flavonoids to become partially dehydrogenated. The formation of *p*-cymene from α-pinene

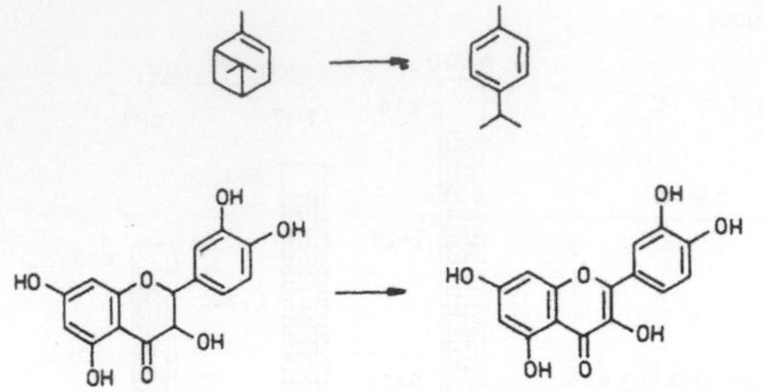

**Fig. 7-18.** Conversion of α-pinene to p-cymene and taxifolin to quercetin during acid sulfite pulping.

and quercetin from taxifolin are well-known reactions of this type (Fig. 7-18). Due to unsaturation, diterpenoids, including the resin acids, are probably partially polymerized to high molecular weight products causing pitch problems in subsequent pulp handling. Delignification of parenchyma cells with a high content of resin remains incomplete.

### 7.2.8 Side Reactions

A considerable part of the hydrogen sulfite ions is consumed in reactions other than the sulfonation of lignin. In the absence of wood, sulfur dioxide solutions are decomposed at elevated temperatures according to the equation:

$$4 \, HSO_3^- \rightarrow S_2O_3^{2-} + 2 \, SO_4^{2-} + 2 \, H^+ + H_2O \qquad (7-9)$$

This disproportionation reaction follows complex kinetics involving formation of polythionates ($S_3O_6^{2-}$, $S_4O_6^{2-}$, and $S_5O_6^{2-}$) as intermediate products. The initial and rather slow decomposition during the induction period is later accelerated by thiosulfate, which functions as an autocatalyst. After a critical thiosulfate concentration has been reached, sulfur is precipitated, and the acidity increases rapidly.

Another mechanism giving rise to thiosulfate formation is the reduction of hydrogen sulfite by wood components. Sugars play an important role in this reduction (see p. 133). Similarly, α-pinene is oxidized to p-cymene, formic acid to carbon dioxide, and taxifolin present in Douglas fir heartwood to quercetin. Sulfite cooking liquors, however, contain less thiosulfate than might be expected on the basis of these side reactions (Table 7-5). This discrepancy can be attributed to continuous consumption of thiosulfate in reactions with lignin, producing thioether cross-links (see p. 129). This type of sulfur, termed "organic excess sulfur," may account for 5–10% of the total

TABLE 7-5. Origins of Thiosulfate Formation during Sulfite Pulping[a]

| Cause | $S_2O_3{}^{2-}$ (g/liter) | Sulfur loss (kg/ton of pulp) |
|---|---|---|
| Aldonic and sugar sulfonic acid formation (5% sugar composition, wood basis) | 4.0 | 23 |
| Carbon dioxide formation (0.3% formic acid, wood basis) | 0.9 | 5 |
| Cymene formation (0.5 kg ptp) | 0.02 | — |
| Sulfate formation (2–3 g/liter) | 1.5 | 9 |
| Total thiosulfate formation | 6.5 | 37 |
| Found as thiosulfate | 0.5–2 | |
| Found as polythionate (expressed as thiosulfate) | 0.5–1.5 | |
| Total thiosulfate found | 1–3.5 | 5–20 |

[a] From Rydholm (1965).

organically bound sulfur. The major portion, 80–90% of the sulfur, exists in the form of sulfonate groups in the lignin, although minor amounts of sulfite are also consumed in the formation of carbohydrate sulfonic acids.

## 7.2.9 Composition of Sulfite Spent Liquors

In addition to lignosulfonates and hemicelluloses and their degradation products, sulfite spent liquors contain small amounts of uronic acids, methyl

TABLE 7-6. Typical Composition of the Sulfite Spent Liquor Resulting from the Acid Sulfite Pulping of Norway Spruce

| Component | Content (% of dry solids) | Composition (% of carbohydrates) |
|---|---|---|
| Lignosulfonates | 55 | |
| Carbohydrates | 28 | |
| Arabinose | | 4 |
| Xylose | | 22 |
| Mannose | | 43 |
| Galactose | | 17 |
| Glucose | | 14 |
| Aldonic acids | 5 | |
| Acetic acid | 4 | |
| Extractives | 4 | |
| Other compounds | 4 | |

glyoxal, formaldehyde, methyl alcohol, furfural, etc. Most of the remaining sulfur dioxide in the liquor is directly titratable, but some of it is liberated slowly under titration or after certain treatments. The α-hydroxysulfonic acids derived from carbonyl compounds are responsible for this "loosely combined sulfur dioxide" (cf. Figs. 2-28 and 7-8). In the liquors from acid sulfite cooking most of the carbohydrates are present in the form of mono-saccharides (Table 7-6, cf. also Table 10-1). After bisulfite and neutral sulfite pulping, however, a large portion of the sugars remains as oligo- and poly-saccharides. Characteristic of the spent liquors from neutral sulfite cooking of hardwood is the high proportion of acetic acid in comparison with the other organic constituents present.

### 7.2.10 Recovery and Conversion of Sulfite Cooking Chemicals

In addition to the organic solids resulting from the degradation and dis-solution of the wood constituents the spent cooking liquors contain the inorganic pulping chemicals, which for the most part have been changed during the pulping reactions. A variety of useful products can be produced from this organic source (see Section 10.2), but most of the solids in the spent liquor are still burned with generation of heat. Beyond the heat econo-my aspects, the combustion of the organic substance is today necessary from the pollution point of view.

Because calcium was long used almost exclusively in the sulfite process as the base, no need existed for the recovery of this comparatively cheap chemical. Interest was therefore only directed toward the relatively simple recovery of excess sulfur dioxide. However, the introduction of the more expensive soluble bases, together with more stringent environmental re-quirements, stimulated the development of methods for the recovery of both heat and inorganic chemicals (base and sulfur).

The concentration of solids in the sulfite spent liquors in the digester after cooking is 11–17%, dropping to 10–15% after pulp washing. For proper ignition and burning, the spent liquors are in most cases concentrated in multiple-effect evaporators to a solids content of 50–65%. Because volatile components, such as acetic and formic acids and furfural, are transferred to the steam condensates upon evaporation, the handling of these dilute li-quors requires specific procedures to avoid water pollution. After combus-tion the resulting ash, consisting of a mixture of calcium sulfate and calcium oxide in roughly equal amounts, is not converted for reuse in pulping; only the dust is collected to reduce air pollution. The stack gases resulting from the combustion of calcium-based spent liquors are especially problematic, because of their high content of sulfur dioxide which can be neither economically recovered nor eliminated.

A furnace similar to the Tomlinson kraft recovery furnace is used for the combustion of magnesium-based sulfite spent liquors. In this case, however, no smelt is obtained; instead the base is completely recovered as magnesium oxide in dust collectors; the sulfur escapes as sulfur dioxide and is absorbed from the combustion gases in scrubber towers. However, because magnesium hydroxide has a very low solubility in water, a complete recovery of sulfur dioxide meets difficulties.

Ammonium-based sulfite spent liquors can be burned in the same type of furnace as the calcium-based liquors. However, during combustion the base is decomposed to form nitrogen and water and the problems with fly ash are thus eliminated. All sulfur escapes to the combustion gases as sulfur dioxide which can be partly absorbed in an ammonia solution.

Sodium-based spent liquors both from the acid and neutral sulfite processes can be burned in a kraft type furnace. A smelt is obtained, consisting of sodium sulfide and sodium carbonate. The sulfur-to-sodium molar ratio is about 1:1 for sulfite spent liquors instead of around 0.15:1 in kraft black liquors, which means that a considerable part of the sulfur escapes as sulfur dioxide. Sodium carbonate from the recovery cycle is suitable for the absorption of the sulfur dioxide, although special problems are encountered because the sulfur dioxide concentration of these gases is low and oxidation to sulfur trioxide must be avoided. However, the greatest problem is the complete removal of sulfide from the smelt and its conversion to pure cook-

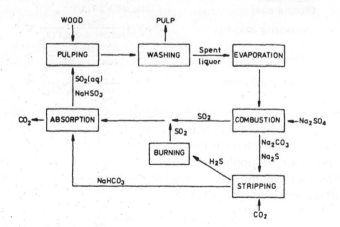

**Fig. 7-19.**  Recovery and conversion cycles for sulfite cooking chemicals.

ing chemicals. Especially for acid sulfite pulping, the cooking chemicals must be very pure, since other sulfur compounds, especially thiosulfate, are detrimental for pulping. Several recovery systems based on the use of a conventional kraft recovery furnace have been developed. In another and a simpler system, which, however, requires a modified furnace construction, all sulfur is first converted to hydrogen sulfide together with partial or complete gasification of the organic constituents. Sodium is thus recovered as pure carbonate. Hydrogen sulfide is finally converted to sulfur dioxide and absorbed. The recovery and conversion cycles for sulfite cooking chemicals are illustrated in Fig. 7-19.

## 7.3   Kraft Pulping

### 7.3.1   Cooking Chemicals and Equilibria

Kraft pulping is performed with a solution composed of sodium hydroxide and sodium sulfide, named "white liquor." According to the terminology the following definitions are used, where all the chemicals are calculated as sodium equivalents and expressed as weight of $NaOH$ or $Na_2O$.

| | |
|---|---|
| Total alkali | All sodium salts |
| Titratable alkali | $NaOH + Na_2S + Na_2CO_3$ |
| Active alkali | $NaOH + Na_2S$ |
| Effective alkali | $NaOH + \frac{1}{2} Na_2S$ |
| Causticizing efficiency | $100 \dfrac{NaOH}{NaOH + Na_2CO_3}$ % |
| Sulfidity | $100 \dfrac{Na_2S}{NaOH + Na_2S}$ % |
| Degree of reduction | $100 \dfrac{Na_2S}{Na_2S + Na_2SO_4}$ % |

In modern pulping chemistry weight units of $NaOH$ are often replaced by molar units, e.g., moles of effective alkali per liter of solution or kilogram of wood. Table 7-7 shows typical conditions for kraft pulping. The charge of effective alkali ($NaOH$) applied is usually 4–5 moles or 16–20% of wood.

The following equilibria are involved in the aqueous solutions containing sodium sulfide and sodium hydroxide:

$$S^{2-} + H_2O \rightleftarrows HS^- + HO^- \tag{7-10}$$

$$HS^- + H_2O \rightleftarrows H_2S + HO^- \tag{7-11}$$

TABLE 7-7.   Alkaline Pulping Methods and Conditions

| Method | pH range | "Base" | Active reagents | Max. temp. (°C) | Time at max. temp. (hr) | Softwood pulp yield (%) |
|---|---|---|---|---|---|---|
| Alkali (soda) | 13–14 | Na+ | HO⁻ | 155–175 | 2–5 | 50–70[a] |
| Kraft | 13–14 | Na+ | HS⁻, HO⁻ | 155–175 | 1–3 | 45–55 |
| Soda–anthraquinone | 13–14 | Na+ | HO⁻, AHQ[b] | 160–175 | 1–3 | 45–55 |
| Prehydrolysis-kraft | | | | | | 35–40 |
| Prehydrolysis stage | 3–4 | | H+ | 160–175 | 0.5–3 | |
| Kraft stage | 13–14 | Na+ | HS⁻, HO⁻ | 155–175 | 1–3 | |

[a] Hardwood.
[b] Anthrahydroquinone.

The equilibrium constants for these reactions are:

$$K_1 = [HS^-][HO^-] / [S^{2-}] \qquad (7\text{-}12)$$

$$K_2 = [H_2S][HO^-] / [HS^-] \qquad (7\text{-}13)$$

Since $K_1 \sim 10$ and $K_2 \sim 10^{-7}$, the equilibrium in equation (7-10) strongly favors the presence of hydrogen sulfide ions and for all practical purposes, sulfide ions can be considered to be absent (Fig. 7-20). The concentration of hydrogen sulfide becomes significant below pH 8 and needs to be considered only in modified kraft pulping processes involving pretreatment at low pH (see Section 7.3.6).

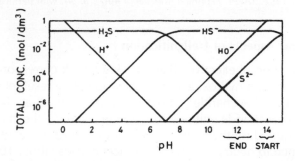

Fig. 7-20.   Equilibrium diagram showing the composition of the kraft white liquor at different pH values. Concentration, 0.2 mole/liter; temp., 25°C. Based on equilibrium constants corresponding to $K_1 = 10$ and $K_2 = 10^{-7}$.

## 7.3.2 Impregnation

In the kraft process thorough impregnation of the chips with cooking chemicals is not as critical as in acid sulfite pulping. The diffusion of chemicals in liquid-saturated wood is controlled by the total cross-sectional area of all the capillaries. In moderately alkaline solutions (pH < 12.5) the *effective capillary cross sectional area* (ECCSA), which is the area of paths available for diffusion, is higher in the longitudinal direction than in the radial and tangential directions (Fig. 7-21). However, because of swelling caused by alkali at pH values above 13, the ECCSA is increased in tangential and radial directions, approaching the same permeability as in the longitudinal direction.

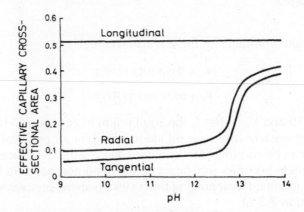

**Fig. 7-21.** Effective capillary cross-sectional area (ECCSA) of aspen wood as a function of pH (Stone, 1957). © 1957. TAPPI. Reprinted from *Tappi* **40**(7), p. 54, with permission.

## 7.3.3 General Aspects of Delignification

The consumption of effective alkali in a kraft cook corresponds to about 150 kilogram sodium hydroxide per ton of wood. As a result of the alkaline degradation of polysaccharides, about 1.6 equivalents of acids are formed for every monosaccharide unit peeled from the chain. Of the charged alkali, 60–70% is required for the neutralization of these hydroxy acids, while the rest is consumed to neutralize uronic and acetic acids (about 10% of alkali) and degradation products of lignin (25–30% of alkali).

Hydrogen sulfide ions react with lignin, but most of the sulfur-containing lignin products are decomposed during the later stages of the cook with formation of elemental sulfur, which combines with hydrogen sulfide ions to form polysulfide. However, kraft lignin still contains 2–3% of sulfur, corresponding to 20–30% of the charge.

As already shown (cf. Fig. 7-4), the selectivity of delignification is rather low for kraft pulping. The carbohydrates are attacked already at a comparatively low temperature (Fig. 7-22). This means that the acetyl groups are completely removed and the primary peeling process is terminated long before the maximum cooking temperature has been attained. The reactivity of the polysaccharides varies depending on their accessibility as well as on their structure. Because of its crystalline nature and high degree of polymerization, cellulose suffers less losses than the hemicelluloses.

The dissolution of lignin can be divided into three phases (Fig. 7-23). The initial phase of delignification takes place at temperatures below 140°C and is controlled by diffusion. Above 140°C, the rate of delignification becomes controlled by chemical reactions and accelerates steadily with increasing temperature. The rate of lignin dissolution remains high during this "bulk delignification" phase, until about 90% of the lignin has been removed. The final slow phase is termed "residual delignification" and can be regulated to some degree by varying the alkali charge and the cooking temperature.

The kinetics of the delignification are of importance especially when considering the control of the pulping process. Since kraft pulping follows simpler kinetics than the sulfite processes, more applications have been adopted for this case. Because of the heterogenity of the system, however,

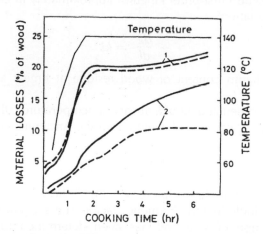

**Fig. 7-22.** Dissolution of carbohydrates (1) and lignin (2) during sulfate (—) and soda (---) pulping of Norway spruce (see Enkvist et al., 1957). Note that the cooking temperature is exceptionally low (140°C), resulting in much less dissolution than in normal pulping.

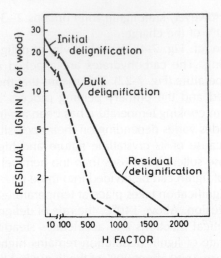

**Fig. 7-23.** Removal of lignin during kraft pulping of pine (—) and birch (---) as a function of the H factor (see Kleppe, 1970). © 1970. TAPPI. Adopted from *Tappi* **53**(1), p. 37, with permission.

pulping reactions are complicated and can therefore not be treated in the same fashion as homogeneous reactions in solution.

The overall rate of the bulk delignification in kraft pulping, during which the variations in hydroxyl and hydrogen sulfide ion concentrations are moderate, follows pseudo-first-order kinetics, approximately in conformity with the following equation:

$$- \frac{dL}{dt} = kL \tag{7-14}$$

where $L$ is the lignin content of wood residue at time $t$ and $k$ the rate constant.

Based on experimental data of $k$ at varying temperatures, the value of the activation energy $E_a$ can be calculated from the Arrhenius equation:

$$\ln k = \ln A - \frac{E_a}{RT} \tag{7-15}$$

where $T$ is the absolute temperature (Kelvin), $R$ the gas constant, and $A$ a further constant including the frequency factor. The activation energy $E_a$ for kraft delignification of softwood has been determined to be 130–150 kJ/mole (31–36 kcal/mole) (bulk phase) and about 50 and 120 kJ/mole (12 and 27 kcal/mole) for the initial and final phases, respectively.

According to a simplified system the net effect of both cooking time and temperature can be expressed by means of a single variable. In this system the rate at 100°C is chosen as unity and rates at all other temperatures are related to this standard. When using a value of 134 kJ/mole (32 kcal/mole) for $E_a$ the rates at any other temperatures can then be expressed by the following equation:

$$\ln \text{(relative reaction rate)} = 43.2 - \frac{16,113}{T} \qquad (7\text{-}16)$$

The time integral of the relative reaction rate is called the *H factor:*

$$H = \int_0^t \exp(43.2 - \frac{16,113}{T}) \, dt \qquad (7\text{-}17)$$

A normal heating period contributes to the *H* factor by 150–200, and 1500–2000 are needed for a complete kraft cook. Within the bulk delignification phase the relative reaction rate is doubled when the temperature is increased by about 8°C. If the hydroxide and hydrogen sulfide ion concentrations vary in a reproducible manner the *H* factor will predict the degree of delignification with sufficient precision.

### 7.3.4  Lignin Reactions

As in sulfite pulping, depolymerization of lignin depends on the cleavage of ether linkages, whereas the carbon-to-carbon linkages are essentially stable. The presence of hydrogen sulfide ions greatly facilitates delignification because of their strong nucleophilicity in comparison with hydroxyl ions. Cleavage of ether linkages, promoted by both hydroxyl and hydrogen sulfide ions, results also in increasing hydrophilicity of lignin because of the liberation of phenolic hydroxyl groups. The degraded lignin is dissolved in the cooking liquor as sodium phenolates. Studies with model substances representing various structural units in lignin have largely clarified the delignification reactions in kraft pulping.

*Etherified Phenolic Structures Containing β-Aryl Ether Bonds*  In etherified *p*-phenolic structures the β-aryl ether linkage is cleaved by hydroxide ions according to the mechanism shown in Fig. 7-24. The reaction proceeds via an oxirane intermediate which is subsequently opened with formation of an α,β-glycol structure. This reaction promotes efficient delignification by fragmenting the lignin and by generating new free phenolic hydroxyl groups.

*Free Phenolic Structures Containing β-Aryl Ether Bonds*  The first step of the reaction involves the formation of a quinone methide from the phenolate anion by the elimination of a hydroxide, alkoxide, or phenoxide ion from

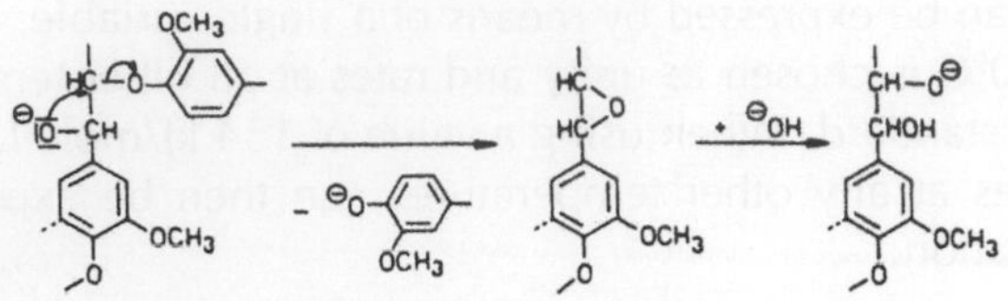

**Fig. 7-24.** Cleavage of β-aryl ether bonds in nonphenolic phenylpropane units during soda pulping (Gierer, 1970).

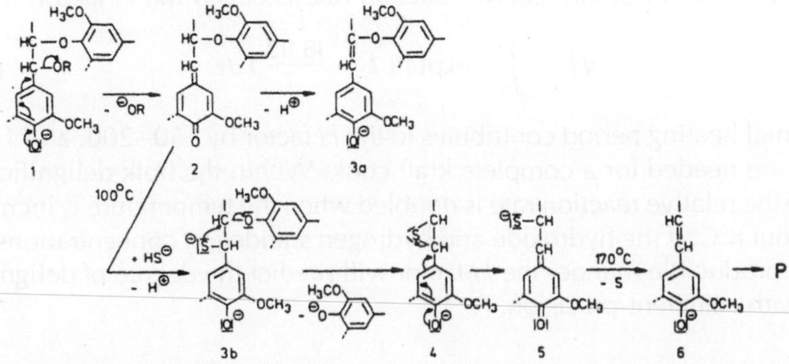

**Fig. 7-25.** Main reactions of the phenolic β-aryl ether structures during alkali (soda) and kraft pulping (Gierer, 1970). R = H, alkyl, or aryl group. The first step involves formation of a quinone methide intermediate (2). In alkali pulping intermediate (2) undergoes proton or formaldehyde elimination and is converted to styryl aryl ether structure (3a). During kraft pulping intermediate (2) is instead attacked by the nucleophilic hydrogen sulfide ions with formation of a thiirane structure (4) and simultaneous cleavage of the β-aryl ether bond. Intermediate (5) reacts further either via a 1,4-dithiane dimer or directly to compounds of styrene type (6) and to complicated polymeric products (P). During these reactions most of the organically bound sulfur is eliminated as elemental sulfur.

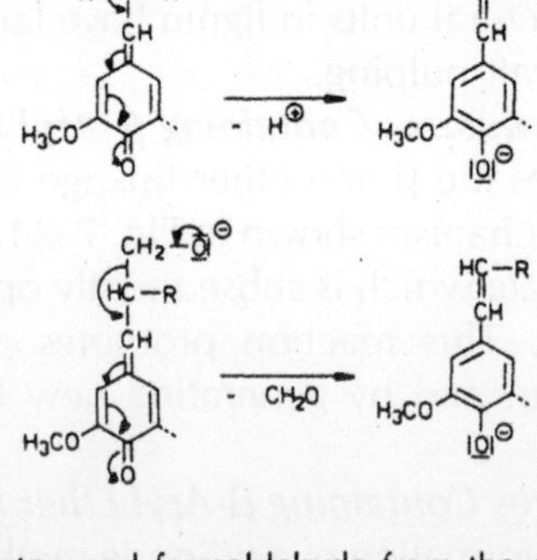

**Fig. 7-26.** Elimination of proton and formaldehyde from the quinone methide intermediate during alkali pulping (Gierer, 1970).

the α-carbon (Fig. 7-25). The subsequent course of reactions depends on whether hydrogen sulfide ions are present or not. In the latter case (soda pulping), the dominant reaction is the elimination of the hydroxymethyl group from the quinone methide with formation of formaldehyde and a styryl aryl ether structure *without* cleavage of the β-ether bond (Fig. 7-26). When hydrogen sulfide ions are present (strong nucleophiles) they react with the quinone methide to form a thiol derivative which is converted to a thiirane structure with simultaneous cleavage of the β-ether bond. The thiirane can be dimerized to a dithiane structure, but this as well as other sulfur-containing intermediates are decomposed, forming elemental sulfur and unsaturated side-chain structures. Competing reaction paths are possible, giving rise to other minor degradation products, such as guaiacol.

*Structures Containing α-Ether Bonds*    The α-ether bonds in phenolic phenylcoumaran (Fig. 7-27) and pinoresinol structures are readily cleaved by hydroxide ions, usually followed by the release of formaldehyde. Only in the case of open α-aryl ether structures does this reaction result in the fragmentation of lignin. In contrast, the α-ether bonds are stable in all etherified structures.

*Methoxyl Groups*    Lignin is partially demethylated by the action of hydrogen sulfide ions forming methyl mercaptan which is convertible to di-

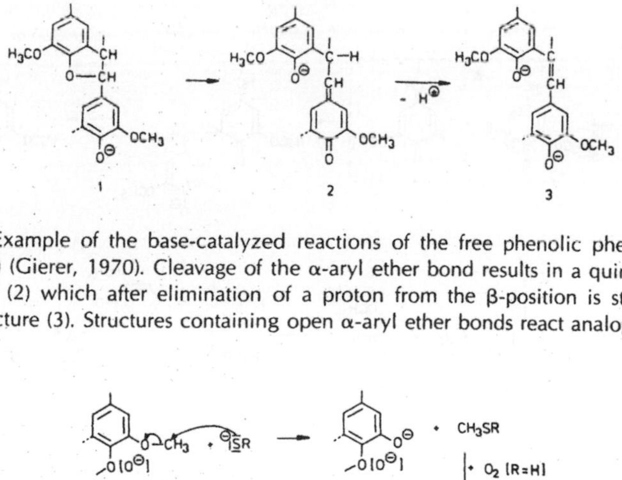

**Fig. 7-27.** Example of the base-catalyzed reactions of the free phenolic phenylcoumaran structures (1) (Gierer, 1970). Cleavage of the α-aryl ether bond results in a quinone methide intermediate (2) which after elimination of a proton from the β-position is stabilized to a stilbene structure (3). Structures containing open α-aryl ether bonds react analogously.

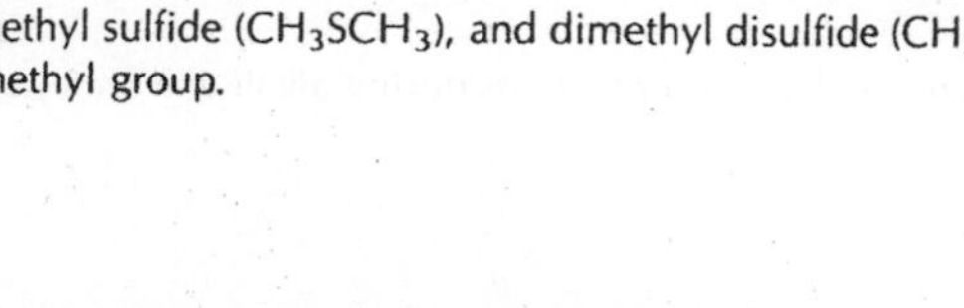

**Fig. 7-28.** Cleavage of methyl aryl ether bonds with simultaneous formation of methyl mercaptan ($CH_3SH$), dimethyl sulfide ($CH_3SCH_3$), and dimethyl disulfide ($CH_3SSCH_3$) during kraft pulping. R = H or methyl group.

methyl sulfide by reaction with another methoxyl group. In the presence of oxygen, methyl mercaptan can be oxidized further to dimethyl disulfide (Fig. 7-28). Because the hydroxide ions are less strong nucleophiles than hydrogen sulfide ions, only small amounts of methanol are formed. Methyl mercaptan and dimethyl sulfide are highly volatile and extremely malodorous, causing an air pollution problem that is difficult to master.

**Condensation Reactions**   A variety of condensation reactions are known to occur in alkaline pulping. Since carbon-to-carbon linkages are formed between lignin entities, it has been proposed that as a result of condensation reactions, lignin dissolution is retarded, particularly during the terminal phases of kraft pulping.

It has been suggested that the major part of condensation processes occurs at the unoccupied C-5 position of phenolic units. Thus, in isolated MWL preparations about half of the C-5 positions are unsubstituted, while in isolated kraft lignins only about one third of these positions remain free. The syringyl units of hardwood lignins cannot, of course, undergo condensation reactions of this type.

Figure 7-29 shows some examples of postulated condensation reactions forming diarylmethane structures of three different types. In the first case (A) a phenolate adds to a quinone methide structure, forming an α-5 linkage. The second case (B) illustrates a similar condensation between the 1- and α-

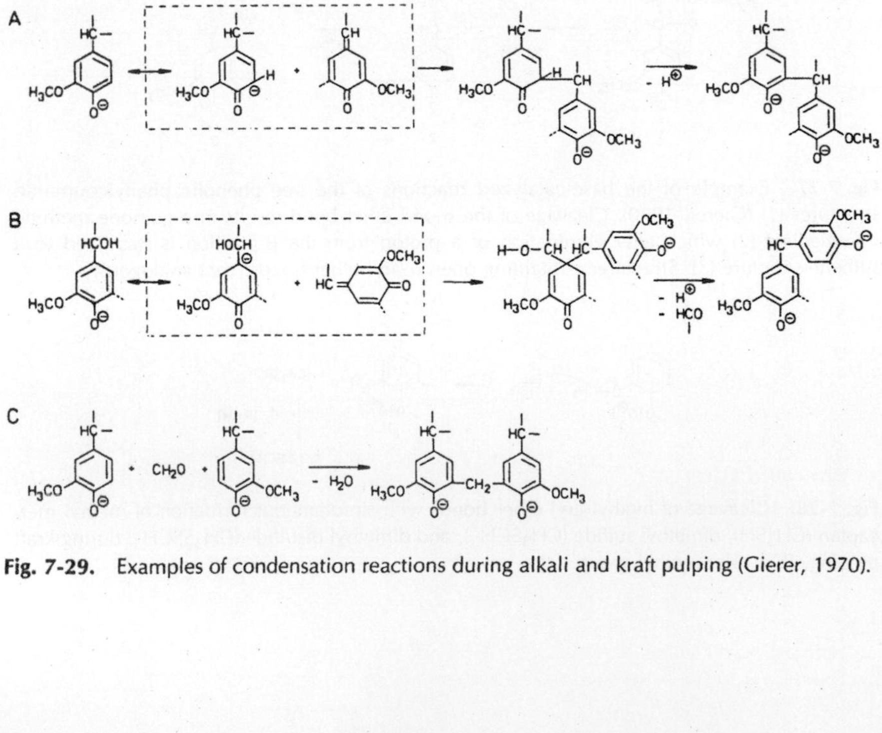

**Fig. 7-29.** Examples of condensation reactions during alkali and kraft pulping (Gierer, 1970).

carbons with simultaneous removal of the propane side chain. The third reaction (C) involves formaldehyde released from the γ-carbinol groups (see Fig. 7-26) and also leads ultimately to a diarylmethane structure.

*Formation of Chromophores* During kraft delignification the color of the chips darkens gradually until a yield level of 60–70% is reached (light absorption coefficient of ca. 40 m²/kg instead of ca. 5–10 m²/kg for the original wood at 457 nm). Further delignification brings about a modest brightening of the pulp. The specific light absorption coefficient of the residual lignin, however, increases continuously, reaching ca. 500 m²/kg at the end of the pulping. For comparison, the corresponding value for wood lignin is 20–40 m²/kg.

The color of unbleached pulps is caused by certain unsaturated structures (chromophores). In addition, leucochromophores, which can be converted into chromophores by air oxidation may be present in the pulp. Most of the chromophores are presumed to be derived from lignin (Fig. 7-30) although some chromophoric groups can also be introduced into the polysaccharides, for example, carbonyl groups.

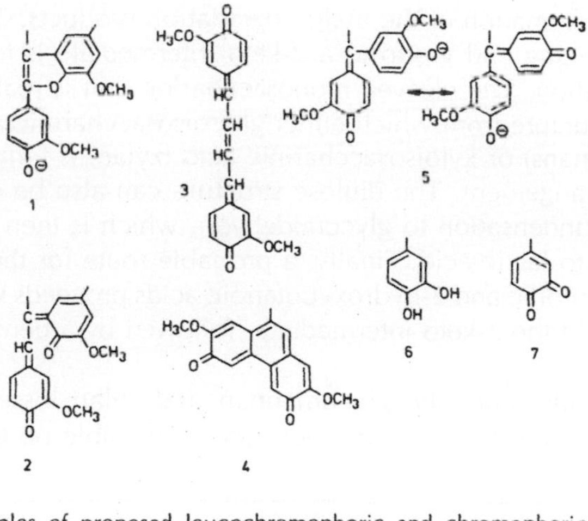

**Fig. 7-30.** Examples of proposed leucochromophoric and chromophoric structures. Aryl-coumarones (1) and stilbene quinones (2) are thought to be formed from stilbenes after oxidation. Butadiene quinones (3) could arise from oxidation of hydroxyarylbutadienes being formed from phenolic pinoresinol structures during kraft or neutral sulfite pulping. Cyclization may yield intermediates which are further oxidized to cyclic diones (4). A resonance-stabilized structure (5) results from the corresponding condensation product formed during pulping. o-Quinoid structures (7) are oxidation products of catechols (6) formed during alkaline or neutral pulping processes.

## 7.3.5    Reactions of the Polysaccharides

Because of the alkaline degradation of polysaccharides kraft pulping results in considerable carbohydrate losses. The acetyl groups are hydrolyzed at the very beginning of the kraft cook (from hardwood xylan and softwood galactoglucomannans). In the earlier stages of cooking the polysaccharide chains are peeled directly from the reducing end groups present (primary peeling). As a result of the alkaline hydrolysis of glycosidic bonds, occurring at high temperatures, new end groups are formed, giving rise to additional degradation (secondary peeling) (cf. Section 2.5.5). As a consequence, the yield of cellulose is always reduced in kraft pulping, although to a lesser extent than that of the hemicelluloses which are degraded more extensively due to their low degree of polymerization and amorphous state (cf. Table 7.4). The peeling reaction is finally interrupted because the competing "stopping reaction" converts the reducing end group to a stable carboxylic acid group.

*Mechanism of the Peeling Reaction*    The carbohydrate material lost in peeling is converted to various hydroxy acids. In addition, formic and acetic acid and small amounts of dicarboxylic acids, are formed as well. Figure 7-31 (lower portion) shows a simplified reaction scheme illustrating the mechanism of formation of the main degradation products. The rearrangement of a reducing end group to a 2-keto intermediate is followed by β-alkoxy elimination. The cleaved monosaccharide unit is rearranged into a 2,3-diulose structure from which either glucoisosaccharinic acid (cellulose and glucomannans) or xyloisosaccharinic acid (xylan) is formed via a benzilic acid rearrangement. The diulose structure can also be cleaved by reversed aldol condensation to glyceraldehyde, which is then converted via methylglyoxal to lactic acid. Finally, a probable route for the formation of 3,4-dideoxypentonic and 2-hydroxybutanoic acids proceeds via formic acid elimination from the 3-keto intermediate, followed by a benzilic acid rearrangement.

Figure 7-32 illustrates the glucomannan and xylan losses during kraft pulping of pine wood. As can be seen, an appreciable portion of the lost

---

**Fig. 7-31.**    Peeling and stopping reactions of polysaccharides (Sjöström, 1977). R = polysaccharide chain and R′ = $CH_2OH$ (cellulose and glucomannans) or H (xylan). Cellulose and glucomannans (R′ = $CH_2OH$): 3-Deoxyhexonic acid end groups (metasaccharinic acid) (1), 2-C-methylglyceric acid end groups (2), 3-deoxy-2-C-hydroxymethylpentonic acid (glucoisosaccharinic acid) (3), 2-hydroxypropanoic acid (lactic acid) (4), and 3,4-dideoxypentonic acid (2,5-dihydroxypentanoic acid) (5). Xylan (R′ = H): 3-Deoxy-2-C-hydroxymethyltetronic acid (xyloisosaccharinic acid) (3), 2-hydroxypropanoic acid (lactic acid) (4), and 2-hydroxybutanoic acid (5). © 1977. TAPPI. Adopted from *Tappi* 60(9), p. 152, with permission.

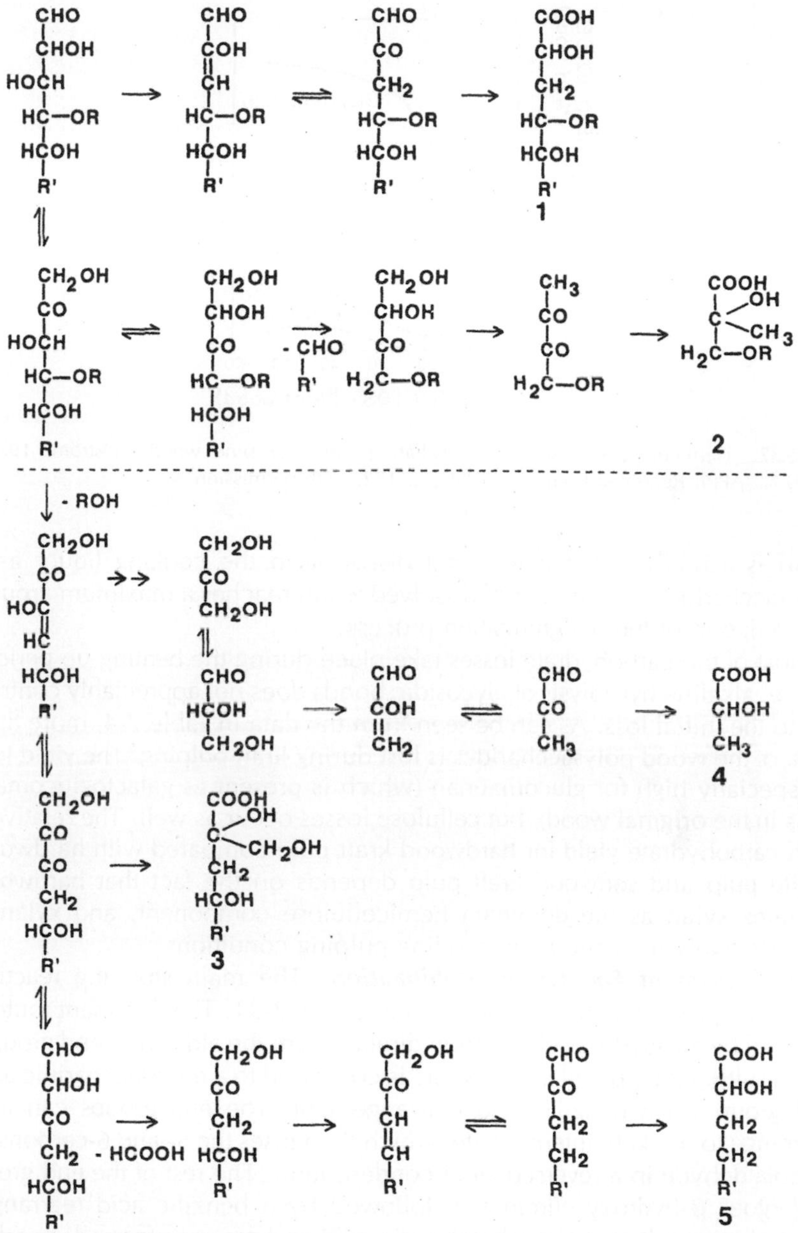

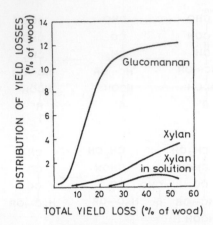

**Fig. 7-32.** Hemicellulose losses during sulfate pulping of pine wood (Sjöström, 1977). © 1977. TAPPI. Reprinted from *Tappi* **60**(9), p. 152, with permission.

xylan is actually not degraded but dissolves in the cooking liquor as a polysaccharide. The amount of dissolved xylan reaches a maximum around the midpoint of the delignification process.

Most of the carbohydrate losses take place during the heating-up period; hence, alkaline hydrolysis of glycosidic bonds does not appreciably contribute to the initial loss. As can be seen from the data in Table 7-4, more than 30% of the wood polysaccharides is lost during kraft pulping. The yield loss is especially high for glucomannan (which is present as galactoglucomannans in the original wood), but cellulose losses occur as well. The relatively high carbohydrate yield for hardwood kraft pulp compared with hardwood sulfite pulp and softwood kraft pulp depends on the fact that hardwood contains xylan as the dominant hemicellulose component, and xylan is comparatively resistant under alkaline pulping conditions.

*Mechanism of End Group Stabilization* The main stopping reaction routes are presented in the upper portion of Fig. 7-31. The dominant route is initiated by a β-hydroxy elimination directly from the aldehydic end groups. The resulting dicarbonyl intermediate is converted to a metasaccharinic acid end group via a benzilic acid rearrangement. The end groups can also rearrange to a 3-keto intermediate, which then loses the 5- and 6-carbons as glycolaldehyde in a reversed aldol condensation. The rest of the end group undergoes β-hydroxy elimination followed by a benzilic acid rearrangement. As a result, a 2-C-methylglyceric acid end group is formed. In addition to the metasaccharinic and 2-C-methylglyceric acid end groups, small amounts of 2-C-methylribonic (glucosaccharinic) and aldonic acid end

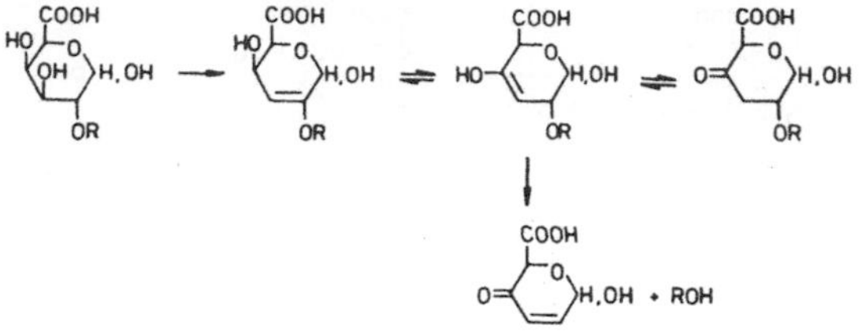

**Fig. 7-33.** Alkaline degradation of 2-O-(α-L-rhamnopyranosyl)-D-galacturonic acid (Johansson and Samuelson, 1977). R represents the rhamnose group in the xylan chain.

groups have been found to be present. The presence of aldonic acid end groups indicate that some oxidative reactions also occur. In the case of pulping with alkali alone (soda process), the oxidation might depend on the presence of dissolved oxygen, whereas the polysulfides generated during kraft pulping can function as oxidants.

Softwood xylan is partially substituted with arabinose at the C-3 position of the xylose units. During the course of peeling, arabinose is easily eliminated from the chain under simultaneous formation of a metasaccharinic acid end group (β-alkoxy elimination) which stabilizes the chain against further alkaline peeling. However, because of its relatively low content in softwood, xylan makes only a small contribution to the total carbohydrate yield in pulping. Much more important is the behavior of hardwood xylan during kraft pulping. In this connection, the detailed structure of the xylan chain is of interest (Fig. 3-17). The terminal xylose unit is rapidly cleaved from the xylan chain, but the remaining galacturonic acid end group is stable against further peeling. Its stability is not permanent, however (Fig. 7-33). After hydroxy elimination at C-3, a 2-enuronic acid group is formed which is decomposed at higher temperatures (> 100°C) after isomerization of the double bond to the C-3-C-4 position. The remaining terminal rhamnose unit is eliminated very easily because its C-3 position is bound to the following xylose unit.

The 4-O-methylglucuronic acid groups prevent the peeling of xylan chains at lower temperatures (< 100°C) but they offer only a partial protection at higher temperatures. Since the 4-O-methylglucuronic acid groups are bound to the C-2 position in the xylose units, no conversion of this carbon atom to a carbonyl group can take place. Instead, HO-3 is eliminated directly (β-hydroxy elimination).

**Significance of Uronic Acid Groups**  The uronic acid content is much lower in the final kraft pulp than in the original wood. In the glucuronoxylan remaining in the pulp it corresponds to a molar ratio of roughly 1:25

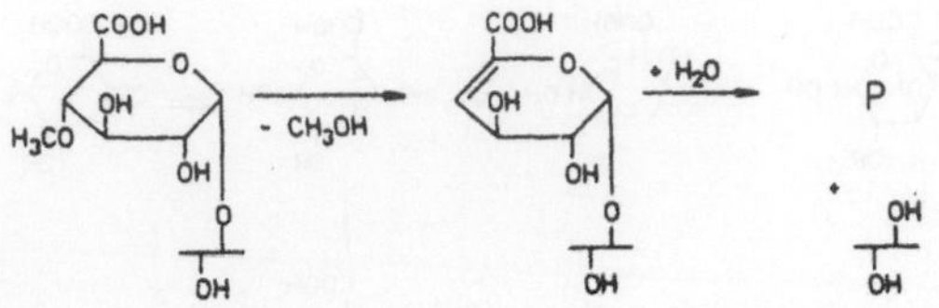

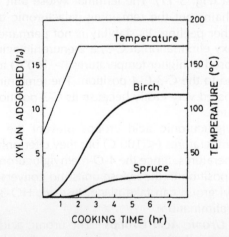

**Fig. 7-34.** Loss of 4-*O*-methylglucuronic acid groups (Johansson and Samuelson, 1977). P denotes fragmentation products formed.

(glucuronic acid/xylose), whereas this ratio for native xylan is about 1:5 (softwood) and 1:10 (hardwood). Since xylan is not evenly substituted by the uronic acid groups it is probable that the fractions of high uronic acid content are preferably dissolved, resulting in a lower uronic acid content of the pulp. Another reason for the decrease in the uronic acid content is the cleavage of these groups from the xylan chain, since the pyranosyluronic acid linkages are more sensitive to alkaline hydrolysis than are the corresponding glycosyl linkages. An additional reaction possibly proceeds via $CH_3O$ elimination at C-4 (β-position to the carboxyl group) followed by H-5 elimination (Fig. 7-34).

*Other Reactions*    In addition to the direct stabilizing effect of the uronic acid groups, the relatively high yield of hardwood kraft pulp is also due to the readsorption of xylan on the fibers (Fig. 7-35). After kraft pulping of softwood, the glucomannan remaining in the pulp still contains traces of

**Fig. 7-35.** Adsorption of xylan on cotton fibers present in the digester during kraft pulping (Yllner and Enström, 1956).

galactose residues, and the xylan has some arabinose residues contrary to sulfite pulping during which these moieties are cleaved completely.

### 7.3.6  Stabilization of Polysaccharides against Alkaline Degradation

The primary peeling of polysaccharides by alkali can be avoided by the elimination of the aldehyde functions from the end groups. The reduction of these groups to alcohols by sodium borohydride inhibits primary peeling, and the carbohydrate yield is thus increased considerably. The end groups can also be stabilized by oxidizing them to carboxyl groups or by conversion to other stable derivatives.

Of the stabilization methods, the polysulfide pulping process is of practical importance. The influence of polysulfides is based on a specific oxidation of the end groups to carboxyl groups via glucosone intermediates (cf. Section 8.1.3). Polysulfides can be prepared by catalytic oxidation of sulfide in the white liquor or by adding elemental sulfur into the kraft cooking liquor:

$$n\,S^{2-} + \frac{n-1}{2}\,O_2 + (n-1)H_2O \rightarrow S_n^{2-} + (2n-2)HO^- \tag{7-18}$$

$$S^{2-} + n\,S \rightarrow S_{n+1}^{2-} \tag{7-19}$$

In the latter method an excess of sulfide is created because of the added sulfur. This must be regenerated to elemental sulfur in order to avoid high sulfidities.

Of the reducing methods, the pretreatment of wood chips with hydrogen sulfide (140°C, pH ~ 7) might be technically feasible. During such a treatment the aldehyde end groups are reduced to thioalditols according to the following equation:

$$R\text{—}CHO \underset{\phantom{xx}}{\overset{H^+,\ HS^-}{\rightleftharpoons}} R\text{—}CH(OH)SH \underset{\phantom{xx}}{\overset{-H_2O,\ H_2S}{\rightleftharpoons}}$$

$$R\text{—}CH(SH)_2 \underset{\phantom{xx}}{\overset{-H_2S}{\rightleftharpoons}} R\text{—}CHS \underset{-S}{\overset{H_2S}{\longrightarrow}} R\text{—}CH_2SH \tag{7-20}$$

The increase in pulp yield may reach 8% on dry wood basis, but requires high pressures (> 1000 kPa) and a large excess of hydrogen sulfide (ca. 10% of wood). Only a fraction of the hydrogen sulfide (1–2% of wood) is consumed and the rest is recoverable. Table 7-8 illustrates the influence of some oxidizing and reducing agents on the carbohydrate yield of kraft pulp. Stabilization with anthraquinone is dealt with in Section 7.3.7.

**TABLE 7-8.    Reductive and Oxidative Stabilization of Softwood Carbohydrates during Kraft Pulping**

| | Yield of polysaccharide (% of wood) | | | |
|---|---|---|---|---|
| Method/addition | Cellulose | Glucomannan | Glucuronoxylan | Total |
| Normal sulfate pulping | 35 | 4 | 5 | 44 |
| Oxidation/polysulfide (4% S) | 36 | 9 | 5 | 50 |
| Reduction/NaBH₄ | 36 | 12 | 4 | 52 |
| Reduction/H₂S | 36 | 9 | 4 | 49 |

## 7.3.7    Sulfur-Free Pulping

In order to be able to reduce the pollution load of the pulp mills, attempts have been made to diminish the use of sulfur chemicals even if their complete elimination is difficult. Sulfur contaminants are easily introduced into the system, e.g., in connection with the use of oil as fuel, and already traces of them give rise to odor.

Alkaline solutions containing oxygen can be used for the removal of lignin from softwoods, but the delignification in this system is quite unselective. Better results have been obtained for hardwoods using pressurized oxygen at low alkalinities, e.g., sodium carbonate and sodium hydrogen carbonate solutions. Interest has also been directed toward a two-stage oxygen pulping system ("soda-oxygen pulping"). Here, wood chips are first subjected to soda pulping and then fiberized mechanically. The final delignification is carried out in the presence of alkali and oxygen.

Sulfide in the kraft cooking liquor can be replaced, at least partly, by anthraquinone (AQ) or similar compounds which possess a marked ca-

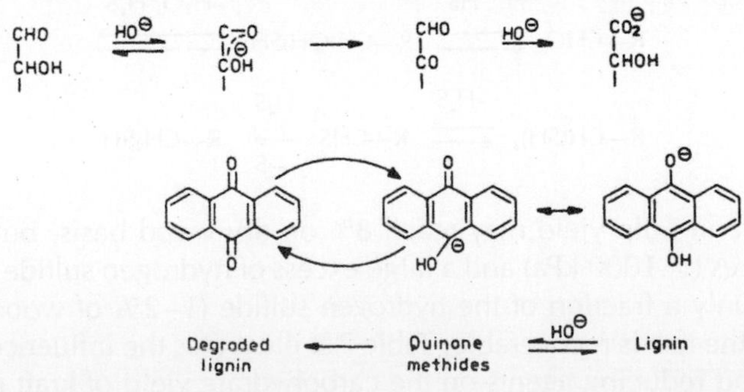

**Fig. 7-36.**    Anthraquinone-anthrahydroquinone reactions with carbohydrates and lignin.

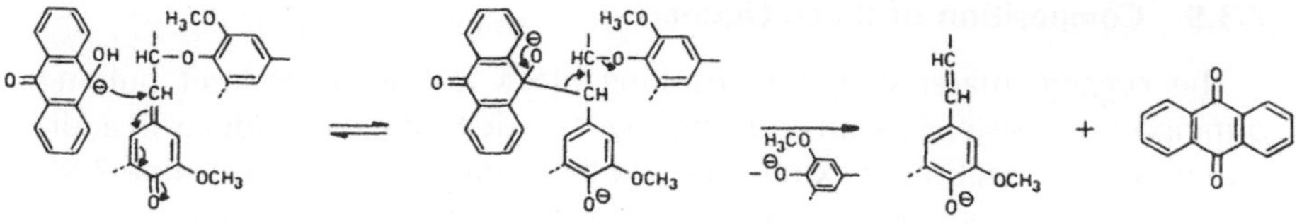

**Fig. 7-37.** Cleavage of β-aryl ether bonds in alkaline media by anthrahydroquinone with regeneration of anthraquinone.

pability of accelerating the delignification while at the same time stabilizing the polysaccharides toward alkaline degradation according to the mechanism illustrated in Fig. 7-36. At moderate temperatures AQ is reduced to anthrahydroquinone (AHQ) by the polysaccharide end groups, which, in turn, are oxidized to alkali-stable aldonic acid groups. The reduced species or AHQ now acts as an effective cleaving agent with regard to the lignin β-aryl ether linkages in free phenolic phenylpropane units and is simultaneously oxidized to AQ (Fig. 7-37). The partly depolymerized lignin is further degraded by sodium hydroxide at elevated temperature. As a result of this reduction-oxidation cycle, additions as low as 0.01% AQ of the dry wood weight markedly improve the delignification. Depending on the wood species, conditions, desired effect, etc., up to 0.5% may be used.

## 7.3.8    Reactions of Extractives

During kraft pulping the fatty acid esters are hydrolyzed almost completely although the waxes are much more stable than the fats. The fatty acids are dissolved together with the resin acids as sodium salts in the cooking liquor. Especially the resin acid soaps are effective solubilizing agents facilitating the removal of sparingly soluble neutral substances such as sitosterol in pine wood and betulinol and betulaprenols in birch wood. Because hardwoods do not contain resin acids, tall soap is usually added to the cook to reduce the content of extractives in the final pulp to a sufficiently low level so that the "pitch problems" can be avoided.

Some of the unsaturated fatty acids and resin acids are partly isomerized at kraft pulping conditions (for structures, see Table 5-2 and Fig. 5-8). Linoleic and pinolenic acids, representing dienoic and trienoic fatty acid types, are converted to the respective isomers with conjugated double bonds at 9,11 and 10,12 positions which are mainly of *cis, trans* configuration. In the case of common resin acids, the principal change is a partial isomerization of levopimaric to abietic acid. The other members of the common resin acids are essentially stable at the kraft pulping conditions.

## 7.3.9 Composition of Black Liquor

The organic material in the resulting black liquors after kraft pulping principally consists of lignin and carbohydrate degradation products in addition to a small fraction of extractives and their reaction products (Table 7-9). The black liquors are extremely complex mixtures containing a huge number of components of diversified constitutions and structures.

The major portion of the lignin fraction constitutes of high-molecular-weight material, which is precipitated when the liquor is acidified (cf. Section 10.2.2). The composition of the "kraft lignins" is complex and it varies depending on the wood species and cooking conditions. However, some characteristic features are known of which examples are given in Table 7-10. Compared with native lignin, kraft lignin typically contains many more phenolic hydroxyl groups and carboxyl groups. Instead of the coniferyl type double bonds, which have been destroyed completely during cooking, new double bonds are present in the kraft lignin in styrene and stilbene struc-

**TABLE 7-9.** Typical Distribution of the Organic Material in Softwood (Pine) Kraft Black Liquors

| Fraction/component | Content (% of dry solids) | Composition of hydroxy acid fraction (% of total acids) |
|---|---|---|
| Lignin[a] | 46 | |
| Hydroxy acids[b] | 30 | |
| Glycolic | | 10 |
| Lactic | | 15 |
| 3,4-Dideoxypentonic | | 10 |
| Glucoisosaccharinic[c] | | 35 |
| 2-Hydroxybutanoic | | 5 |
| 3-Deoxypentonic | | 5 |
| Xyloisosaccharinic | | 5 |
| Others | | 15 |
| Formic acid | 8 | |
| Acetic acid | 5 | |
| Extractives | 7 | |
| Other compounds | 4 | |

[a] About 90% of this fraction constitutes of polymeric material; minor amounts are monomeric and dimeric fragments (cf. Table 7-10 and Fig. 7-38).

[b] For structures, see Fig. 10-8.

[c] The ratio of threo (β) and erythro (α) forms is about 2.5.

**TABLE 7-10.** Characteristics of Softwood Kraft Lignin
Compared with Milled Wood Lignin (MWL)

| Characteristic | Kraft lignin | MWL |
|---|---|---|
| Molecular weight ($\bar{M}_n$) | 3000–5000 | 8000–10000 |
| Polydispersity ($\bar{M}_w/\bar{M}_n$) | 3–4 | 2–3 |
| Sulfur content (%) | 1–3 | — |
| Functional groups[a] | | |
| Hydroxyl groups, total | 120 | 120 |
| Guaiacyl OH | 60 | 30 |
| Catechol OH | 10 | — |
| Aliphatic OH | 50 | 90 |
| Carboxyl groups | 13 | — |
| Carbonyl groups | 3 | 10–15 |

[a] Per 100 $C_6C_3$ units.

tures, which are formed during pulping. Many of the typical linkages originally present in native lignin have been extensively modified.

Although the low-molecular-weight degradation products represent only a
minor portion of the total lignin fraction, several hundreds of single components have been identified. Of the monomeric compounds in softwood
black liquors some are shown in Fig. 7-38. To the identified compounds

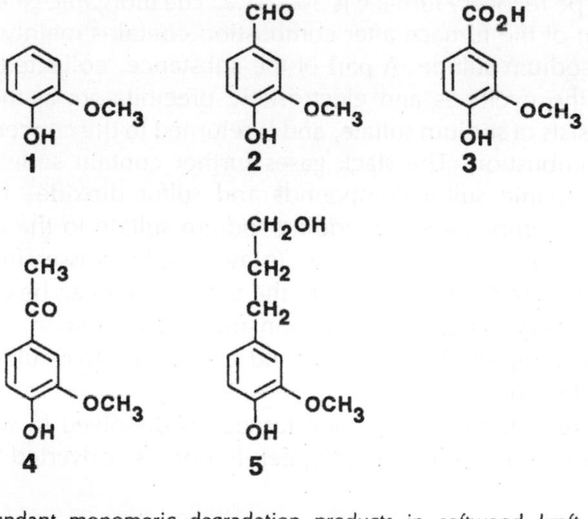

**Fig. 7-38.** Abundant monomeric degradation products in softwood kraft black liquors.
Guaiacol (1), vanillin (2), vanillic acid (3), acetovanillone (4), and dihydroconiferyl alcohol (5).

belong also a number of hydroxylated monomeric arylalkanoic acids and dimeric hydroxy acids having the stilbene structure. Noteworthy is also that hydroxyalkanoic acids containing a guaiacyl or syringyl nucleus are present in softwood and hardwood black liquors. They are obviously condensation products between lignin fragments and degradation intermediates of carbohydrates.

The carbohydrate degradation products in black liquors consist of aliphatic carboxylic acids of which the hydroxy monocarboxylic acids are the dominant components.

## 7.3.10   Recovery and Conversion of Kraft Cooking Chemicals

In the kraft process large amounts of comparatively expensive cooking chemicals are used which has necessitated the development of an advanced technology for the recovery of these chemicals in combination with the generation of process energy. The total content of solids in pine kraft black liquor leaving the diffusers of filter washers is 15–20%. It contains most of the degraded and dissolved wood material together with the inorganic chemicals. Most of the base has been consumed for the neutralization of the organic acids; sulfur is still predominantly present as hydrogen sulfide ions. After partial evaporation, the tall oil skimmings are recovered and treated separately (see Section 10.2.2). The content of solids in the concentrated black liquor coming from the multiple-effect evaporators and entering the Tomlinson-type recovery furnace is 50–70%. The inorganic smelt remaining at the bottom of the furnace after combustion contains mainly sodium carbonate and sodium sulfide. A part of the substance, collected as dust and fumes from the cyclones and electrostatic precipitators at the top of the furnace, consists of sodium sulfate, and is returned to the concentrated black liquor for combustion. The stack gases further contain sodium sulfate in addition to organic sulfur compounds and sulfur dioxide. The losses of chemicals are compensated by adding sodium sulfate to the concentrated black liquor prior to its combustion. To avoid sulfur losses from the stack gases and also to reduce air pollution, the black liquor can be oxidized with air or gaseous oxygen before evaporation and/or combustion. The oxidation converts remaining sulfides to sulfates and mercaptans to disulfides and their further oxidation products.

The smelt leaving the combustion furnace is dissolved in water and the sodium carbonate in the resulting "green liquor" is converted to hydroxide by lime:

$$CaO + H_2O \rightarrow Ca(OH)_2 \tag{7-21}$$

$$Ca(OH)_2 + Na_2CO_3 \rightleftarrows 2\,NaOH + CaCO_3 \tag{7-22}$$

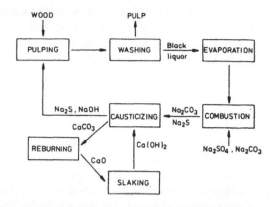

**Fig. 7-39.** Recovery and conversion of the kraft cooking chemicals.

In this procedure, calcium carbonate (lime sludge) is precipitated and separated from the liquid. The remaining solution, consisting mainly of sodium sulfide and sodium hydroxide ("white liquor"), is used as such for cooking. After washing and drying the lime sludge is reburned to give new calcium oxide.

Figure 7-39 illustrates the recovery and conversion of the kraft cooking chemicals. In the case of sulfur-free pulping (soda process, soda-oxygen process, and anthraquinone-alkali pulping) only sodium carbonate is recovered and chemical losses are compensated by adding sodium carbonate. The hydroxide-carbonate and lime cycles are the same as for the kraft process.

# 7.4 Special Features of High-Yield Pulping

Although the lignin and carbohydrate reactions discussed in the previous sections are applicable to high-yield pulping, the chemistry associated with this type of processes deserves special attention.

*Material Losses* In high-yield pulping processes, including CTMP, a treatment with alkaline sulfite solution is the most usual prestage prior to the final mechanical defibration. The material losses are low, but the handling of dilute effluents to avoid pollution is a special problem. Most of the resulting organic material consists of lignin and carbohydrate degradation products in addition to components from extractives.

The average molecular weight of the dissolved lignin fraction has been found to be very low. Recent studies have revealed that fragments from the end group structures of lignin are released when softwood is treated at relatively mild conditions with neutral or alkaline sulfite solutions. The major monomeric fragments consist of coniferyl alcohol and coniferyl aldehyde, of which only the latter is sulfonated. These observations suggest that sulfonation of coniferyl aldehyde structures is very fast in comparison with other lignin structures.

Of the carbohydrates mainly hemicelluloses are degraded to soluble fragments present in the effluents from high-yield pulping. Some of this material has not been degraded completely and it still exists as polysaccharides, but, in addition, various monomeric degradation products are present. This material is mainly composed of aliphatic carboxylic acids, among which hydroxy carboxylic acids and acetic acid dominate. The hydroxy carboxylic acids are typical alkaline degradation products of polysaccharides. The extent of this type of degradation depends on the alkalinity and is still rather insignificant during high-yield pulping in comparison to kraft pulping (see Section 7.3.5).

Most of the acetic acid originates from hemicelluloses, that is, either from softwood galactoglucomannans or from hardwood glucuronoxylans. Since the acetyl groups are extensively hydrolyzed already after a relatively short time of treatment, acetic acid is a prominent component particularly in effluents from high-yield pulping of hardwood.

A direct mechanical processing of wood and defibration after its chemical pretreatment also results in some yield losses. Most of this material consists of finely divided fiber fragments ("fines") and colloidal particles, but some of the substance is dissolved. On prolonged washing more lignosulfonates are also leached out of the fibers. Of the hemicelluloses xylans are preferentially dissolved, and fragments of a surprisingly high degree of polymerization are present in effluents resulting from mechanical treatment of pulps.

*Acidic Groups*   The wood and the fibers liberated from it either by mechanical means or by chemical treatments carry a negative charge because of the presence of certain functional groups. These groups exhibit an acidic character provided that they can be ionized at ambient conditions. The positively charged counter ions ($X^+$), maintaining the balance of electroneutrality, are exchangeable to other cations ($Y^+$) present in the surrounding solution:

$$\text{Fiber-}X^+ + Y^+ \rightleftarrows \text{Fiber-}Y^+ + X^+ \tag{7-23}$$

This ion exchange ability of wood and pulps leads to important consequences, which must be taken into consideration in operations such as pulp washing. The ion exchange capacity of the high-yield pulps is higher than that of chemical pulps, particularly because they contain much lignin carrying sulfonic acid groups.

TABLE 7-11.   Type of Acidic Groups in Wood[a]

| Acidic group | Structure[b] | $pK_a$ (25°C) | Degree of ionization at pH 7 (%) |
|---|---|---|---|
| Carboxylic | $R-CO_2H$ (minor) | 4–5 | 99–99.9 |
| | $R-CH(OR')CO_2H$ | 3–4 | 99.9–99.99 |
| Phenolic | $R-\overset{\overset{O}{\|}}{C}$-⟨⟩-OH (minor) | 7–8 | 10–50 |
| | $R$-⟨⟩-OH | 9.5–10.5 | 0.03–0.3 |
| Alcoholic | $R-CH(OH)-R'$ (minor) | 15–17 | $10^{-8}-10^{-6}$ |
| | $R-CH(OR')CH(OH)-R''$ | 13.5–15 | $10^{-6}-3\cdot10^{-5}$ |
| Hemiacetalic | ⟨O⟩-OH$\atop$OR | 12–12.5 | $10^{-3}-3\cdot10^{-4}$ |

[a] From Sjöström (1989).
[b] R, R', R'' = H, alkyl, or aryl.

The acidic groups favorably affect the fiber properties. The polyelectrolytic and colloidal properties of the pulp, regulating the extent of swelling as well as the brightness stability, are more or less, and often even markedly, influenced by these charges. The acidic groups are indeed important factors contributing to the hydrodynamic properties of the fiber suspensions as well as to the physical properties of fibers and forces between them.

Various ionizable groups are present in the polymeric constituents of wood (Table 7-11). Of these, however, only the carboxyl groups are ionized in neutral or weakly acidic conditions. Ionization of phenolic hydroxyl groups demands rather alkaline conditions, and the very weak alcoholic hydroxyl groups are ionized only in the presence of strong alkali. However, due to the inductive effect of certain substituents that are present in wood constituents, the acidity of the hydroxyl groups can increase considerably.

**TABLE 7-12.   Carboxyl Groups in Some Wood Species (mmol/100 g)[a]**

| Species | Methylglucuronic acid content | Carboxyl group content | | |
|---|---|---|---|---|
| | | | Accessible to ion exchange | |
| | | Total | Untreated | Hydrolyzed[b] |
| Picea abies | 7 | 15–25 | 7 | 13 |
| Pinus sylvestris | 8 | 15 | 5 | 9 |
| Betula verrucosa/pubescens | 15 | 25–35 | 6 | 17 |

[a] From Sjöström (1989).
[b] The sample has been treated to hydrolyze the carboxylic acid lactones and esters.

At usual conditions of papermaking the carboxyl groups are obviously the only type of functional groups giving rise to the generation of charged sites on the fibers. Both in native wood and in mechanical pulps the majority of the carboxyl groups are of uronic acid type mainly attached to the xylan. Some of them originate from pectic substances, enriched in the compound middle lamella region. Native lignin contains few, if any, carboxyl groups. An additional source of the carboxyl groups is the fraction of free fatty acids and resin acids, but their contribution is relatively small.

Some of the carboxyl groups are located at inaccessible regions in the wood cell wall or they are blocked through esterification or lactonization. These groups are liberated during alkaline pretreatment (Table 7-12). As a result, the solvation and swelling of the material are increased, leading to improved bonding of the fibers.

When the pretreatment is made in the presence of sulfite, as usual, lignin is sulfonated to a degree depending on the treatment conditions. The strongly acidic sulfonic acid groups introduced are ionized even at a low pH when they are in a free hydrogen ion form. However, their contribution to swelling and other material properties depends also on the particular counter ion.

Whether or not it is possible to improve the properties of high-yield pulps by more proper treatments leading to a smooth liberation of the fibers and by increasing the concentration of acidic material on the fiber surfaces are open questions.

# Chapter 8

# PULP BLEACHING

The light absorption (color) of pulp is mainly associated with its lignin component. To reach an acceptable brightness level the residual lignin should thus either be removed from the pulp or, alternatively, freed from strongly light-absorbing groups (chromophores) as completely as practicable. Accordingly, the following two alternatives are possible: (1) *lignin-removing* (*delignifying*) or (2) *lignin-preserving* (*retaining*) bleaching.

Delignifying bleaching, which results in both high and reasonably permanent brightness, is applicable to chemical pulps. It is performed in a bleaching *sequence* comprising several treatment stages with bleaching chemicals and alkali (sodium hydroxide). Common bleaching chemicals today are chlorine, chlorine dioxide, and oxygen. Because of environmental reasons, gaseous chlorine will probably be completely replaced within a few years, first with chlorine dioxide. However, in the long run, the chlorine-free chemicals (oxygen, hydrogen peroxide, and ozone) will probably take over the role of chlorine dioxide (Fig. 8-1).

The purpose of bleaching is also to improve the cleanliness of the pulp through removing of extractives and other contaminants including inorganic impurities and bark residues. In the case of paper-grade pulps hemicellulose losses should be avoided, whereas the presence of hemicelluloses detrimentally affects the quality of dissolving pulps.

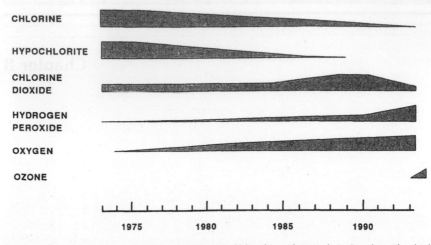

**Fig. 8-1.** Development and forecast for the use of bleaching chemicals in Sweden. The thickness of the lines represents relative proportions (Grundelius, personal communication, 1992).

Lignin-preserving (retaining) bleaching, which usually gives only a moderate brightness increase, is the appropriate method for high-yield pulps of semichemical, chemimechanical, and mechanical types. The most common lignin-preserving bleaching chemicals are sodium dithionite and hydrogen peroxide.

## 8.1 Lignin-Removing Bleaching

### 8.1.1 Background

Chlorine was discovered by Scheele in 1774 and its bleaching properties by C. L. Berthollet in 1784, but it took further more than a hundred years or until the late 1880s before the technology according to which chlorine is produced electrolytically from sodium chloride was ready for industrial production. From this time it took about fifty years further or until the early 1930s for bleaching with chlorine gas to be introduced into the pulp industry. However, calcium (or sodium) hypochlorite was used much earlier or soon after the beginning of the pulp industry. Its use was based on the discovery of S. C. Tennant, who about twenty years after Scheele's discovery observed that chlorine gas reacts with anhydrous lime, resulting in a dry bleaching powder. This finding was of great practical importance because of the effortlessness in handling and transporting of sodium and calcium hypochlorite compared with that of chlorine gas. In the long run, however, the technology and materials improved, making the industrial use of even chlorine gas possible and common.

In the early 1930s advances in bleaching technology had resulted in a process including several stages, starting first with chlorine and then followed with an alkaline treatment (sodium hydroxide) and ending with a hypochlorite stage. This multistage bleaching was rather successful in the case of sulfite pulps, but conditions necessary to obtain high brightness of kraft pulps resulted in depolymerization of cellulose and loss in the fiber strength.

A decisive breakthrough particularly in the bleaching of kraft pulps took place because of the introduction of chlorine dioxide in the middle of 1940s. Although the potential of chlorine dioxide as a bleaching agent had been discovered much earlier, time was needed for the development of integrated processes and systems suitable for production and safe handling of this explosive and corrosive gas within the mill site. This was the start of a new era of steadily growing production of bleached kraft pulp and a decline in sulfite pulp production.

Innovations in the process technology resulted in new types of bleaching systems, of which so-called displacement bleaching and gas-phase bleaching deserve mention. The advantages are that the diffusion of chemicals into the pulp phase is increased, resulting in a shorter bleaching time and a higher solids concentration of the spent liquors. However, the success has not been the same as for continuous cooking, and few bleaching plants based on these principles have been installed.

Around the beginning of 1970s the bleaching technology was again at the entrance of great changes. Chlorinated and toxic organic compounds, of which some responded positively in standard mutagenicity tests, were detected in the spent liquors from bleaching. Of a great importance for the period ahead was the discovery of the method according to which the lignin content of the pulp can be decreased considerably by using oxygen together with alkali. From the laboratory scale this technique was soon introduced into a common industrial practice. Today there are several modifications, but usually the unbleached pulp is first treated in an oxygen reactor. After this delignification ("oxygen bleaching") prestage the pulp is bleached in an appropriate bleaching sequence, but the consumption of chlorine or chlorine-based chemicals is much lower.

When the analytical methods successively improved it was possible to detect chlorinated compounds present in extremely low concentrations. In the 1980s such sensitive methods disclosed the presence of chlorinated compounds including highly toxic dioxins at part per billion (ppb) levels even in papers made from chlorine-bleached pulps.

Because of environmental concerns the limits allowed for the discharges from the pulp mills have been set today to a very low level and are to be even more stringent in the future. These regulations concern not only chlorinated compounds, but also easily degradable organic materials and suspended solids. Because of changes in the bleaching practices during the last few years, these discharges have decreased markedly. However, the bleaching technique involving chlorine-free chemicals is still under vigorous development in order to minimize the discharges in the future. New parameters are also needed for more adequate testing of the toxicity and other detrimental effects of the effluents after bleaching. Hundreds of individual compounds have been identified in these effluents. Section 8.1.7 deals with the composition of the spent liquors after bleaching with regard for environmental effects.

Table 8-1 illustrates the changes in the bleaching systems from the early beginning up to the present time, and examples of the conditions of modern bleaching sequences are given in Table 8-2.

Even if other agents, such as peroxyacetic acid and sodium chlorite, are applied in a laboratory scale as delignification agents, they are not useful for industrial pulp bleaching. In addition to chemical agents, even the use of

**TABLE 8-1.  Progress in the Bleaching of Chemical Pulps: Typical Sequences through Early Periods until Modern Times[a]**

| Period (decades) | Sequence | Comments[b] |
|---|---|---|
| 1880 | H | |
| 1910 | HEH | |
| 1930 | CEH, CEHEH | |
| 1950 | CEHDED, CEDED | |
| 1960 | (C + D)EHDED | |
| 1970 | (C + D)(E + P)HD(E + P)D | $D$ = 0–15% |
| 1980 | O(C + D)(E + P)D(E + P)D | $D$ = 0–15% |
| 1985 | O(DC)(EO)D(E + P)D | $D$ = 20–60% |
| 1990 | O(DC)(EPO)D(E + P)D | $D$ = 40–90% |

[a] Nomenclature: C, chlorination ($Cl_2$), H, hypochlorite (NaOCl), E, extraction (NaOH), D, chlorine dioxide ($ClO_2$), $O_2$, oxygen ($O_2$ + NaOH), P, peroxide ($H_2O_2$ + NaOH). A + sign between the chemicals (letters) means that the respective chemicals are added simultaneously. When no sign, the chemicals are added one after another.

[b] The percentage dosage $D$ of chlorine dioxide given here refers to (C + D) and (DC) stages. Addition of P in all the E stages is low.

TABLE 8-2. Examples of Bleaching Sequences and Conditions (Softwood Kraft Pulp)[a]

| Sequence[b] | Stage | Chemicals | Charge (kg/ton pulp)[c] | Consistency (%) | Temperature (°C) | Time (min) |
|---|---|---|---|---|---|---|
| A | C | $Cl_2$ | 55 | 4 | 30 | 60 |
| | E | NaOH | 37 | 10 | 60 | 120 |
| | H | NaOCl | 12 | 8 | 40 | 150 |
| | D | $ClO_2$ | 15 | 10 | 70 | 180 |
| | E | NaOH | 7 | 10 | 65 | 120 |
| | D | $ClO_2$ | 5 | 10 | 70 | 240 |
| B | O | $O_2$ + NaOH | 15 + 20 | 10(30) | 100 | 60 |
| | D40C60 | $ClO_2$, $Cl_2$ | 12 + 18 | 4 | 70 | 45 |
| | EO | NaOH + $H_2O_2$ | 15 + 5 | 10 | 70 | 120 |
| | D | $ClO_2$ | 17 | 10 | 70 | 180 |
| | E + P | NaOH + $H_2O_2$ | 8 + 3 | 10 | 70 | 90 |
| | D | $ClO_2$ | 5 | 10 | 70 | 180 |
| C | O | $O_2$ + NaOH | 16 + 18 | 10 | 100 | 60 |
| | X[d] | e.g., EDTA | 2 | 4 | 50 | 30 |
| | E + P | NaOH + $H_2O_2$ | 14 + 19 | 10 | 80 | 180 |
| | D | $ClO_2$ | 32 | 10 | 70 | 180 |
| | E + P | NaOH + $H_2O_2$ | 11 + 2 | 8 | 70 | 90 |
| | D | $ClO_2$ | 8 | 10 | 70 | 180 |

[a] Data from Grundelius (1991).

[b] Sequence A was typical in the mill practice during the 1960s. Sequence B represents 1990s bleaching practice and sequence C bleaching technology still under development.

[c] NaOCl and $ClO_2$ as active chlorine.

[d] This stage is a slightly acidic washing stage in the presence of complexing agents for removal of heavy metals from the pulp.

enzymes such as ligninases and other type of biocatalysts has received interest. Among many obstacles in using enzymes are, however, their restricted accessibility and tendency to be adsorbed on the fiber surfaces. The recycling of the enzymes is difficult and their use thus expensive.

Another type of approach involves the use of enzymes capable of cleaving linkages between lignin and polysaccharides. This type of specific treatment would, indeed, be ideal provided that lignin-carbohydrate bonds really represent the main obstacle. It has been shown that treatment of hardwood pulps with hemicellulase mixtures rich in endo-β-xylanase makes these pulps more susceptible toward chemical bleaching agents. However, it is not known whether this effect depends on the cleavage of lignin–xylan bonds or is to be ascribed to other factors. One disadvantage is that the pulp yield is decreased because of the xylan losses.

Instead of enzymes, so-called biomimetic catalysts have been suggested as bleaching agents. Experiments have been made using hemoglobin as an oxygen carrier, but this type of approach seems so far even more unrealistic than the application of enzymes.

### 8.1.2   Bleaching Chemicals

Even if chlorine and hypochlorite are losing importance as bleaching agents, the technology of bleaching is based on the use of these chemicals. The reactions of the chlorine-based chemicals with the wood substance have also for decades been subjected to extensive research, giving a necessary background for further studies of new systems and other bleaching agents. Of these, oxygen is especially important and the use of hydrogen peroxide is increasing. Ozone is a potential newcomer.

*Chlorine and Hypochlorite*   Molecular chlorine is a greenish and highly reactive gas. On dissolution of chlorine in water, the following equilibria are almost instantly attained:

$$Cl_2 + H_2O \rightleftharpoons H^+ + Cl^- + HOCl \tag{8-1}$$

$$HOCl \rightleftharpoons H^+ + ClO^- \tag{8-2}$$

The equilibrium constants relating to equations (8-1) and (8-2), respectively, are:

$$K_1 = [H^+][Cl^-][HOCl]/[Cl_2] = 3.9 \times 10^{-4} \quad (pK_1 = 3.4) \tag{8-3}$$

$$K_2 = [H^+][ClO^-]/[HOCl] = 2.9 \times 10^{-8} \quad (pK_2 = 7.5) \tag{8-4}$$

The $K$ values given here have been measured at 25°C for 0.1 $M$ solution, and they increase with temperature and also vary with concentration. An equilibrium diagram based on these values is shown in Fig. 8-2. Starting from the acidic side the concentration of chlorine is gradually decreased and is practically zero at around pH 5.5 where the maximum concentration of undissociated hypochlorous acid has been reached. After this point the degree of dissociation of the hypochlorous acid is gradually increased, and at a pH above 10 practically only hypochlorite ions are present.

At certain conditions some dichlorine monoxide can be formed from hypochlorous acid according to the following reaction:

$$2 HOCl \rightleftharpoons Cl_2O + H_2O \tag{8-5}$$

for which the equilibrium constant is:

$$K_3 = [Cl_2O]/[HOCl]_2 < 10^{-2} \quad (pK_3 < 2) \tag{8-6}$$

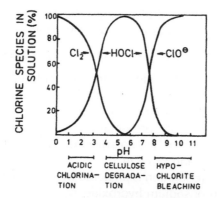

**Fig. 8-2.** Composition of aqueous chlorine solution (0.1 $M$) as a function of pH. Based on equilibrium constants of $K_1 = 3.9 \times 10^{-4}$ and $K_2 = 2.9 \times 10^{-8}$.

This equilibrium is attained rather slowly, but because dichlorine monoxide is extremely reactive, its presence is to be considered within the pH region below 7.5.

In the acidic chlorination stage of pulp bleaching, most of the chlorine is consumed rapidly within 5–10 min. This initial phase is diffusion controlled, and effective mixing of the pulp slurry facilitates the reaction. Chlorine can react either as a molecular species or after being decomposed to chlorine radicals. A spontaneous decomposition may take place under the influence of light, followed by combination with chloride ions:

$$Cl_2 \rightarrow 2\ Cl\cdot \tag{8-7}$$

$$Cl\cdot + Cl^- \rightarrow Cl_2^-\cdot \tag{8-8}$$

Another possibility is the abstraction of a hydrogen atom from an organic substrate by chlorine, for example (R is an organic residue):

$$Cl_2 + RH \rightarrow Cl\cdot + R\cdot + HCl \tag{8-9}$$

Although this process is slow, the chlorine radicals once formed will initiate a rapidly accelerating chain reaction:

$$RH + Cl\cdot \rightarrow R\cdot + HCl \tag{8-10}$$

$$R\cdot + Cl_2 \rightarrow RCl + Cl\cdot \tag{8-11}$$

The chlorination stage of bleaching is usually carried out in the presence of chlorine dioxide, which is either added first, followed somewhat later by chlorine (DC bleaching) or, alternatively, chlorine and chlorine dioxide are added simultaneously (D + C bleaching). The presence of chlorine dioxide is important because it acts as a radical scavenger:

$$Cl\cdot + ClO_2\cdot + H_2O \rightarrow Cl^- + ClO_3^- + 2H^+ \qquad (8\text{-}12)$$

It seems that the reactions of chlorine and hypochlorous acid with carbohydrates proceed mainly via radical mechanisms because these are retarded by chlorine dioxide and other radical scavangers. The corresponding reactions with lignin instead are presumably more of ionic character, taking place via the positive ends of the polarized chlorine and hypochlorous acid molecules ($^{\delta-}Cl\text{-}Cl^{\delta+}$ and $^{\delta-}HO\text{-}Cl^{\delta+}$).

Sodium hypochlorite solution is prepared by introducing chlorine gas into an aqueous solution of sodium hydroxide:

$$2NaOH + Cl_2 \rightarrow NaOCl + NaCl + H_2O \qquad (8\text{-}13)$$

On prolonged storage, the hypochlorite ions disproportionate gradually into chloride and chlorate. Hypochlorite solutions may also be decomposed to chloride and oxygen in the presence of heavy metal ions. When hypochlorite is reduced to chloride, two oxidation equivalents are consumed. For example, 1 $M$ sodium hypochlorite solution contains $2 \times 35.5 = 71.0$ g/liter of "active" chlorine.

***Chlorine Dioxide*** Chlorine dioxide can be made from sodium chlorate in the presence of reducing agents, e.g., sulfur dioxide:

$$2NaClO_3 + H_2SO_4 + SO_2 \rightarrow 2ClO_2 + 2NaHSO_4 \qquad (8\text{-}14)$$

Notwithstanding the comparatively high price, chlorine dioxide has almost completely replaced sodium hypochlorite in the modern bleaching technique because of its superior delignification selectivity. Moreover, the formation of chlorinated and harmful reaction products is virtually eliminated, which is also the reason why the current tendency is to substitute more and more of the chlorine with chlorine dioxide in the first bleaching stage.

At slightly acidic conditions (pH 3–5) chlorine dioxide can be reduced to chloride ions:

$$ClO_2 + 4H^+ + 5e^- \rightarrow Cl^- + 2H_2O \qquad (8\text{-}15)$$

In this reaction five oxidation equivalents can be released. Therefore, 1 $M$ chlorine dioxide solution contains $5 \times 35.5$ g/liter of active chlorine.

In alkaline media chlorine dioxide is reduced to chlorite ions involving a change of only one oxidation equivalent:

$$ClO_2 + e^- \rightarrow ClO_2^- \qquad (8\text{-}16)$$

Even at the usual (slightly acidic) bleaching conditions both chlorite ions and hypochlorous acid can be generated from chlorine dioxide in its reactions with pulp components (lignin). Chlorite is not highly reactive, but it can disproportionate to hypochlorous acid and chlorate ions:

$$2ClO_2^- + H^+ \rightarrow HOCl + ClO_3^- \tag{8-17}$$

Of course, hypochlorite is in equilibrium with chlorine [cf. equation (8-1)]. Contrary to hypochlorite and chlorine, chlorate is rather unreactive and it mainly remains in the bleaching effluent as such. Another reason for generation of chlorate is the oxidation of chlorite with chlorine or hypochlorite:

$$ClO_2^- + Cl_2 + H_2O \rightarrow ClO_3^- + 2\ Cl^- + 2\ H^+ \tag{8-18}$$

$$ClO_2^- + HOCl \rightarrow ClO_3^- + Cl^- + H^+ \tag{8-19}$$

However, chlorite is reoxidized to chlorine dioxide:

$$2\ ClO_2^- + Cl_2 \rightarrow 2\ ClO_2 + 2\ Cl^- \tag{8-20}$$

$$2\ ClO_2^- + HOCl \rightarrow 2\ ClO_2 + Cl^- + HO^- \tag{8-21}$$

Since a number of species, including chlorine, hypochlorous acid, chlorite, and chlorate, are formed as intermediates, the oxidation pathways of chlorine dioxide bleaching are complex.

*Oxygen and Hydrogen Peroxide*   The chemistry and interaction of oxygen with organic materials have been subjected to intensive studies, particularly in conjunction with biological processes. Oxygen has also long been used for a variety of industrial applications, and its bleaching effect on cotton has been known since historical times. In spite of that, oxygen was not introduced to the pulp bleaching technology until the 1970s. The reasons still limiting the use of oxygen are associated with the difficulties of directing its action more specifically toward lignin without degrading carbohydrates.

Hydrogen peroxide is used for lignin-preserving bleaching (cf. Section 8.2), but in recent years its use in the bleaching of chemical pulps has increased considerably. The reactions of oxygen and hydrogen peroxide bleaching have common features because in both cases the medium is alkaline and oxygen is partly converted to hydrogen peroxide and vice versa. A number of reactive intermediates are formed in the reactions of oxygen and peroxides with the substrate. The resulting reaction pattern is thus extremely complex and so far imperfectly understood or based on speculations.

In its ground state (*triplet state*) the oxygen molecule has two unpaired electrons in its outer shell, but it can be excited to higher energy levels (*singlet state*) characterized by two paired or unpaired electrons with an antiparallel spin (Fig. 8-3). When oxygen reacts with the substrate, it may be reduced to water by one-electron transfer in four successive stages giving

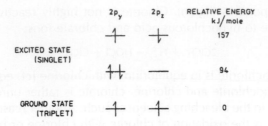

**Fig. 8-3.** Energy states of oxygen molecule.

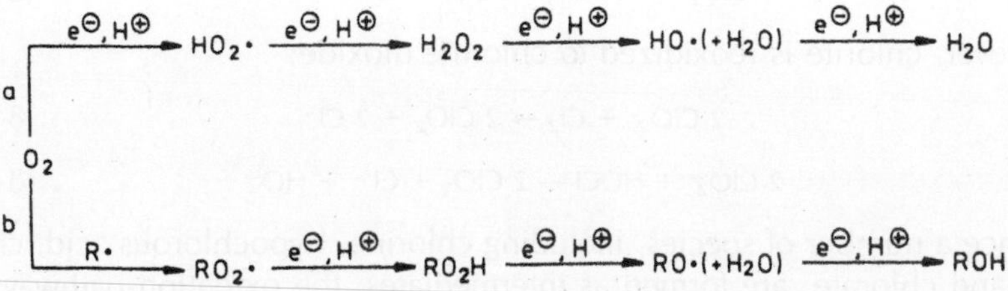

**Fig. 8-4.** Reduction of oxygen by one-electron transfer mechanism. (a) Each of the four reaction steps involves addition of one electron and one proton, giving rise to hydroperoxy radicals, hydrogen peroxide, and hydroxyl radicals as intermediate products. (b) The corresponding organic intermediates are formed when oxygen reacts with an organic radical R·.

rise to intermediate products, namely, hydroperoxy radicals ($HO_2\cdot$), hydrogen peroxide ($H_2O_2$), and hydroxyl radicals ($HO\cdot$), as well as their organic counterparts (Fig. 8-4). Provided the pH of the medium is sufficiently high, ionization will take place:

$$HO_2\cdot + H_2O \rightarrow O_2\cdot^- + H_3O^+ \qquad (pK_a \sim 4.8) \qquad (8\text{-}22)$$

$$HO\cdot + H_2O \rightarrow O\cdot^- + H_3O^+ \qquad (pK_a \sim 11.9) \qquad (8\text{-}23)$$

$$H_2O_2 + H_2O \rightarrow HO_2^- + H_3O^+ \qquad (pK_a \sim 11.6) \qquad (8\text{-}24)$$

The reactive intermediates may be formed as a result of radical reactions typified by phases of initiation, propagation, and termination (R is the organic residue):

*Initiation:*

$$RO^- + O_2 \rightarrow RO\cdot + O_2\cdot^- \qquad (8\text{-}25)$$

or

$$RH + O_2 \rightarrow R\cdot + HO_2\cdot^- \qquad (8\text{-}26)$$

*Propagation:*

$$R\cdot + O_2 \rightarrow RO_2\cdot \tag{8-27}$$

$$RO_2\cdot + RH \rightarrow RO_2H + R\cdot \tag{8-28}$$

*Termination:*

$$RO\cdot + R\cdot \rightarrow ROR \tag{8-29}$$

The reaction is initiated by the attack of oxygen molecules resulting in abstraction of either electrons from the ionized phenolic hydroxyl groups [equation (8-25) and Fig. 8-5)] or certain hydrogen atoms linked to carbon atoms elsewhere in the substrate [equation (8-26)]. The reaction is propagated through combination of oxygen with the initiated organic radical, which is afterward liberated under simultaneous formation of hydroperoxides ($RO_2H$) [equations (8-27) and (8-28)]. The reaction chain is terminated by coupling of radicals [equation (8-29)].

The pulps always contain traces of ions of transition metals (mainly of Fe, Co, Cu, and Mn, possibly present in the form of finely divided colloidal particles, complexes, or bound to the acidic groups in fibers). They act as effective catalysts decomposing peroxides to hydroxyl radicals:

$$H_2O_2 + M^{n+} \rightarrow HO\cdot + HO^- + M^{(n+1)+} \tag{8-30}$$

or

$$HO_2^- + M^{n+} \rightarrow O\cdot^- + HO^- + M^{(n+1)+} \tag{8-31}$$

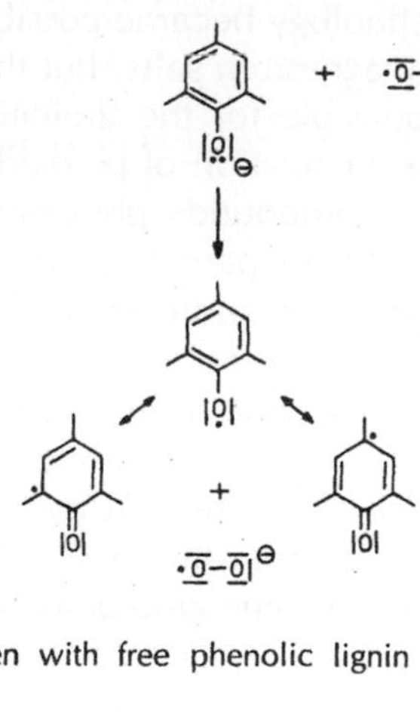

**Fig. 8-5.** Reaction of oxygen with free phenolic lignin structures leading to resonance-stabilized phenoxy radicals.

The redox system involved in this decomposition implies that the catalyst is reduced to its initial valence state. It is possible that superoxide or hydrogen peroxide anions act as the reducing agents:

$$O_2 \cdot ^- + M^{(n+1)+} \rightarrow O_2 + M^{n+} \tag{8-32}$$

or

$$HO^- + HO_2^- + M^{(n+1)+} \rightarrow O_2 \cdot ^- + H_2O + M^{n+} \tag{8-33}$$

Another mechanism not needing intervention of heavy metal ion catalysts is the homolytic decomposition of hydroperoxides to hydroxyl radicals that may take place at higher temperatures:

$$RO_2H \rightarrow RO \cdot + HO \cdot \tag{8-34}$$

In a neutral aqueous solution the hydroxyl radical is an electrophile with very strong oxidizing properties (standard redox potential $E° \sim 2.3$ V), but after ionization at higher pH values it looses its electrophilic character and the redox potential of the resulting anion ($O \cdot ^-$) is substantially lower ($E° \sim 1.4$ V). Judging from the acidity constant ($pK_a \sim 4.8$), the hydroperoxide radical ($HO_2 \cdot$) is ionized even in a weakly acidic medium. The ionized superoxide ($O_2 \cdot ^-$) is a weak oxidant ($E° \sim 0.41$), thus having more selective properties, and it can also act as a reducing agent. The hydrogen peroxide anion ($HO_2^-$) is a rather weak oxidant ($E° \sim 0.87$).

The system is even further complicated by the interaction of the reactive species. For example, in alkaline medium combination of ionized hydroxyl radicals with oxygen may generate ozonide anion radicals:

$$O \cdot ^- + O_2 \rightarrow O_3 \cdot ^- \tag{8-35}$$

The oxygen bleaching technology became possible after the discovery of the inhibiting properties of magnesium salts, but the severe degradation of polysaccharides is still responsible for the limitations. The most harmful reaction is obviously the decomposition of peroxides to hydroxyl radicals. Both inorganic and organic compounds possessing a more or less pronounced ability either to stabilize peroxides or inhibit the formation of radicals are known. However, such additives usually also retard the delignification, and so far the magnesium compounds are the best and the only inhibitors used for oxygen bleaching. Even if magnesium is added as a water-soluble salt, e.g., as sulfate, it is readily precipitated as magnesium hydroxide in the presence of alkali. Several mechanisms have been suggested, but it seems clear that the transition metals play a decisive role and that they are deactivated through some kind of association with magnesium hydroxide.

*Ozone*   Although various ways are possible for preparation of ozone, its industrial production is based on the use of so-called "ozonator discharge" in which a stream of oxygen gas or air is passed through an electrical discharge. The oxygen atoms formed are combined with oxygen molecules to form ozone. Although on a large scale the production costs of ozone are lower, the consumption of electrical energy for the process is high. No other reaction products are formed when ozonizing pure and dry oxygen, but the concentration of ozone in the gas obtained is relatively low. If the air is used, small amounts of nitrogen oxides are generated as impurities. Ozone is much more soluble in water than oxygen. However, its low partial pressure at ordinary temperatures and conditions limits its solubility to a few milligrams per liter.

Ozone is a very strong oxidant ($E° \sim 2.07$ V). Its reactions with organic substrates are complicated because it is decomposed in water to other reactive species including superoxide and hydroxyl radicals. The ozone molecule may exist in different mesomeric structures:

$$\tag{8-36}$$

The equilibrium and proportions of these resonance hybrids depend on the conditions, such as solvent and pH, but the positively charged terminal oxygen atoms, having only a sextet of electrons, are believed to be the reactive electrophilic sites. However, in some reactions ozone is considered to behave as a nucleophile. When ozone reacts with an organic substrate it is reduced to oxygen or hydrogen peroxide. These species, in turn, will give rise to their typical reactions.

### 8.1.3   General Aspects of Bleaching

Although the last residues of lignin can be removed from the pulp much more selectively with bleaching agents than by cooking, the bleaching chemicals are expensive and the elimination of water-polluting wastes at reasonable costs meets great difficulties. The delignification in the cooking stage should therefore be extended as far as possible but there is a certain limit (Fig. 8-6). In the case of a normal kraft cook of softwood the point when the cook should be interrupted corresponds to a residual lignin content of about 5%. When the pulp is then bleached in a conventional bleaching

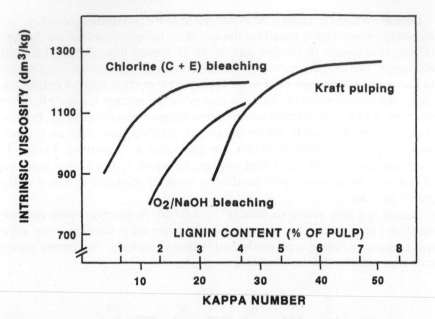

**Fig. 8-6.**   Limits of delignification for kraft pulping and bleaching. The viscosity is a measure for the degree of polymerization of cellulose. Note that these curves only show tendencies; the actual shapes vary considerably, depending on the pulping and bleaching conditions.

sequence including chlorine in the first stage, only a moderate depolymerization of cellulose takes place as indicated by the viscosity measurements. The carbohydrate losses are also moderate. However, when oxygen/alkali is applied as the first delignification stage, a sharp viscosity drop occurs rather early. Not only is the depolymerization of cellulose severe but also the carbohydrate losses are considerable. Because of these reasons oxygen delignification cannot be extended to low lignin contents. Only about 50% of the residual lignin in the unbleached pulp can be removed in the oxygen delignification stage; the remaining portion is to be removed by using other bleaching agents.

The action of the bleaching agents is only fragmentarily understood. In addition to the chemical reactions, physical and morphological factors play an important role. Much of the present knowledge of the chemical reactions, and especially their mechanisms, is based on experiments with simplified systems with model compounds representing structures that are believed to be characteristic of the residual lignin. Sometimes conclusions have been drawn from experiments far from realistic bleaching conditions.

**TABLE 8-3.** Example of the Decrease in the Lignin Content after Various Stages during Bleaching of Sulfate and Sulfite Pulp[a]

| Sulfate pulp | | Sulfite pulp | |
|---|---|---|---|
| Stage | Lignin content[b] (% of pulp) | Stage | Lignin content (% of pulp) |
| Unbleached | 2.4 | Unbleached | 2.4 |
| C | 2.0 | C | 0.7 |
| E | 0.5 | E | 0.2 |
| C | 0.3 | D | <0.05 |
| H | <0.05 | E | <0.05 |
| D | <0.05 | D | <0.05 |
| E | <0.05 | | |

[a] cf. Sjöström and Enström (1966).
[b] Lignin content of unbleached pulp is lower than normally.

Examples of these are studies in organic solvents in the presence of which the reactions may proceed quite differently compared with those in aqueous solution.

In recent years much attention has been focused on the characterization of residual lignins. An insight into their properties and detailed structures, including lignin–carbohydrate bonds, will undoubtedly contribute to a better understanding of delignification processes and to the choice of more adequate process conditions.

An example of the differences in the bleaching response between sulfite and kraft pulps is given in Table 8-3. As can be seen, the residual lignin can be removed much more easily from the sulfite pulp than from the kraft pulp. This difference is especially pronounced in the chlorination (first) bleaching stage. Only more or less speculative explanations can be offered to explain this difference. However, the low reactivity of the kraft pulp lignin is obviously associated with its condensed structure and with cross-links to polysaccharides. During sulfite pulping lignin is less drastically modified, but strongly acidic sulfonic acid groups are introduced to its structure. These groups present in the residual sulfite pulp lignin are dissociated over the entire pH range, thus contributing to the hydrophilicity (and solubility) also in the acidic chlorination stage. The phenolic hydroxyl groups and carboxyl groups are the only type of acidic groups in kraft pulps, and they contribute markedly to the solubility only in the alkaline stage when they are dissociated and converted into the salt form.

### 8.1.4    Lignin Reactions

*Chlorine and Alkali*    The reactions of chlorine with an organic substrate are either of ionic or radical type. Chlorine reacts with aromatic lignin structures mainly by substitution and by oxidation, but some chlorine can also be introduced by addition into the double bonds present in the side chain. Chlorine or the reactive species generated from it (cf. Section 8.1.2) are electrophiles and they preferentially attack negatively charged sites in the substrate. Such partially charged sites are generated by the influence of substituents. For example, the methoxyl groups as well as the free·and etherified hydroxyl groups activate *ortho* and *para* positions of the aromatic nucleus. Likewise, the double bonds in the side chains conjugated to the aromatic ring, e.g., in stilbene structures, are susceptible to an electrophilic attack (Fig. 8-7).

Chlorination results in a partial substitution of the hydrogen atoms in the aromatic nucleus under simultaneous formation of hydrogen chloride. Another consequence of this electrophilic attack is displacement of the side chain and a partial depolymerization of lignin (Figs. 8-8 and 8-9). During normal chlorination (bleaching) conditions at least one-half of the charged amount of chlorine is rapidly consumed in substitution reactions. The other half or less is mainly consumed for oxidation reactions which also proceed quickly at the initial phase. Chlorine oxidation results in an extensive cleavage of the aromatic rings, generating unsaturated (and chlorinated) dicarboxylic acid structures of muconic acid type in the lignin polymer, which is almost completely demethylated under liberation of methanol (Fig. 8-9). Even chlorinated quinoid and catechol structures are formed during chlorination, but the latter are readily oxidized. Even if an oxidative cleavage of the ether bonds seems to be a plausible mechanism, it is possible that another mechanism, involving an initial hydrolytic cleavage of these bonds and a subsequent oxidation of the resulting catechols, is operating.

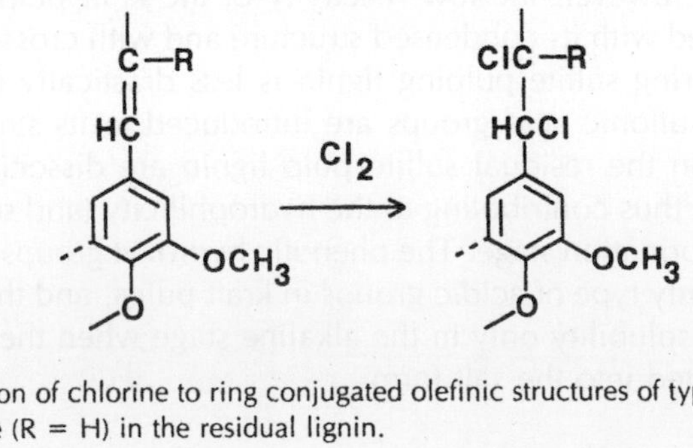

**Fig. 8-7.** Addition of chlorine to ring conjugated olefinic structures of type stilbene (R = aryl group) or styrene (R = H) in the residual lignin.

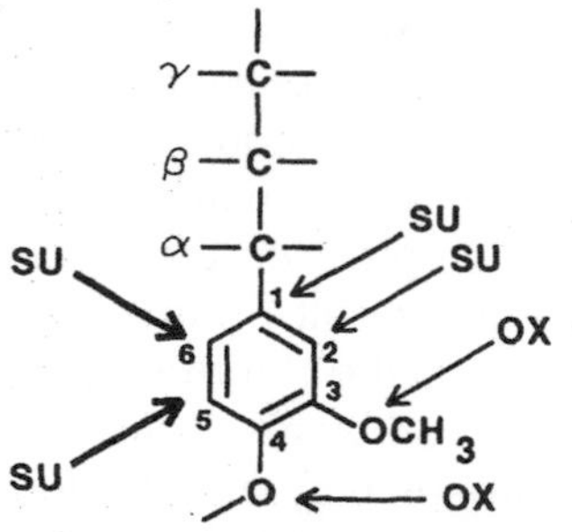

**Fig. 8-8.** Possible reaction sites of chlorine in a guaiacylpropane unit. *Substitution* (SU) which occurs mainly at C-6 and C-5 (uncondensed) positions leads to chlorinated lignin products under liberation of hydrogen chloride. Substitution at C-1 results in the displacement of the side chain and lignin fragmentation. *Oxidative* attack (OX) cleaves ether bonds, generating methanol and lignin fragmentation products. *Addition* of chlorine to double bonds, present in the residual lignin mainly in stilbene and styrene structures, should also be considered even if the frequency of these bonds is relatively low (cf. Fig. 8-7).

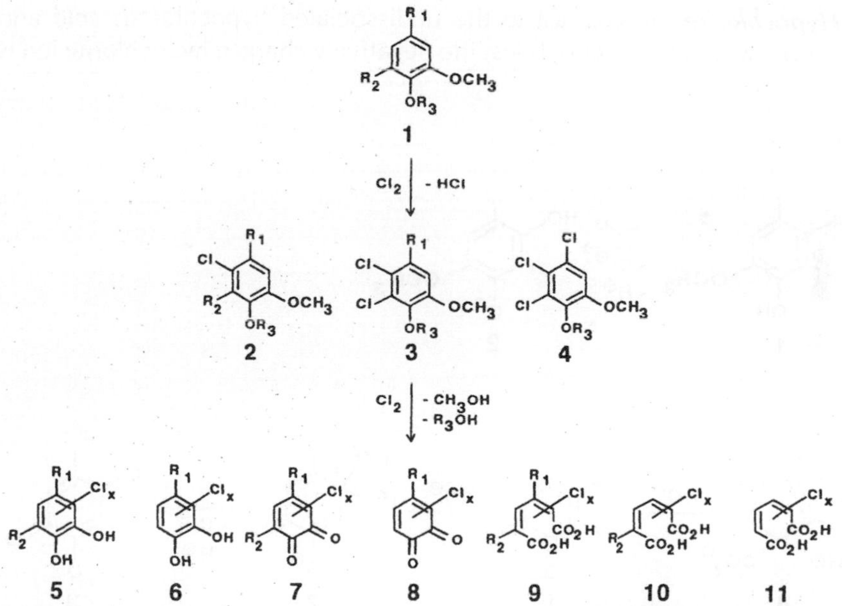

**Fig. 8-9.** Examples of reaction products of residual lignin after substitution and oxidation reactions of chlorine. A number of chlorinated products of types 2–4 are formed after substitution. These are oxidized to a variety of products of which those having catechol (5–6), o-quinone (7–8), and muconic acid (9–11) structures are typical. The chlorine content also greatly varies ($x \sim 1–3$).

The type of comprehensive oxidative action of chlorine just described is typical irrespective of whether the phenolic hydroxyl groups are free or etherified. However, although the aromatic structure is extensively destroyed, the linkages between the monomeric units are cleaved only partially, and most of the reaction products are still in a polymeric form.

The alkaline extraction stage results in an effective dissolution of chlorinated lignin. About two-thirds of the chlorine substituted in lignin during the previous chlorination is removed by alkali under liberation of phenolic hydroxyl groups. Carboxyl groups may also be formed from the o-quinoid structures (Fig. 8-10). After neutralization with alkali the acidic groups are effectively contributing to the solubility of the lignin fragments.

Although normally at least 80% of the lignin is removed in the prebleaching stages by chlorine and alkali, the brightness of the pulp is low (35–30% ISO), and the light absorption coefficient $k$ at 457 nm is 20–40 m²/kg. The corresponding $k$ value for the residual lignin is at least 2000 m²/kg, indicating a substantial increase of strong chromophoric structures ($k$ can be approximated to be of the order of 10–20 for native lignin).

*Hypochlorite* In contrast to the undissociated hypochlorous acid and chlorine, which are electrophiles, the negatively charged hypochlorite ion is

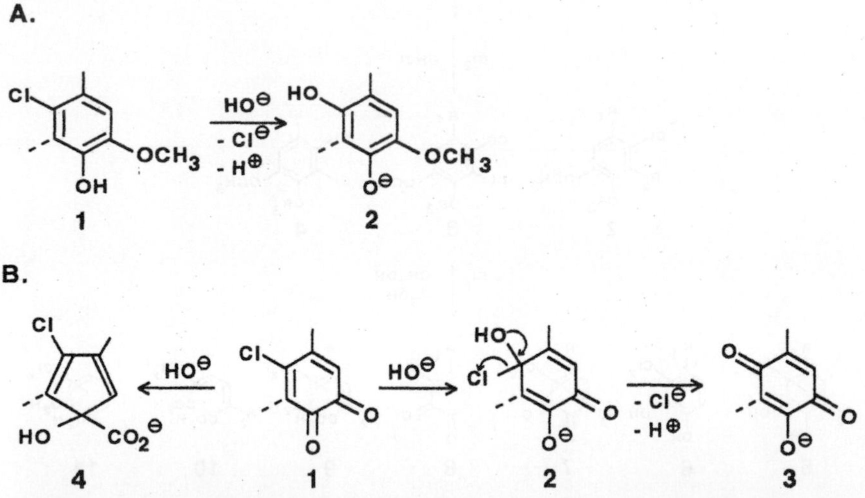

**A.**

**B.**

Fig. 8-10. Reactions of chlorinated lignin with alkali. (A) Nucleophilic substitution of chlorine by a hydroxyl ion. (B) Nucleophilic addition of a hydroxyl ion to a chlorinated o-quinone structure (1 → 3). A benzilic acid type rearrangement is an alternative reaction (1 → 4) resulting in a furanosidic acid structure. Note that introduction of acidic groups of phenolic, carboxylic acid, or other types result in an increased alkali solubility of the lignin fragments. The hydroxyl group present in the quinone structure 3 is exceptionally acidic (pK 4–5).

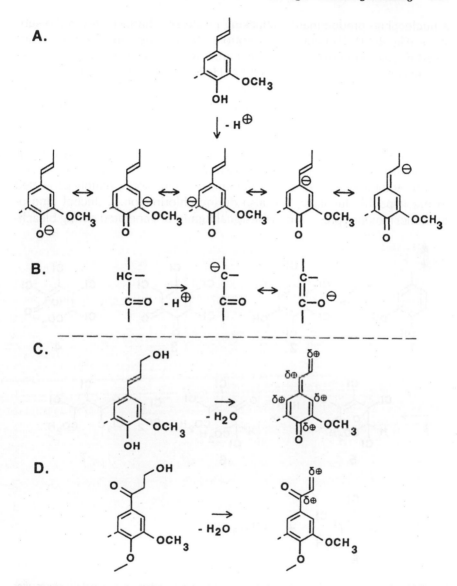

**Fig. 8-11.** Formation of negatively and positively charged sites in the lignin structure. Free phenolic (A) or carbonyl (B) structures (above the broken line) are ionized in alkali resulting in mesomeric carbanions. Below the broken line are examples of structures which can be converted by alkali (or acid) to conjugated structures with positively charged sites. See also Gierer and Imsgard (1977).

a nucleophile predominantly attacking positively charged sites in the substrate (Fig. 8-11). Particularly susceptible sites are the carbonyl and conjugated double bond structures, which are few in native lignin but generated during kraft pulping and prebleaching. Hypochlorite ion is a strong oxidant and it cleaves carbon–carbon bonds in such structures. The lignin residues rich in carboxylic acid groups are fragmented further, and most of them are dissolved. As a result the brightness of the pulp is markedly increased.

A large number of low-molecular-weight reaction products, most of them chlorinated, are formed during chlorine and hypochlorite bleaching according to reactions routes which have been only partially clarified (cf. Section 8.1.7). Among the detrimental compounds is chloroform, mainly generated in the hypochlorite stage, but also during chlorination. Catechol (and o-quinone) moieties present in the residual lignin, or generated during bleach-

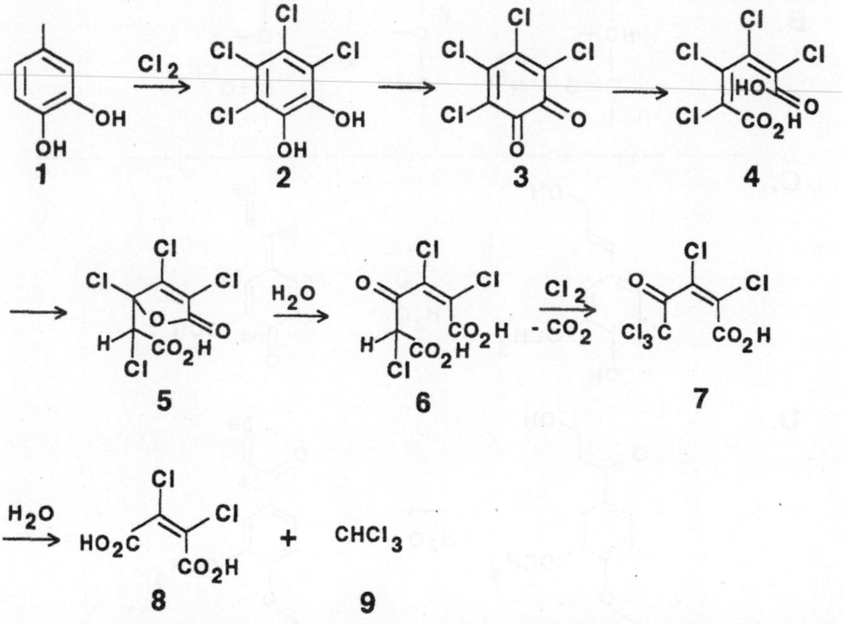

**Fig. 8-12.** Example of the formation of chloroform from uncondensed catechol structures after chlorination of residual lignin. According to this reaction sequence a possible precursor is a demethylated guaiacylpropane unit (1). It reacts to form a chlorinated muconic acid (4), which is then lactonized to structure 5. After hydrolysis of the lactone, the resulting chlorinated muconic acid (6) is decarboxylated and further chlorinated to yield chloromethyl ketone (7). This is the final precursor, which can be hydrolyzed according to the haloform reaction to chlorinated muconic acid (8) and chloroform (9). (Adapted from Hrutfiord and Negri, 1990.)

ing, have been suggested as the most probable sources for chloroform formation (Fig. 8-12).

*Chlorine Dioxide*   In a modern bleaching technique chlorine dioxide has almost totally replaced sodium hypochlorite. The tendency is also to add increasing amounts of chlorine dioxide in the first bleaching stage instead of chlorine. Chlorine dioxide is an electrophile and a rather selective oxidant, attacking predominantly aromatic rings with free phenolic hydroxyl groups, which indeed are abundant in the residual lignin after kraft pulping (Fig. 8-13). The reaction products are similar muconic acid structures as produced by chlorination, but only slightly chlorinated. However, this type of oxidation alone cannot explain the delignification effectivity of chlorine dioxide. It is known that hypochlorous acid (and chlorine in equilibrium with it) is generated from chlorine dioxide as reaction products (see Section 8.1.2). These reagents are probably contributing to the delignification and dissolution of residual lignin as already described.

The bleaching sequence is usually also terminated by a chlorine dioxide stage. When the pulp is bleached to a high ISO brightness, 90–92%, the light absorption coefficient at 457 nm decreases to a value less than 0.3 $m^2/kg$.

*Oxygen*   The bleaching with oxygen is normally performed at a rather high pH. The reactive species generated from oxygen include hydrogen

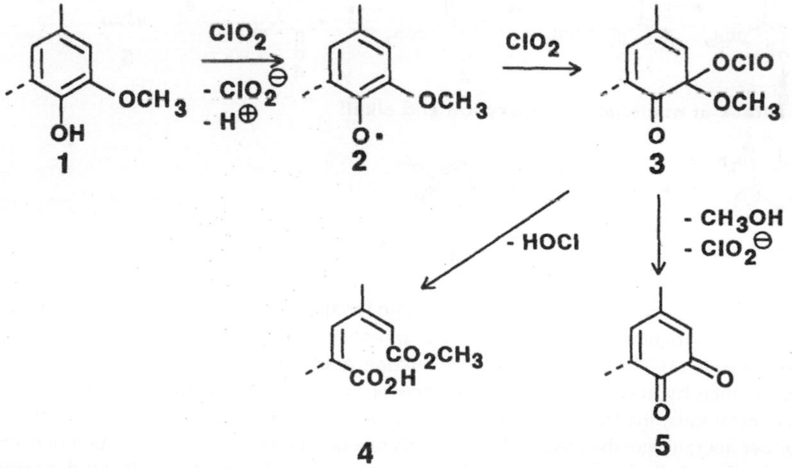

**Fig. 8-13.**   Examples of reactions of residual lignin with chlorine dioxide. A free phenolic acid structure (1) gives rise to the formation of radical (2), which is then oxidized to intermediate (3). This is either cleaved to a muconic acid structure (4) or converted to an o-quinone structure (5). See also Gellerstedt et al. (1991).

peroxide anion ($HO_2^-$) and dissociated oxide and superoxide radicals ($O\cdot^-$, $O_2\cdot^-$), as well as undissociated and highly reactive hydroxyl radical ($HO\cdot$) (see Section 8.1.2). These reactive species primarily attack aromatic rings substituted with free phenolic hydroxyl groups and cleave them to unsaturated dicarboxylic acid structures of muconic acid type.

In alkaline conditions lignin structures are converted to carbanions and conjugated carbonyl structures (Fig. 8-11) giving rise to electron attracting

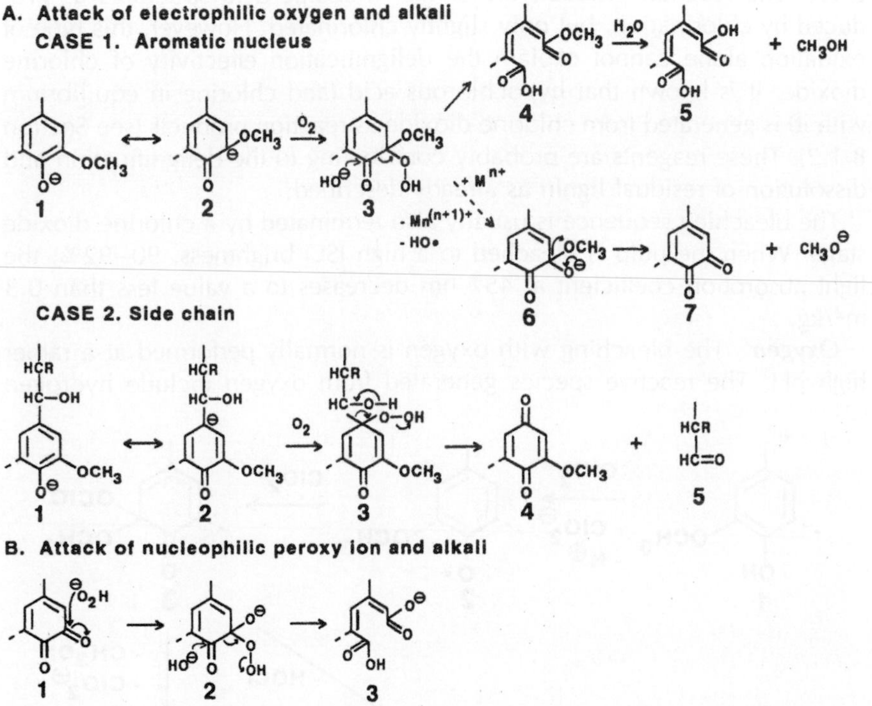

**A. Attack of electrophilic oxygen and alkali**

**CASE 1. Aromatic nucleus**

**CASE 2. Side chain**

**B. Attack of nucleophilic peroxy ion and alkali**

**Fig. 8-14.** Susceptible sites in the residual lignin for attack of oxygen (A) and peroxy anion (B). (A) *Case 1.* Carbanion (2) in resonance with phenoxy anion (1) reacts with oxygen to form a hydroperoxide (3). After decomposition, a muconic acid methyl ester derivative (4) is obtained, which is then hydrolyzed (5) under liberation of methanol. Alternatively, in the presence of heavy metal ions, the hydroperoxide (3) can be decomposed via 6 to an *o*-quinone derivative (7) under liberation of the methoxyl group. This reaction gives rise to the formation of hydroxyl radicals. *Case 2.* Carbanion (2) reacts with oxygen to a hydroperoxide (3). This is decomposed to a *p*-quinone derivative (4) under cleavage of the side chain (5). (B) The peroxy anion attacks a *o*-quinone structure (1) resulting in a hydroperoxide (2), which is decomposed to a muconic acid derivative (3).

and repelling positions. As an electrophilic reagent, oxygen (and hydroxyl radicals) prefers negatively charged positions, whereas the nucleophilic hydrogen peroxide anion ($HO_2^-$) like hypochlorite anion reacts with positively charged positions. An important reaction of oxygen and hydrogen peroxide is to form hydroperoxides with organic compounds, which are then cleaved to muconic acid and quinone structures and giving rise to the formation of hydroxyl radicals (Fig. 8-14).

Although the hydrophilicity of the lignin is increased, the degradation and depolymerization taking place are rather limited resulting only in a partial solubility. Prolonged and a more drastic treatment even in the presence of inhibitors (magnesium salts) is not possible because cellulose is depolymerized (see Section 8.1.5). Many attempts have been made to increase the reactivity of lignin toward oxygen so that an increased degree of delignification would be possible. For example, pretreatment with nitrogen dioxide prior to oxygen bleaching has given promising results, but this type of process has so far not been introduced into industrial practice.

**Hydrogen Peroxide** The conditions for hydrogen peroxide bleaching are similar to those for oxygen bleaching, and the same reactive species, although in different proportions, are generated. The nucleophilic hydrogen peroxide ion ($HO_2^-$) itself is a mild oxidant reacting mainly with carbonyl groups, but this type of attack is insufficient to prompt delignification (cf. Fig. 8-14). The hydroxyl radicals formed after decomposition of peroxide are powerful lignin oxidants, but the problem is that they are nonspecific, attacking carbohydrates as well. To avoid or limit the decomposition of peroxide, which is catalyzed by transition metal ions, the bleaching must be performed under carefully controlled conditions by using suitable sequestering and stabilizing agents. Peroxide bleaching is only a complementary technique to be used in combination with other bleaching chemicals for production of fully bleached pulps.

**Ozone** The conditions under ozone bleaching are acidic and ozone reacts as an electrophile (see Section 8.1.2). As from oxygen, reactive radical species are generated from ozone during bleaching. Their attack is directed both toward aromatic rings and to double bonds in the side chains (Fig. 8-15). This attack results in an effective delignification under formation of muconic acid, o-quinoid, and partially hydroxylated structures. Methanol is formed from the methoxyl groups. Even if ozone is an effective delignification agent, it is like most of the oxidants nonspecific creating alkali-labile glycosidic bonds through introduction of carbonyl groups in the polysaccharide chains (see Section 8.1.5). This is a serious limitation for its use, but it has potential especially when combined with other bleaching agents.

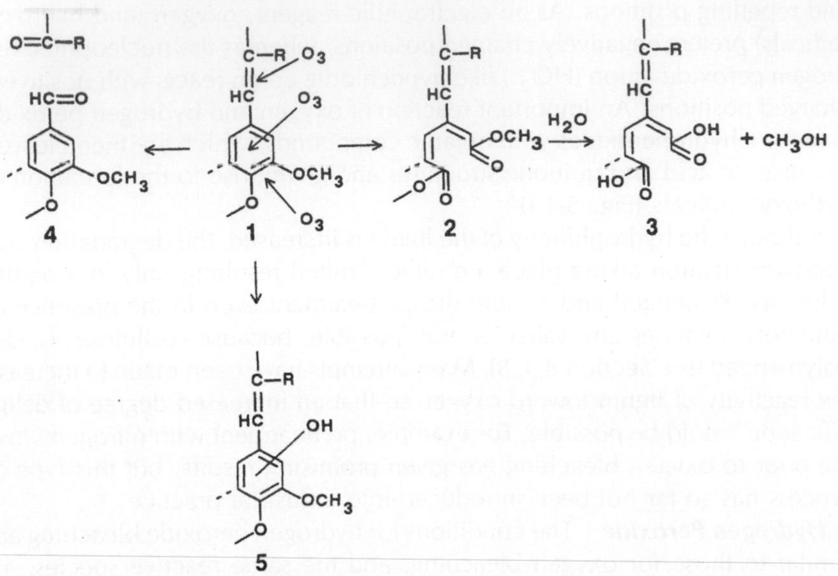

**Fig. 8-15.** Examples of reactions of ozone with stilbene (R = aryl group) and styrene (R = H) structures (1). The attack of ozone is directed as indicated by the arrows. Cleavage of the aromatic ring results in the formation of muconic acid structures (2, 3) and methanol. The other reactions illustrated are cleavage of the side chain (4) and introduction of hydroxyl groups to the aromatic ring (5).

## 8.1.5   Carbohydrate Reactions

Among all commercial bleaching chemicals, chlorine dioxide is the most selective lignin oxidant and it reacts very slowly with polysaccharides. The oxidation of polysaccharides in the chlorination stage is also prevented in the presence of chlorine dioxide because it acts as a radical scavanger, but hypochlorous acid is a strong oxidant. Because of generation of reactive radicals, it is very difficult to avoid degradation of polysaccharides especially in oxygen-alkali bleaching. Decomposition of ozone leads to the same radical species as is formed from oxygen and hydrogen peroxide. However, because the conditions during chlorine and ozone bleaching are acidic, the oxidized intermediate structures are susceptible to degradation mainly in the following alkaline extraction stage.

The main oxidizing attack of bleaching agents occurs within the polysaccharide chains, but it can also be directed to the end groups (Fig. 8-16). Although these end groups in kraft pulps are predominantly of carboxylic acid type, an aldehydic end group is formed after each cleavage of the

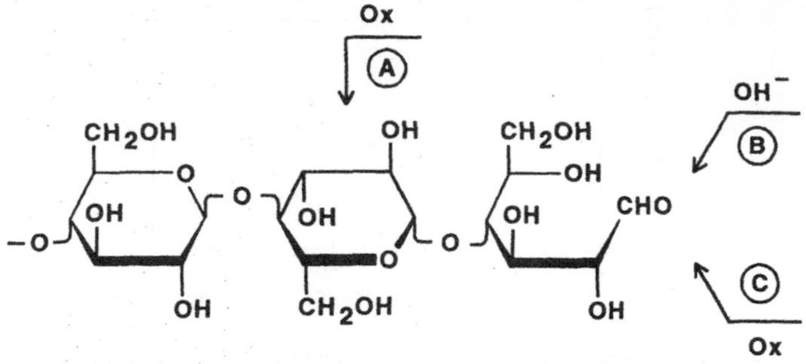

**Fig. 8-16.** Susceptible sites in cellulose to attack of oxidative agents (Ox) and alkali. (A) Oxidation of any position within the chain units (C-2, C-3, or C-6) to carbonyl groups generates alkali labile glycosidic linkages. (B) Peeling reaction starts from the reducing end group. (C) Oxidation of aldehyde end groups to carboxyl groups prevents the peeling reaction.

glycosidic linkage. The aldehydic groups can then either be oxidized to carboxylic acid groups or, if the conditions are alkaline, the peeling of the polysaccharide chains will start from them.

**Reactions within the Polysaccharide Chains**   The most harmful reaction in any bleaching system is oxidation of C-2, C-3, or C-6 positions of the monomeric sugar units to carbonyl or carboxyl groups. Hypochlorous acid oxidizes these positions predominantly to the carbonyl state, whereas in the presence of hypochlorite ions the reaction proceeds generating mainly carboxyl groups. In other words, buffering the chlorine solution to different pH values significantly changes the oxidation pattern (cf. Fig. 8-2).

The presence of carbonyl groups gives rise to alkali-labile glycosidic bonds. For example, if carbonyl groups have once been generated during

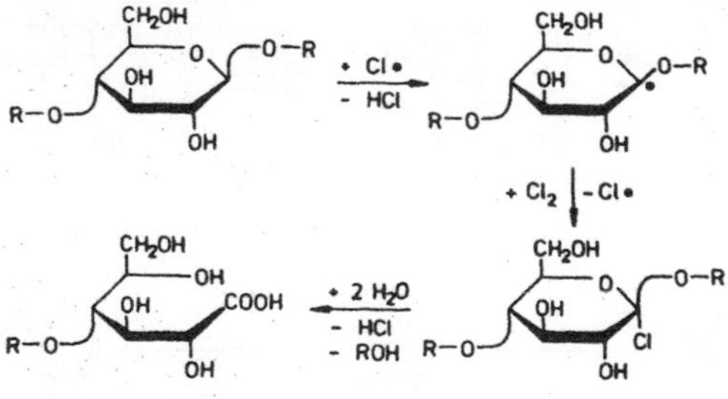

**Fig. 8-17.** Example of the cleavage of a glycosidic bond in cellulose directly after attack of a chlorine radical. R' denotes cellulose chain. After oxidation at C-1, the glycosidic bond is cleaved with formation of a gluconic acid end group.

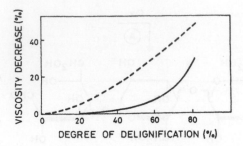

**Fig. 8-18.** Depolymerization of cellulose (viscosity decrease) during oxygen bleaching of pine kraft pulp. Dotted line, no inhibitor added; full line, magnesium salt added as inhibitor.

unfavorable chlorination conditions, the glycosidic bonds of cellulose and hemicelluloses are easily cleaved in the subsequent alkali extraction stage according to the β-alkoxy elimination mechanism (cf. Section 2.5.5). In addition to this type of depolymerization, glycosidic bonds may also be cleaved directly, for example, after a chlorine radical attack at C-1, which position is then oxidized to an aldonic acid end group (Fig. 8-17).

In the presence of magnesium salts about 50% of the lignin can be removed from the pulp by oxygen and alkali, but on further delignification cellulose is gradually depolymerized concomitant with loss of fiber strength (Fig. 8-18). Also in this case the primary reason for depolymerization is formation of carbonyl groups giving rise to alkali-labile glycosidic bonds. As illustrated in Fig. 8-19, the most common type of depolymerization of cellulose occurs after oxidation of the C-2 position. A corresponding oxidation at C-3 leads to the same result because of the migration of the carbonyl group to C-2. The 2-ulose thus formed is easily degraded by β-alkoxy elim-

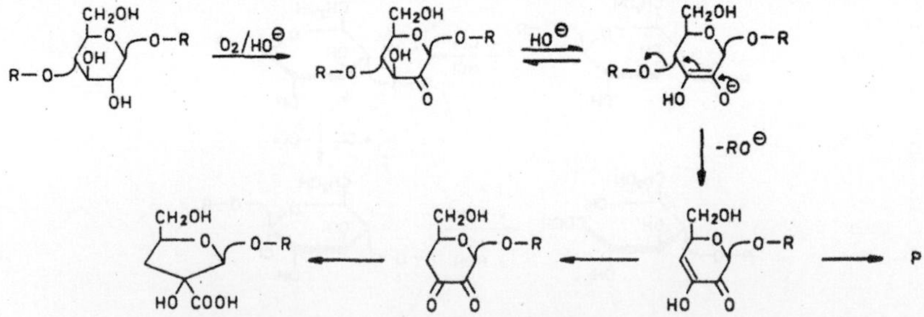

**Fig. 8-19.** Cleavage of a glycosidic bond after formation of a carbonyl group. R is cellulose chain and P denotes the reaction products. For explanations, see the text.

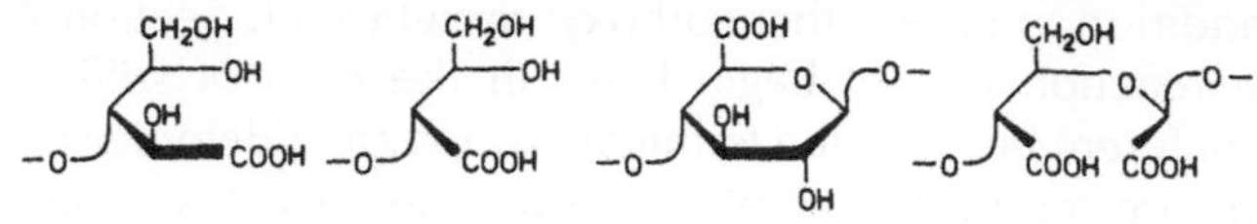

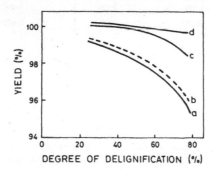

**Fig. 8-20.** Examples of carboxyl structures formed in cellulose by oxidative action of bleaching agents. Types: arabinonic acid (1), erythronic acid (2), glucuronic acid (3), and dicarboxylic acid structure (4).

ination at C-4, resulting in chain cleavage and formation of a new reducing end group. Although chain cleavage at C-1 is also possible (after oxidation at C-3), this does not seem to occur. Position C-2 and C-3 can also be oxidized simultaneously. The resulting dicarbonyl derivative is converted to a furanosidic acid unit or directly degraded. Some oxidation of C-6 may also occur, giving rise to chain cleavage at C-4.

In addition to carbonyl groups, carboxyl groups can be introduced into cellulose and hemicelluloses by bleaching agents (Fig. 8-20). Even though the carboxyl groups, unlike carbonyl groups, do not render polysaccharides extremely sensitive toward alkaline degradation, they affect detrimentally the brightness stability of the final (bleached) pulp.

**Peeling Reaction**   Depolymerization of cellulose according to the mechanism already described is not directly responsible for carbohydrate yield losses in oxygen bleaching. However, the new reducing end groups generated as a result of the cleavage of glycosidic bonds will give rise to the peeling of the cellulose (or hemicellulose) chains. Most of the terminal units in the polysaccharides of kraft pulps are alkali-stable metasaccharinic acid

**Fig. 8-21.** Pulp yield in oxygen-alkali bleaching. Full line, inhibitor added; dotted line no inhibitor addition. (a) Spruce sulfite pulp; (b and c) pine kraft pulp; (d) spruce sulfite pulp after reduction of aldehydic end groups to alcohol groups by treatment with $NaBH_4$.

groups in addition to some other carboxyl moieties (cf. Section 7.3.5), and the peeling reaction cannot begin here. In the case of sulfite pulps the situation is different because the terminal groups are aldehyde sugar units. If these groups are reduced to alcohol groups, the yield losses are considerably decreased (Fig. 8-21). It is thus clear that the peeling reaction is contributing to the yield losses especially in the case of sulfite pulps.

The peeling reaction is to some extent counteracted by oxidation of the end groups to aldonic acid end groups (Fig. 8-22). The initial oxidative step involves formation of an aldos-2-ulose (glycosone) end group, which, after a benzilic acid rearrangement, is transformed to an epimeric pair of aldonic acids—in the case of cellulose and glucomannan mainly to mannonic acid. Cleavage of carbon–carbon bonds occurs as well, resulting predominantly in the formation of arabinonic and erythronic acid end groups. However, before this type of oxidative stabilization has occured, several sugar units are peeled off from the reducing end of the polysaccharide chain (Fig. 8-23). Instead of the usual alkaline peeling products formed at nonoxidative condi-

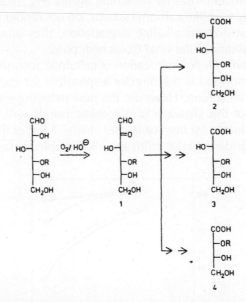

**Fig. 8-22.** Oxidative stabilization of cellulose through formation of aldonic acid end groups. R is cellulose chain. The glucosone intermediate (1) undergoes a benzilic acid rearrangement resulting in a hexonic acid end group (mannonic acid (2) dominates instead of gluconic acid). In addition, cleavage of C-1 to C-2 and C-2 to C-3 carbon bonds gives rise to the formation of arabinonic acid (3) and erythronic acid (4) end groups. (In this simplified scheme only the main products are shown.)

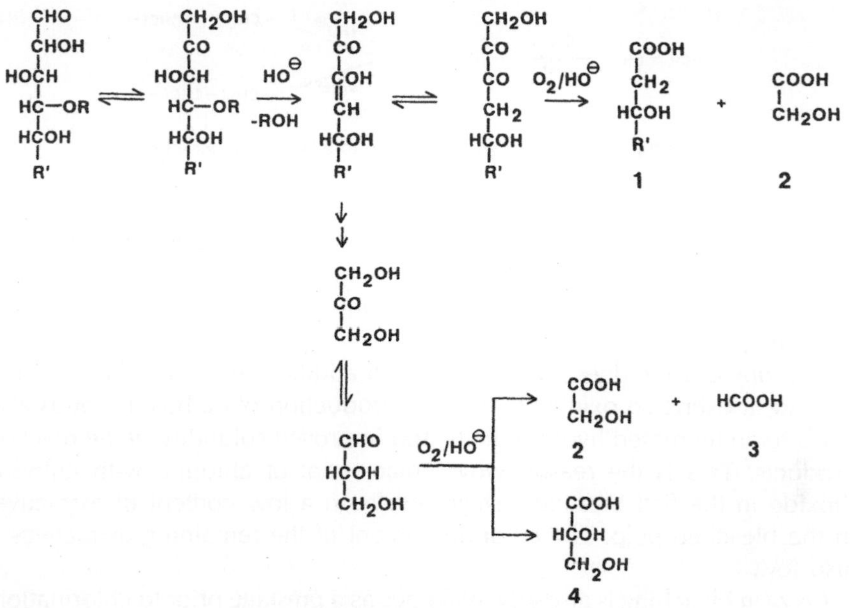

$$
\begin{array}{l}
\text{CHO} \\
\text{CHOH} \\
\text{HOCH} \\
\text{HC}-\text{OR} \\
\text{HCOH} \\
\text{R'}
\end{array}
\;\rightleftharpoons\;
\begin{array}{l}
\text{CH}_2\text{OH} \\
\text{CO} \\
\text{HOCH} \\
\text{HC}-\text{OR} \\
\text{HCOH} \\
\text{R'}
\end{array}
\;\;\text{HO}^{\ominus}\;\;
\xrightarrow{-\text{ROH}}
\begin{array}{l}
\text{CH}_2\text{OH} \\
\text{CO} \\
\text{COH} \\
\text{CH} \\
\text{HCOH} \\
\text{R'}
\end{array}
\;\rightleftharpoons\;
\begin{array}{l}
\text{CH}_2\text{OH} \\
\text{CO} \\
\text{CO} \\
\text{CH}_2 \\
\text{HCOH} \\
\text{R'}
\end{array}
\;\;\text{O}_2/\text{HO}^{\ominus}\;\;
\xrightarrow{\;\;}
\begin{array}{l}
\text{COOH} \\
\text{CH}_2 \\
\text{HCOH} \\
\text{R'} \\
\mathbf{1}
\end{array}
\;+\;
\begin{array}{l}
\text{COOH} \\
\text{CH}_2\text{OH} \\
\;\\
\mathbf{2}
\end{array}
$$

$$
\downarrow\downarrow
$$

$$
\begin{array}{l}
\text{CH}_2\text{OH} \\
\text{CO} \\
\text{CH}_2\text{OH}
\end{array}
$$

$$
\Updownarrow
$$

$$
\begin{array}{l}
\text{CHO} \\
\text{HCOH} \\
\text{CH}_2\text{OH}
\end{array}
\;\;\text{O}_2/\text{HO}^{\ominus}\;
\xrightarrow{\;\;}
\left\{
\begin{array}{l}
\text{COOH} \\
\text{CH}_2\text{OH} \quad + \quad \text{HCOOH} \\
\quad\;\mathbf{2}\qquad\qquad\quad\mathbf{3} \\[1ex]
\text{COOH} \\
\text{HCOH} \\
\text{CH}_2\text{OH} \\
\quad\;\mathbf{4}
\end{array}
\right.
$$

**Fig. 8-23.** Peeling reaction of polysaccharides during oxygen-alkali bleaching. *Cellulose* and *glucomannans* (R' = CH$_2$OH): 3,4-dihydroxybutanoic acid (1), glycolic acid (2), formic acid (3), and glyceric acid (4). *Xylan* (R' = H):3-hydroxypropanoic acid (2-deoxyglyceric acid) (1), glycolic acid (2), formic acid (3), and glyceric acid (4).

tions (cf. Section 7.3.5), the diulose structures are converted to 3,4-dihydroxybutanoic and glycolic acids in the presence of oxygen. Correspondingly, glyceraldehyde is oxidized to glycolic and glyceric acids.

### 8.1.6 Reactions of Extractives

The extractives remaining in the pulps, especially if they are chlorinated, are detrimental for the pulp quality. It is therefore important to perform the bleaching in conditions resulting in a lowest possible content of extractives. It is equally important to avoid reactions generating toxic or otherwise harmful resinous products, which are carried to the liquors after bleaching. However, the system is complex and difficult to master because of the diversity of compounds involved, of which some are harmful already when present in trace amounts.

*Chlorine* typically participates in addition reactions with unsaturated compounds such as fatty acids present as stable esters (waxes) in hardwood

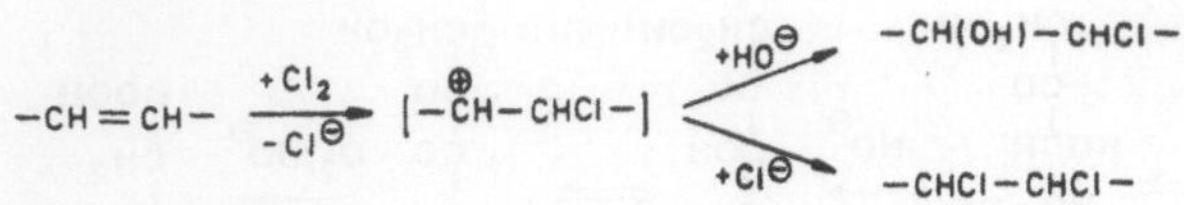

$$-CH=CH- \xrightarrow[-Cl^\ominus]{+Cl_2} \quad [-\overset{\oplus}{C}H-CHCl-] \begin{array}{c} \xrightarrow{+HO^\ominus} \quad -CH(OH)-CHCl- \\ \\ \xrightarrow{+Cl^\ominus} \quad -CHCl-CHCl- \end{array}$$

**Fig. 8-24.**  Addition of chlorine to double bonds in extractives.

pulps (Fig. 8-24). Especially the extractives left in hardwood pulps after chlorination contains dichlorinated compounds. It is extremely difficult to remove such compounds completely from the pulp in the subsequent bleaching stages.

*Chlorine dioxide* does not participate in addition reactions like chlorine; instead, it exerts an oxidizing action. Introduction of carboxyl groups also leads to an increased hydrophilicity and improved solubility of the reaction products. This is the reason why replacement of chlorine with chlorine dioxide in the first bleaching stage results in a low content of extractives in the bleached pulp. The chlorine content of the remaining extractives is also low.

*Oxygen* bleaching is also advantageous as a prestage prior to chlorination, particularly because it counteracts the formation of chlorinated extractives.

### 8.1.7  Spent Liquors after Bleaching

Disposal of spent liquors from bleach plants represents a serious environmental problem which is today fully recognized. Reduction of the effluents from the production of unbleached pulp has been much easier by different measures, including introduction of more closed pulp washing systems. Even if the load of the bleach plant effluents to the receiving waters has been reduced markedly during the last decade, today these wastes are still responsible for the major part of the total pollution load of the pulp industry.

The load from bleach plants can be at least reduced by the following arrangements: (1) Extended delignification, e.g., through an improved cooking liquor circulation system. (2) Oxygen bleaching, enabling the effluents from this stage to be combined with cooking spent liquors and thus entering the chemical regeneration system. (3) Countercurrent flow system, reducing the total volume of effluents and increasing the solids concentration.*

---

*According to a closed-cycle system of Rapson and Reeve the water requirement is minimized by countercurrent washing through the bleaching stages, and the resulting effluent is used instead of fresh water for washing of the unbleached pulp. This effluent, which contains the solids both from bleaching and cooking, is introduced into the kraft chemical recovery system. Because of recycling, the bleaching chemicals are accumulated in the system. To

(4) Replacement of the chlorine by chlorine dioxide. (5) Improvement of the final bleaching stages by introduction of oxygen, hydrogen peroxide, and ozone. (6) Improved external processes for the purification of the effluents from bleaching as well as more effective activated sludge treatment system.

Most of these arrangements are already in use in modern pulp mills, but further development is needed. Particularly important is to find new ways for improving the selectivity of oxygen delignification.

An adequate characterization of bleach plants effluents with respect to both the chemical composition and biological effects is an important and necessary target for a meaningful assessment of the risks caused by such discharges to the receiving waters. More traditional parameters for evaluating these risks are the biological oxygen demand (BOD), chemical oxygen demand (COD), total suspended solids (TSS), color, and acute toxicity. BOD is primarily a measure for the easily degradable organic components.

Measured by conventional parameters (BOD and TSS) the discharges from the mills in both North America and Scandinavia have decreased more than 50% during the last two decades even if the production has doubled. Acute toxicity tests with fish, invertebrates, and algae were commonly incorporated into pulp mill permits in the 1980s. Some years later, however, the concerns were focused on emissions of persistent and bioaccumulable nature, i.e., on compounds resulting in chronic (long-term) toxic effects already when present at relatively low concentrations. As a new parameter TOCl (total organic chlorine) was introduced, which was later replaced by AOX (adsorbable organically bound halogen [X]). AOX is a sum parameter comprising all the chlorinated organic material, not taking into consideration the specific effects of the individual compounds or fractions. However, the most detrimental products with serious ecological effects are associated with the low-molecular-weight and lipophilic compounds present at low concentrations and corresponding only to a very small fraction of the AOX. In addition to the organic material, chlorate ions generated in intermediate reactions from chlorine dioxide (cf. Section 8.1.2) are also contributing with serious effects on brown algae. Also, since the presence (or absence) of detrimental compounds in the liquors from bleaching greatly varies depending on the chemicals and conditions, the toxicity cannot be measured simply by means of single parameters. It is therefore not surprising that the introduction of AOX as an universal parameter has met much criticism.

---

overcome these difficulties chlorine, chlorine dioxide, and sodiun hydroxide with given proportions are used for bleaching (mainly chlorine dioxide) to result in the formation of sodium chloride predominantly, which is continuously removed from the white liquor by evaporation and crystallization and reconverted to bleaching chemicals.

Apart from these objections, specified load limits for chlorinated material, measured as AOX, have either already been introduced or will be valid in the middle of 1990s for pulp mills in many pulp-producing countries including Scandinavia, Germany, Canada, and Australia. These limits vary somewhat, but are in the range of 1–3 kg AOX/ton of pulp. However, in the United States and Canada stringent regulations have been introduced specifically on strongly toxic chlorinated compounds, dioxins and furans. Fulfillment of these regulations means considerable modifications of the present bleaching practise.

The overall composition of the spent liquors from bleaching can be characterized only by dividing the material into certain groups of constituents because only a small fraction of the total material constitutes of distinctly

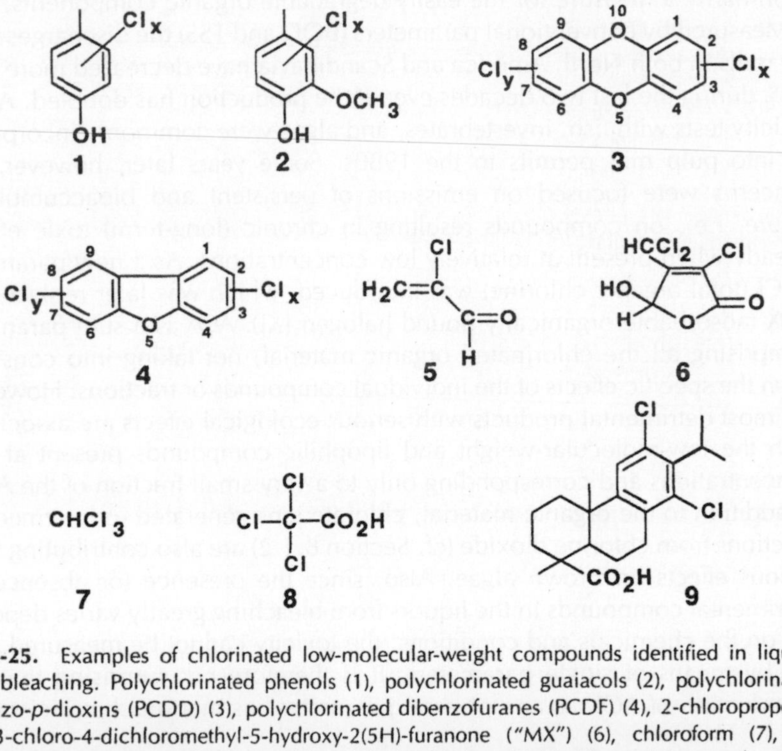

**Fig. 8-25.** Examples of chlorinated low-molecular-weight compounds identified in liquors from bleaching. Polychlorinated phenols (1), polychlorinated guaiacols (2), polychlorinated dibenzo-p-dioxins (PCDD) (3), polychlorinated dibenzofuranes (PCDF) (4), 2-chloropropenal (5), 3-chloro-4-dichloromethyl-5-hydroxy-2(5H)-furanone ("MX") (6), chloroform (7), trichloroacetic acid (8), and 12,14-dichlorodehydroabietic acid (9). The toxicity greatly varies depending on the degree of chlorination and the structure. Of the dioxins (3 and 4) the 2,3,7,8-tetrachlorinated compounds are highly toxic of which 2,3,7,8-TCDD represents the most toxic type. The analytical data found for various PCDD and PCDF are often related to the corresponding weight of 2,3,7,8-TCDD (TCDD-equivalents or dioxin value).

TABLE 8-4.    Effluent Characteristics of Alternative Bleaching Sequences[a]

| Alternative sequences | COD (kg) | BOD (kg) | AOX (kg) | Chlorinated phenolics[b] (g) | TCDD (μg) |
|---|---|---|---|---|---|
| A.  C90 + D10EHDED | 100 | 25 | 8 | 90 | 5–30 |
| B.  O(C85 + D15)(EO)DED | 65 | 15 | 3 | 25 | <1 |
| C.  O(D30C70)(EPO)D(E + P)D | 50 | 12 | 2 | 11 | <0.1 |
| D.  Seq.C + aerated lagoon | 30 | 3 | 1.3 | 9 | <0.1 |
| E.  O(D70C30)(EPO)D(E + P)D | 50 | 12 | 1.4 | <0.5 | <0.1 |
| F.  Seq.C + aerated lagoon + UF[c] | 15 | <2 | 0.9 | <9 | <0.1 |
| G.[d]  OXPD(E + P)D | 40 | 10 | 0.5 | <1 | <0.1 |
| H.[e]  OD100(EPO)DD | 40 | 10 | <1 | <1 | <0.1 |
| I.  Seq.H + aerated lagoon | 25 | 2 | <0.7 | <1 | <0.1 |

[a] All data are expressed per ton of pulp. Softwood kraft pulp, bleaching to ISO brightness of 90% (personal communication from Haglund; data from internal reports). Sequences A–E are being used for full-scale production of market pulp. The technology for Sequences F–I is still not fully adequate for production of acceptable pulp qualities.

[b] Three to five Cl/aromatic ring.

[c] UF, ultrafiltration (technique not fully developed for this application).

[d] Sign X denotes an acidic washing stage in the presence of complexing agents for removal of heavy metals.

[e] A modified kraft pulp.

definable compounds. In addition to methanol, a relatively small group of miscellaneous products, probably mostly of carbohydrate origin, are contributing to the biological oxygen demand (BOD) of the spent liquors from bleaching. However, the dominant fraction is composed of chlorinated products from lignin. This is the most heterogeneous fraction of components with respect to both molecular weight and chemical structure. Most of it is polymeric material with molecular weights exceeding 1000. This material cannot be characterized by usual methods simply because it cannot be divided into subfractions representing homogeneous constituents. The high-molecular-weight fraction derived from residual lignin is biologically inert and thus of less concern. It is also improbable that this "chlorolignin" would give rise to low-molecular-weight degradation products, such as dioxins. It is therefore obvious that the small group of low-molecular-weight products generated during bleaching are mainly responsible for the toxic effects. Application of modern and sensitive analytical methods, especially gas chromatography-mass spectrometry, has made it possible to identify more than 300 individual compounds, of which about 200 chlorinated, in the liquors from pulp bleaching. Examples of typical members possessing strong toxic properties at trace concentrations are given in Fig. 8-25.

Modern bleaching sequences, in which mainly oxygen, chlorine dioxide, and hydrogen peroxide are used, have markedly decreased the amounts of both toxic and oxygen-consuming substances in the effluents from bleaching (Table 8-4). As the technology, including ozone bleaching, develops, it is to be expected that the bleaching effluents will be virtually free from chlorinated and toxic constituents.

## 8.2   Lignin-Preserving Bleaching

The formation of various chromophoric structures giving rise to absorption of visible light cannot be avoided during pulping. Even if such structures are mainly generated by the action of pulping chemicals, mechanical defibration also results in the formation of chromophors. The relatively low brightness of the high-yield pulps is the most serious limitation for their use. Although much research has been devoted to find better methods for bleaching of this type of pulps, the results so far look not very promising. Another problem in this connection is the permanence (stability) of the brightness.

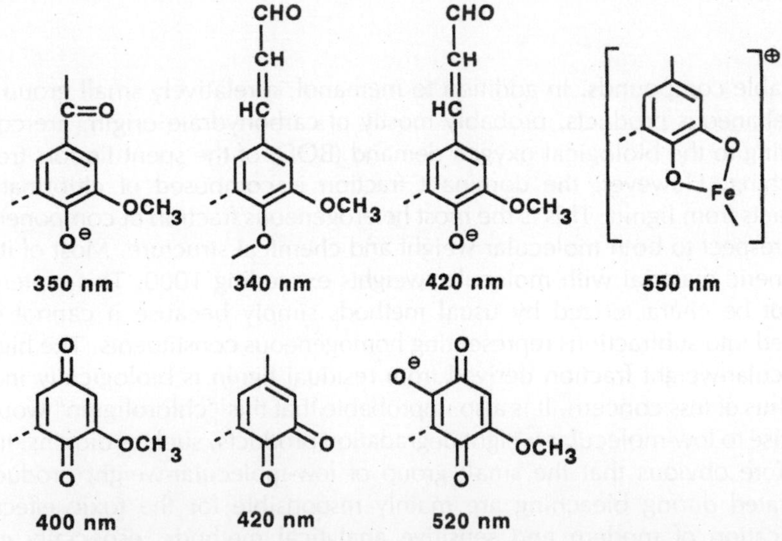

**Fig. 8-26.**   Examples of chromophoric structures. The data refer to the light absorption maxima at the respective wavelengths.

The chromophores and potential chromophores (so-called leucochromophores) are derived from a large variety of structural types such as phenols, catechols, and quinones in combination with unsaturated systems of styrene, diphenylmethane, and butadiene structures (cf. Figs. 7-30 and 8-26). Principally, such unsaturated structures can be eliminated either by reduction ("reducing bleaching") or by oxidation ("oxidative bleaching").

**Reducing Bleaching**   The effect of the reducing agents depends on the redox potential of the system, for example:

$$H_2SO_3 + H_2O \rightarrow SO_4^{2-} + 4H^+ + 2e^- \qquad (E° = -0.20 \text{ V}) \qquad (8\text{-}37)$$

$$S_2O_4^{2-} + 4HO^- \rightarrow 2SO_3^{2-} + 2H_2O + 2e^- \qquad (E° = 1.12 \text{ V}) \qquad (8\text{-}38)$$

$$BH_4^- + 8HO^- \rightarrow B(OH)_4^- + 4H_2O + 8e^- \qquad (E° = 1.24 \text{ V}) \qquad (8\text{-}39)$$

The respective standard redox potentials ($E°$) for the quinone structures in the residual lignin are probably of the order of 0.7–0.9 V, and they can thus be reduced to the corresponding hydroquinones by sodium dithionite and sodium borohydride, but not with sulfurous acid (cf. Fig. 8-27). However, the hydroquinones are readily reoxidized to quinones in the presence of oxygen and light. This is the reason why this type of bleaching does not result in a permanent brightness of the pulp.

Sodium borohydride is too expensive an agent to be used for bleaching, but sodium dithionite is the most common chemical applied for this purpose.

According to equation (8-38) dithionite ions are converted to sulfite ions during bleaching, which means consumption of hydroxyl ions and decrease of pH. Some of the dithionite is consumed for undesirable side reactions. For example, disproportionation of dithionite gives rise to the formation of thiosulfate and hydrogen sulfite:

$$2S_2O_4^{2-} + H_2O \rightarrow S_2O_3^{2-} + 2HSO_3^- \qquad (8\text{-}40)$$

In the presence of air (oxygen), dithionite is oxidized:

$$S_2O_4^{2-} + H_2O + O_2 \rightarrow HSO_3^- + HSO_4^- \qquad (8\text{-}41)$$

A maximal bleaching effect of sodium dithionite is theoretically obtained at a pH between 8 and 9 at temperatures of 20–60°C, but because it is

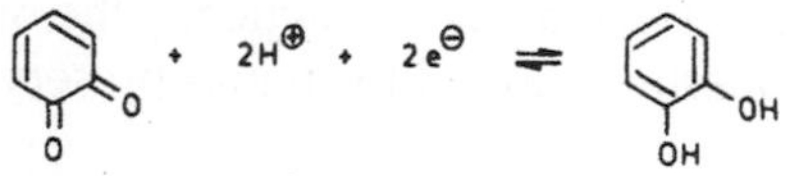

**Fig. 8-27.**   Reduction of o-quinone to catechol.

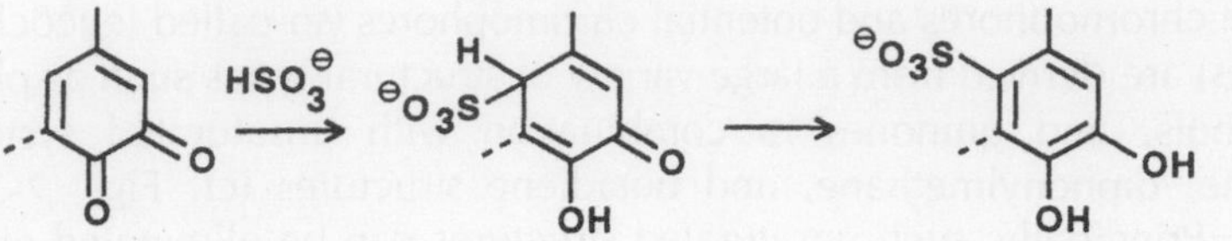

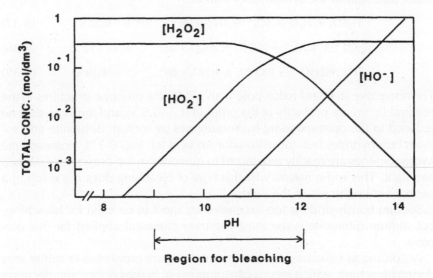

**Fig. 8-28.** Conversion of o-quinone structures to nonchromophoric structures by hydrogen sulfite.

**Fig. 8-29.** Dissociation of hydrogen peroxide.

difficult to exclude oxygen completely from the system, a pH of 5–6 is used in practise to avoid decomposition of dithionite and alkali-induced darkening of the lignin.

Even though sulfurous acid (or its sodium salts) are weak oxidants, sodium hydrogen sulfite* is sometimes used because it can give some bleaching effect. This effect may be associated with the addition of nucleophilic hydrogen sulfite (or sulfite) ions to quinonoid double bonds and subsequent conversion of the resulting intermediates to aromatic sulfonic acid structures (Fig. 8-28).

*Oxidative Bleaching*   The main and practically the only bleaching chemical with oxidative properties applied for lignin-preserving bleaching is hy-

---

*Instead of sodium hydrogen sulfite, which does not exist in solid form, its anhydride, sodium pyrosulfite, is commonly used ($Na_2S_2O_5 + H_2O = 2NaHSO_3$).

drogen peroxide or its sodium salt (sodium peroxide). Hydrogen peroxide is a weak acid [cf. equation (8-24), p. 174, and Fig. 8-29] and the active bleaching species is the nucleophilic peroxide anion ($HO_2^-$), which attacks carbonyl structures converting them to less chromophoric systems without an extensive degradation and dissolution of lignin (see Section 8.1.4). Sodium peroxide is a more effective bleaching agent than sodium dithionite.

In practice, the peroxide bleaching is performed at 50–60°C at an initial pH value of about 11, which drops toward the end to about pH 9. In order to avoid decomposition of peroxide, which takes place in the presence of heavy metal ions, it is necessary to add stabilizers, such as magnesium silicates and sequestering agents. At optimized conditions an ISO brightness increase of about 25% can be obtained, but the bleaching response of different pulp types varies to a great extent.

## 8.3   Yellowing of High-Yield Pulps

The high-yield pulps can be bleached to relatively high, but not to permanent, brightness levels. The loss of brightness, i.e., the "yellowing" tendency, is an inherent property of the lignin-rich pulps, and much research has been devoted to improve the brightness stability. So far no satisfactory or practical means have been found to prevent this yellowing, which is the most serious defect and limitation of the use of high-yield pulps, such as GW, TMP, and CTMP.

Although the carbohydrates and extractives contribute to the yellowing, the main reason is attributable to the lignin component. The yellowing of pulp can take place in the absence of light, but the yellowing in light, which is induced by wavelengths at the near-UV region below 400 nm, is more important. Like the yellowing in dark, the yellowing in light is accelerated by oxygen, but it is less dependent on variations in humidity, unlike the yellowing in dark, which is strongly accelerated in a humid atmosphere.

The complex phenomena of pulp yellowing are so far imperfectly understood, but it is clear that those structures in lignin which absorb light at 300–400 nm are strongly contributing to the light-induced yellowing. Examples of these reactions are given in Fig. 8-30. According to the first example (Case A) a direct hydrogen abstraction can take place in the presence of oxygen and light from conjugated structures carrying free phenolic hydroxyl groups (1). Further reactions of the phenoxy radicals (2) and peroxy radicals result in the formation of hydroperoxides and hydroxyl radicals, which in turn are generating quinoid and other chromophoric structures according to similar mechanisms as described under Section 8.1.4 (Oxygen).

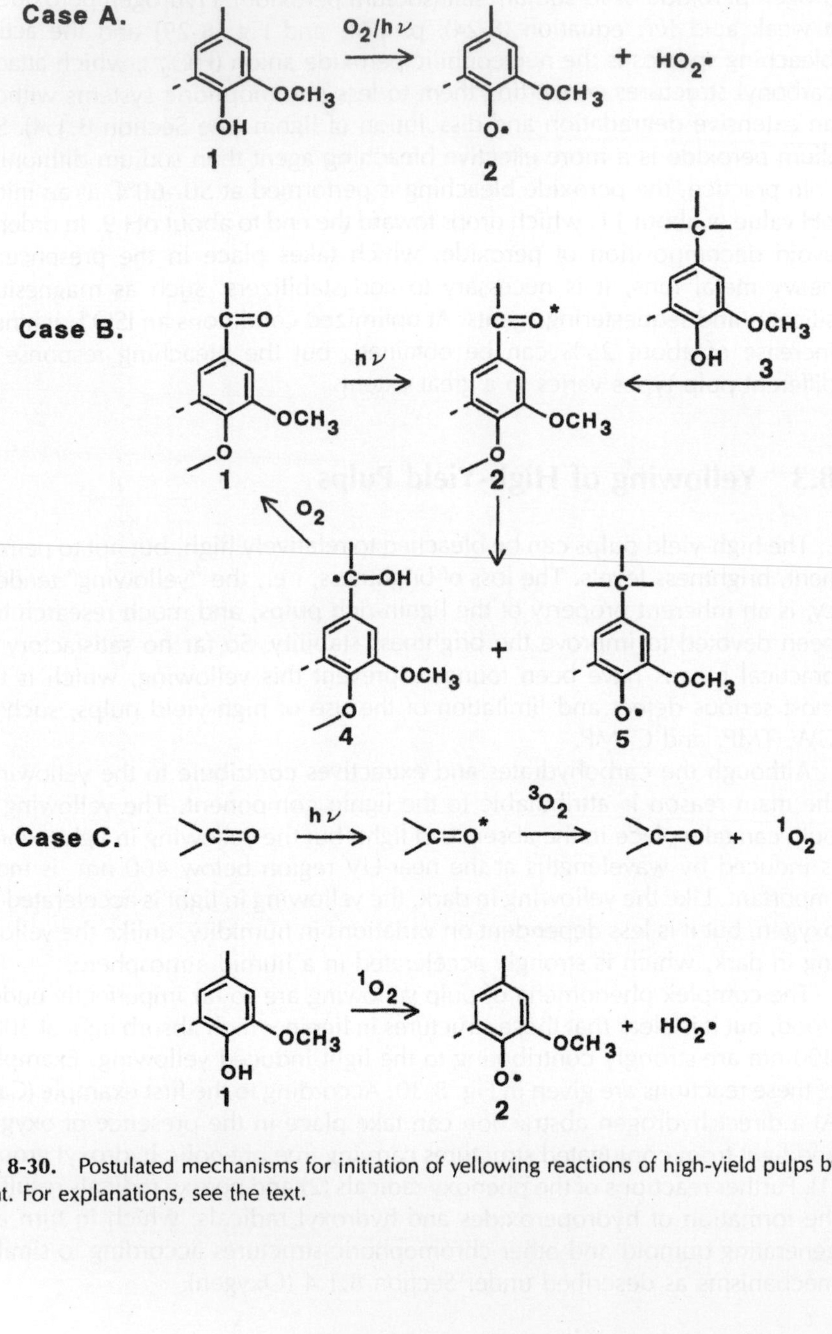

**Case A.**

**Case B.**

**Case C.**

**Fig. 8-30.** Postulated mechanisms for initiation of yellowing reactions of high-yield pulps by light. For explanations, see the text.

It has been postulated that the α-carbonyl groups in the lignin side chain play a key role for yellowing. According to Case B in Fig. 8-30 an α-carbonyl group in a free phenolic or etherified (as in the example) lignin structure (1) is irradiated to an excited state (2). This reacts then with a structure carrying a free phenolic hydroxyl group (3), resulting in a benzyl alcohol radical (4) and a phenoxy radical (5). The former is finally reoxidized by oxygen to the original carbonyl structure. According to this mechanism the carbonyl-carrying structures are thus acting as photosensitizers. Finally, Case C is a modified pathway according to which the excited carbonyl group transfers its energy to an oxygen molecule in its ground (triplet) state. The generated singlet oxygen then abstracts a hydrogen atom from a phenolic hydroxyl group (1). The further reactions of the phenoxy radical (2) leading to yellowing can proceed in the presence of the triplet oxygen.

The brightness stability of high-yield pulps is considerably improved when the carbonyl groups in lignin are reduced to alcohol groups and the phenolic hydroxyl groups are simultaneously either etherified or esterified. However, this type of modification cannot be applied for practice. Other alternatives include the use of various types of stabilizing substances preventing the influence of light and oxygen, i.e., UV screens and antioxidants. However, although it has been established that a number of such additives improve the brightness stability of high-yield pulps, so far all of these, in relation to their effect, are either too expensive or otherwise unsuitable for practical applications.

# Chapter 9

# CELLULOSE DERIVATIVES

## 9.1 Reactivity and Accessibility of Cellulose

Each β-D-glucopyranose unit within the cellulose chain has three reactive hydroxyl groups, two secondary (HO-2 and HO-3) and one primary (HO-6). Although several factors are involved, one prerequisite for etherification of cellulose is the ionization of the hydroxyl groups. Owing to the inductive effects of neighboring substituents the acidity and the tendency for dissociation is increased according to the series: HO-6 < HO-3 < HO-2. It can therefore be understood why HO-2 is, as a rule, most readily etherified in comparison with the other hydroxyl groups. After substitution of HO-2 the acidity of HO-3 is usually increased which results in its higher reactivity. As concerns esterification, the primary hydroxyl group (HO-6) possesses the highest reactivity.

Another important factor to be considered in the reactions of cellulose concerns the accessibility, which means the relative ease by which the hydroxyl groups can be reached by the reactants. For instance, being least sterically hindered, HO-6 groups show higher reactivity toward bulky substituents than do the other hydroxyl groups.

**TABLE 9-1.   Degree of Crystallinity of Some Cellulose Samples Measured by Various Techniques[a]**

| Technique | Cotton | Mercerized cotton | Wood pulps | Regenerated cellulose |
|---|---|---|---|---|
| **Physical** | | | | |
| X-ray diffraction | 0.73 | 0.51 | 0.60 | 0.35 |
| Density | 0.64 | 0.36 | 0.50 | 0.35 |
| **Adsorption and chemical swelling** | | | | |
| Deuteration or moisture regain | 0.58 | 0.41 | 0.45 | 0.25 |
| Acid hydrolysis | 0.90 | 0.80 | 0.83 | 0.70 |
| Periodate oxidation | 0.92 | 0.90 | 0.92 | 0.80 |
| Iodine sorption | 0.87 | 0.68 | 0.85 | 0.60 |
| Formylation | 0.79 | 0.65 | 0.75 | 0.35 |
| **Nonswelling chemical methods** | | | | |
| Chromic acid oxidation | 0.997 | 0.66–0.60[b] | — | — |
| Thallation | 0.996 | 0.69–0.42[b] | — | — |

[a] From Wadsworth and Cuculo (1978).
[b] Mercerization followed by solvent exchange.

The morphology of cellulose has a profound effect on its reactivity. The hydroxyl groups located in the amorphous regions are highly accessible and react readily, whereas those in crystalline regions with close packing and strong interchain bonding can be completely inaccessible. The degree of crystallinity depends on the origin of the cellulose preparation (Table 9-1). A preswelling of the cellulose is necessary in both etherifications (alkali) and esterifications (acids).

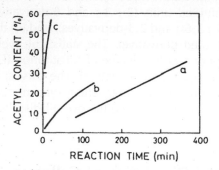

**Fig. 9-1.**  Effect of crystallinity and hydrogen bonding on the acetylation of cotton fibers (Demint and Hoffpauir, 1957). (a) Original fibers. (b) Crystallinity has been destroyed by ethylene amine treatment. Subsequent drying has resulted in the formation of hydrogen bonds. (c) Crystallinity has been destroyed as above but because drying has been omitted no hydrogen bonds have been formed.

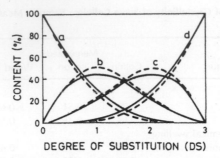

**Fig. 9-2.** Theoretical distribution of unsubstituted (a), monosubstituted (b), disubstituted (c), and trisubstituted (d) glucose units in cellulose ethers (Timell, 1950). —, Ratio of rate constants 1:1:1 (HO-2:HO-3:HO-6); ---, ratio of rate constants 5:1:2.

The effect of the fine structure on the reactivity of cellulose is demonstrated in Fig. 9-1. When samples of cotton yarn are treated with strong swelling agents to destroy crystallinity, the reactivity toward acetylation is substantially increased especially if drying is omitted. Drastic drying conditions cause extensive interchain bonding, thus reducing the accessibility of the hydroxyl groups.

In the strongly swollen or soluble state of cellulose all the hydroxyl groups are accessible to the reactant molecules. However, owing to the random nature of the reaction a homogeneous product is obtained only by complete substitution of the hydroxyl groups (degree of substitution, DS = 3). At any degree of substitution lower than 3 the reaction leads to random sequences of units consisting of the following components: (1) unreacted glucose units, (2) three monosubstituted units (2-; 3-; and 6-derivatives), (3) three disubstituted units (2,6-; 3,6-; and 2,3-derivatives), and (4) the fully substituted units (2,3,6-trisubstituted derivative). The statistical overall distribution of substituents as a function of the degree of substitution (DS) calculated theoretically is shown in Fig. 9-2 (methylation). The curves, which are in good agreement with experimental findings, show that moderate differences in reactivity do not essentially influence the overall distribution of the substituents.

## 9.2 Swelling and Dissolution of Cellulose

Cellulose swells in different solvents. The extent of swelling depends on the solvent as well as on the nature of the cellulose sample. In the case of native cellulose with fibrous structure more or less drastic morphological changes take place depending on whether the swelling is *interfibrillar* or *intrafibrillar*. More generally the differences caused are distinguished in terms of *intercrystalline* and *intracrystalline* swelling. In the former case the swelling agent enters only into the disordered (amorphous) regions of the cellulose microfibrils and between them, whereas in the latter case the ordered (crystalline) regions are penetrated.

When bone-dry cellulose fibers are exposed to humidity, they adsorb water and the cross section of the fibers is increased because of swelling. At a 100% relative humidity this swelling corresponds roughly to a 25% increase in the fiber diameter. An additional 25% increase in swelling takes place when the fibers are immersed in water. In the longitudinal direction the dimensional change is very small.

The water retention of cellulose fibers at a given relative humidity varies depending on whether the equilibration has taken place by desorption or adsorption (hysteresis). The water uptake also continuously decreases after repeated drying and moistening of the fibers. Additional factors influencing the ability of pulp fibers to swell are their chemical composition, such as their hemicellulose and lignin content.

Cellulose swells in electrolyte solutions because of the penetration of hydrated ions which require more space than the water molecules.

Intracrystalline swelling can be accomplished by concentrated solutions of strong bases or acids and also of some salts. This type of swelling can be either *limited* or *unlimited*. In the former case the swelling agent combines with the ordered cellulose in certain stoichiometric proportions but does not completely destroy the interfibrillar bonding. The latter type refers to cases where the swelling agent is bulky and forms complexes with cellulose thus resulting in breakage of the adjacent bonds and separation of the chains so that gradual dissolution occurs.

The ability of inorganic salt solutions to swell and even dissolve cellulose is usually related to the lyotropic series for the solvated ions but the mechanism is complicated.

Alkali is not capable of dissolving native cellulose. Only depolymerized cellulose fragments with a low degree of polymerization are alkali soluble. Certain quaternary ammonium compounds are more effective resulting in full solubility. A mixture of dimethyl sulfoxide and paraformaldehyde (DMSO-PF) has interesting properties as a cellulose solvent. However, its effect depends at least partly on the formation of a hydroxymethylcellulose derivative. The most important cellulose solvents are metal complexes of organic bases; common are cupriethylenediamine (CED) and cadmium ethylenediamine (Cadoxen) (see Table 9-2). Even if other mechanisms are operating, their dissolving ability entails formation of a complex with the two

TABLE 9-2.    Properties of Common Cellulose Solvents

| Abbreviation | Formula | Properties |
|---|---|---|
| Schweizer's solution (Cuoxam) | $[Cu(NH_3)_4](OH)_2$ | Dark blue. Extensive depolymerization of cellulose in the presence of oxygen. |
| CED (Cuen) | $[Cu(en)_2](OH)_2{}^a$ | Dark blue. Depolymerization of cellulose in the presence of oxygen. |
| Cadoxen | $[Cd(en)_3](OH)_2$ | Colorless, useful for optical measurements. Cellulose shows good stability in this solvent. |
| EWNN | $[FeT_3]Na_6{}^b$ | Greenish. Cellulose shows good stability in this solvent. |

[a] en is ethylenediamine.
[b] T is tartrate.

secondary hydroxyl groups in cellulose and with breaking of hydrogen bonds. These solvents are used in connection with the studies of cellulose polymer properties, such as viscosity measurements. Because the solutions are alkaline, cellulose can be depolymerized in the presence of oxygen.

## 9.3   Swelling Complexes—Alkali Celluloses

When the penetrating agent causes intracrystalline swelling, the X-ray diagram of the cellulose is changed indicating the formation of a cellulose-swelling agent complex. This complex is formed only at a given concentration of the swelling agent. Although extensive swelling can be achieved in solutions of various acids and salts, evidence of definite complexes is often lacking.

The most important swelling complexes of cellulose are those with sodium hydroxide although corresponding addition compounds are formed also with other inorganic and organic bases. The alkali celluloses are compounds with given stoichiometric relations between alkali and cellulose. They are hence usually classified as addition compounds even if their reactions refer to hydrates of alkoxides. For instance, alkali cellulose reacts with alkyl halides in analogy with alkoxides during the Williamson type of ether formation (cf. Section 9.6).

Alkali celluloses are extremely important intermediates because they exhibit a markedly enhanced reactivity compared with original cellulose. The reagents can penetrate more easily into the swollen cellulose structure and thus react with the hydroxyl groups. For instance, preparation of alkali cellulose, named mercerization after its inventor John Mercer (1844), is an

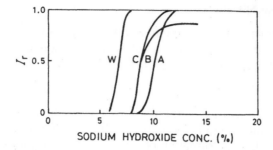

**Fig. 9-3.** Transition of cellulose I ($I_r = 0$) to cellulose II ($I_r = 1$) during cold alkali treatment of wood cellulose (W), cotton cellulose (C), bacterial cellulose (B), and animal cellulose (A) (Rånby, 1952).

important step when producing cellulose xanthate, from which viscose fibers and cellophane are prepared.

When cellulose fibers are mercerized in 12–18% sodium hydroxide solution, the original cellulose (cellulose I) is transformed into cellulose II and the unit cell dimensions are changed. This transformation, taking place at somewhat different concentrations depending on the origin of the sample, can be followed by X-ray measurements (Fig. 9-3).

So-called inclusion compounds can also be prepared from cellulose. According to this technique water-swollen cellulose is dried via a series of solvents and the final solvent used is trapped inside the structure on drying. Although the embedded organic solvent has no stoichiometric relation to the cellulose, it has a property of activating cellulose for organic reactions. Such inclusion compounds can be prepared from water-swollen cellulose after successive replacement of a series of solvent with decreasing polarity and solubility. As entrapping and activating organic solvents both highly volatile liquids (hexane, benzene, toluene, etc.) and less volatile compounds (e.g., fatty acids) can be used.

## 9.4 · Esters of Inorganic Acids

Cellulose is esterified with certain inorganic acids such as nitric acid, sulfuric acid, and phosphoric acid. A prerequisite is that the acids used can bring about a strong swelling thus penetrating throughout the cellulose structure. The esterification can be considered as a typical equilibrium reaction in which an alcohol and acid react to form ester and water. Of the inorganic esters cellulose nitrate is the only important commercial product.

## 9.4.1   Cellulose Nitrate

Cellulose nitrate is usually prepared in nitrating acid mixtures containing besides nitric acid sulfuric acid as a catalyst. The first reaction step involves generation of the nitronium ion ($NO_2^{\oplus}$):

$$HONO_2 + 2\ H_2SO_4 \rightleftarrows NO_2^{\oplus} + H_3O^{\oplus} + 2\ HSO_4^{\ominus} \qquad (9\text{-}1)$$

This reaction is an acid-base equilibrium in which sulfuric acid is the acid and the weaker nitric acid is the base, so that this kind of dissociation can take place instead of the formation of usual $H^{\oplus}$ and $NO_3^{\ominus}$ ions. In the next step the electrophilic nitronium ion attacks the hydroxyl groups of the cellulose:

$$NO_2^{\oplus} + HO-Cell \overset{\longrightarrow}{\rightleftarrows} NO_2-\overset{\oplus}{OH}-Cell \overset{\longrightarrow}{\rightleftarrows} NO_2-O-Cell + H^{\oplus} \qquad (9\text{-}2)$$

Esterification is retarded by the formation of water which must be removed from the system to force the reaction to completion.

The nitric acid concentration in the nitration acid mixture is usually 20–25%. The degree of nitration can be regulated by changes in the water content. Examples of the solubility and use of cellulose nitrates are given in Table 9-3. As a by-product in the nitration process some cellulose sulfate is also formed (see below). Because this results in instability of the cellulose nitrate, the sulfate groups must be removed by various treatments and the sulfuric acid formed removed by washing.

**TABLE 9-3.   Commercial Grades of Cellulose Nitrate**[a]

| DS | Solvents | Applications |
|---|---|---|
| 1.9–2.0 | Ethanol | Plastics |
| 1.9–2.3 | Esters, ethanol, ether-alcohol | Lacquers |
| 2.0–2.3 | Esters | Films, cements |
| 2.4–2.8 | Acetone | Explosives |

[a] From Rånby and Rydholm (1956).

## 9.4.2   Cellulose Sulfate

Cellulose sulfates can be prepared by using a variety of reagent combinations (Table 9-4). The active agent is sulfur trioxide ($SO_3$), present in fuming sulfuric acid or generated according to the following acid-base equilibrium between sulfuric acid molecules:

$$2\ H_2SO_4 \rightleftarrows H_3O^{\oplus} + HSO_4^{\ominus} + SO_3 \qquad (9\text{-}3)$$

**TABLE 9-4.** Suitable Reagents for Preparing Cellulose Sulfate

Sulfuric acid/ethanol, propanol, butanol
Fuming sulfuric acid/sulfur trioxide
Sulfur trioxide/sulfur dioxide, dimethylformamide, carbon disulfide
Chlorosulfonic acid/sulfur dioxide, pyridine

The strongly electrophilic sulfur trioxide is then added to the hydroxyl group and the intermediate oxonium ion is decomposed to sulfate ester and proton:

$$\text{Cell—OH} + SO_3 \rightleftarrows \left[ \text{Cell—}\overset{H}{\underset{\oplus}{O}}\text{—}SO_3^{\ominus} \right] \rightleftarrows \text{Cell—O—}SO_3^{\ominus} + H^{\oplus} \qquad (9\text{-}4)$$

An alternative mechanism is:

$$\text{Cell—OH} + H_2SO_4 \rightleftarrows \left[ \text{Cell—}\overset{H}{\underset{\oplus}{OH}} \right] + HSO_4^{\ominus} \rightarrow \text{Cell—O—}SO_3^{\ominus} + H_3O^{\oplus} \quad (9\text{-}5)$$

Because only one valence of sulfur is occupied for ester formation, the product is acid. Cellulose sulfates are water soluble and can be used as thickening agents.

### 9.4.3 Other Inorganic Cellulose Esters

Considerable interest has been directed to the preparation of cellulose phosphates because of their flame retarding properties and potential use in textiles. Phosphorylation can be accomplished in several ways, e.g., by heating cellulose at high temperatures with molten urea and phosphoric acid. Other phosphor-containing esters of cellulose include phosphites, phosphinates, and phosphonites. In addition, boric acid esters have been prepared.

## 9.5 Esters of Organic Acids

### 9.5.1 Cellulose Acetate

Cellulose acetate has replaced cellulose nitrate in many products, for example, in safety-type photographic films. When a solution of cellulose acetate in acetone is passed through the fine holes of a spinneret and the

**TABLE 9-5.    Commercial Grades of Cellulose Acetate[a]**

| DS | Solvents | Applications |
|---|---|---|
| 1.8–1.9 | Water–propanol–chloroform | Composite fabrics |
| 2.2–2.3 | Acetone | Lacquers, plastics |
| 2.3–2.4 | Acetone | Acetate rayon |
| 2.5–2.6 | Acetone | X-ray and safety films |
| 2.8–2.9 | Methylene chloride–ethanol | Insulating foils |
| 2.9–3.0 | Methylene chloride | Fabrics |

[a] From Rånby and Rydholm (1956).

solvent evaporates, solid filaments are produced. *Acetate rayon* is prepared from threads of these filaments. Some applications and solvents of commercial cellulose acetate grades are summarized in Table 9-5.

Because acetylation of cellulose proceeds in a heterogeneous system, the reaction rate is controlled by the diffusion of the reagents into the fiber structure. The quality of the cellulose raw material used for acetate rayon is of great importance. Although cotton linters fulfill high quality requirements, most cellulose acetate today is produced from wood pulps because of their favorable price and constant availability. Both sulfite and prehydrolyzed kraft pulps are used. Some quality requirements are shown in Table 9-6.

Cellulose acetate is usually produced by the so-called solution process with exception of the fully acetylated end product (triacetate). In the solution process the pulp is first pretreated with acetic acid in the presence of a catalyst, usually sulfuric acid. The purpose of this activation step is to swell the fibers and increase their reactivity as well as to decrease the DP to a suitable level. Acetylation is then performed after addition of acetic anhydride and catalytic amounts of sulfuric acid in the presence of acetic acid. After full acetylation the final triacetate obtained is dissolved. This

**TABLE 9-6.    Typical Specifications for Acetylation Grade Pulp[a]**

| | |
|---|---|
| α-Cellulose (%) | >95.6 |
| Pentosans (%) | <2.1 |
| Intrinsic viscosity (dm³/kg) | 550–750 |
| Ether extractable (%) | <0.15 |
| Ash (%) | <0.08 |
| Iron (mg/kg) | <10 |

[a] From Malm (1961).

"primary" acetate is usually partially deacetylated in aqueous acetic acid solution to obtain a "secondary" acetate with a lower DS of about 2 to 2.5

The fibrous acetylation process is performed in the presence of a suitable liquid, such as benzene, in which the reaction product is insoluble and which thereby retains the fiber form. For fibrous acetylation vapor-phase treatment with acetic anhydride can also be used. Besides sulfuric acid, perchloric acid and zinc chloride have been used as catalysts.

The acid-catalyzed acetylation of cellulose proceeds according to the following reaction formula:

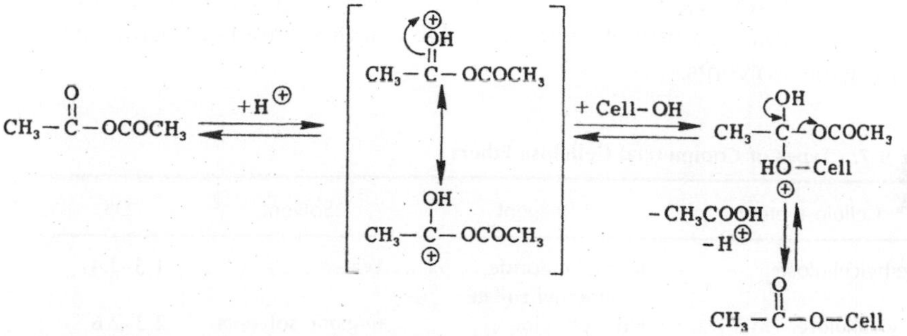

$$
\begin{array}{cc}
\underset{\substack{\|\\ \text{O}}}{\text{CH}_3-\text{C}-\text{OCOCH}_3} \;\; \overset{+\text{H}^{\oplus}}{\rightleftharpoons} \;\;
\left[
\begin{array}{c}
\underset{\substack{\|\\ \text{O}}}{\overset{\overset{\oplus}{\text{OH}}}{\text{CH}_3-\text{C}-\text{OCOCH}_3}} \\
\updownarrow \\
\underset{\substack{|\\ \oplus}}{\overset{\text{OH}}{\text{CH}_3-\text{C}-\text{OCOCH}_3}}
\end{array}
\right]
& \begin{array}{c}
\overset{+\text{Cell}-\text{OH}}{\rightleftharpoons} \;\; \underset{\substack{|\\ \text{HO}-\text{Cell}\\ \oplus}}{\overset{\text{OH}}{\text{CH}_3-\text{C}-\text{OCOCH}_3}} \\
\\
\begin{array}{c} -\text{CH}_3\text{COOH} \\ -\text{H}^{\oplus} \end{array} \; \updownarrow \\
\\
\underset{\substack{\|\\ \text{O}}}{\text{CH}_3-\text{C}-\text{O}-\text{Cell}}
\end{array}
\end{array}
\tag{9-6}
$$

After protonation of the acetic anhydride the electrophilic carbonium ion formed is added to a nucleophilic hydroxyl oxygen atom of the cellulose. This intermediate is then decomposed into cellulose acetate and acetic acid with liberation of a proton.

## 9.5.2   Other Esters of Organic Acids

A number of various esters of cellulose are known, for example, propionate and butyrate and mixed esters such as acetate-butyrate, propionate-isobutyrate, and propionate-valerate. The mixed esters have found use in plastic composites when good grease- and water-repelling properties are required.

Cellulose can also be esterified by aromatic acids. However, derivatives of any importance are only the cellulose cinnamic and salicylic acid esters. A number of nitrogen-containing esters are also known, for example, cellulose dialkyl diaminoacetate, cellulose acetate-N,N-dimethylaminoacetate, and cellulose propionate-3-morpholine butyrate. Because of the presence of basic substituents these derivatives, although water insoluble, can be dissolved in acidic solutions. Such derivatives have found use as surface coatings in photographic films and in tablets for pharmaceutical purposes.

## 9.6    Ethers

Cellulose ethers can be prepared by treating alkali cellulose with a number of various reagents including alkyl or aryl halides (or sulfates), alkene oxides, and unsaturated compounds activated by electron-attracting groups. A variety of products of considerable commercial importance has been developed for different uses (Table 9-7). Most of the cellulose ethers are water soluble and they generally possess similar properties, but because of specific characteristics they complete rather than compete with each other. The water solubility is generally attained at very low degrees of substitution. With hydrophobic substituents and a high DS cellulose ethers become soluble in organic solvents.

**TABLE 9-7.    Types of Commercial Cellulose Ethers**

| Cellulose ether | Reagent | Solvent | DS |
|---|---|---|---|
| Methylcellulose | Methyl chloride, dimethyl sulfate | Water | 1.5–2.4 |
| Ethylcellulose | Ethyl chloride | Organic solvents | 2.3–2.6 |
| Carboxymethylcellulose | Sodium chloroacetate | Water | 0.5–1.2 |
| Hydroxyethylcellulose | Ethylene oxide | Water | 1.3–3.0[a] |
| Cyanoethylcellulose | Acrylonitrile | Organic solvents | 2.0 |

[a] Molar substitution.

### 9.6.1    Alkyl Ethers

The simplest representatives of cellulose ethers are the corresponding alkyl derivatives. The most common representatives manufactured industrially are methyl- and ethylcellulose. Methylcellulose is soluble in cold water when the DS is 1.4 to 2.0, whereas nearly completely substituted products (DS 2.4–2.8) are insoluble in water but soluble in organic solvents.

Alkali cellulose from cotton linters or wood pulp, usually prepared in a way similar to the first step in the viscose process (see Section 9.7), is used as raw material. Alkylation is carried out by using alkyl chlorides. The reaction proceeds according to the $S_N2$ mechanism (bimolecular nucleophilic substitution):

$$Cell\text{—}OH + HO^\ominus \rightleftarrows Cell\text{—}O^\ominus + H_2O \tag{9-7}$$

$$Cell\text{—}O^\ominus + R\text{—}Cl \rightleftarrows Cell\text{—}OR + Cl^\ominus$$
$$(R = CH_3 \text{ or } C_2H_5) \tag{9-8}$$

As a by-product methanol or ethanol is formed:

$$RCl + HO^\ominus \rightleftarrows ROH + Cl^\ominus \tag{9-9}$$

which then reacts with alkyl chloride to form dimethyl or diethyl ether:

$$ROH + HO^\ominus \rightleftarrows RO^\ominus + H_2O \tag{9-10}$$

$$RO^\ominus + R\!-\!Cl \rightleftarrows ROR + Cl^\ominus \tag{9-11}$$

Methylcellulose solutions generally form gels at higher temperatures. The gelation temperature is increased when hydroxyethyl or hydroxypropyl groups are introduced into the methylcellulose (cf. Section 9.6.2). Hydroxyethylmethylcellulose and hydroxypropylmethylcellulose are prepared industrially by the reaction of alkali cellulose first with ethylene oxide or propylene oxide and then with methyl chloride. Similarly, hydroxyethylethylcellulose is prepared by consecutive ethylene oxide and ethyl chloride treatments. Cellulose ethers with both methyl and ethyl groups have also been manufactured.

At a viscosity range exceeding 600 mN·sec·m$^{-2}$ (cP) methylcellulose solutions are pseudoplastic which means that the apparent viscosity decreases with increasing shear rate. Solutions of low viscosities again tend to be thixotropic, resulting in a decreased viscosity with increasing shear times.

Alkyl ethers of cellulose are used as additives for a variety of products. These applications include agricultural products (thickening and dispersing seeds and powders), food products (stabilizer and thickening agents), ceramics (agents improving viscosity and shrink resistance), technochemical products (additives improving wet-rub resistance and flow, etc.), pharmaceutical preparations (tablets, suspensions, emulsions, etc.), cements (control of the setting time), textiles (sizing and coating), paper products, and plywood.

### 9.6.2  Hydroxyalkyl Ethers

Hydroxyalkyl celluloses are obtained in the reaction of cellulose with alkene oxides or their corresponding chlorohydrins. The reaction is a base-catalyzed $S_N2$-type substitution, and the reaction rate is proportional to the product [epoxide] [Cell—O$^\ominus$]. The commercial preparations include hydroxyethyl- and hydroxypropylcellulose for which ethylene oxide and propylene oxide are used as reagents. Hydroxyethylcellulose is formed according to the following equation:

$$\text{Cell}-\text{O}^{\ominus} + \underset{\text{O}}{\overset{\text{H}_2\text{C}-\text{CH}_2}{\triangle}} \longrightarrow \text{Cell}-\text{O}-\text{CH}_2\text{CH}_2\text{O}^{\ominus} \qquad (9\text{-}12)$$

Ethylene oxide can also react with hydroxide ions resulting in the formation of ethylene glycol:

$$\text{HO}^{\ominus} + \underset{\text{O}}{\overset{\text{H}_2\text{C}-\text{CH}_2}{\triangle}} \longrightarrow \text{CH}_2\text{OH}-\text{CH}_2\text{O}^{\ominus} \qquad (9\text{-}13)$$

In addition, ethylene oxide is polymerized to polyethylene oxide. The terminal primary hydroxyl group of the substituent reacts with additional epoxide to form pendant oxyethylene chains:

$$\text{Cell}-\text{O}-\text{CH}_2\text{CH}_2\text{OH} + n\ \underset{\text{O}}{\overset{\text{H}_2\text{C}-\text{CH}_2}{\triangle}} \xrightarrow{\text{HO}^{\ominus}} \text{Cell}-(\text{O}-\text{CH}_2\text{CH}_2)_{n+1}\text{OH} \qquad (9\text{-}14)$$

Because of this reaction the degree of substitution (DS) of hydroxyalkylcelluloses is lower than the molar substitution (MS). The ratio MS/DS is a measure of the relative length of side chains. Usually only half of the ethylene oxide reacts with cellulose, the other half is consumed for side reactions.

Water soluble hydroxyethylcelluloses have molar substitutions ranging from 1.5 to 2.5. Hydroxyethylcellulose is used as a thickener for latex paints, for emulsion polymerization of polyvinylacetate, for paper sizing, and for improving the wet strength of paper (together with glyoxal), in ceramic industry, etc.

Hydroxypropylcellulose is applied for similar purposes as hydroxyethylcellulose although its use is more limited. Because hydroxypropyl substitution improves thermoplasticity and solubility in organic solvents, hydroxypropylcellulose can also be used as a thickener for organic solutions. Hydroxypropylation of cellulose does not result in water and alkali soluble products until the MS is close to 4.

### 9.6.3  Carboxymethylcellulose

Carboxymethylcellulose (CMC) is the most widely used water-soluble derivative of cellulose. It is prepared from alkali cellulose with sodium chloroacetate as reagent:

$$\text{Cell}-\text{O}^{\ominus} + \underset{\text{Cl}}{\text{CH}_2\text{COO}^{\ominus}} \longrightarrow \text{Cell}-\text{O}-\text{CH}_2\text{COO}^{\ominus} + \text{Cl}^{\ominus} \qquad (9\text{-}15)$$

CMC is manufactured in the DS range from 0.4 to about 1.4. The DP varies from 200 to 1000. The water solubility of CMC is increased when the DS increases. At DS values of 0.6–0.8 good water solubility is attained, whereas preparations with a DS of 0.05–0.25 are soluble only in alkali. Because of the carboxyl groups CMC is a polyelectrolyte. Its $pK_a$ value varies from about 4 to 5, depending on the DS.

CMC can be used in a variety of products such as detergents, foods (as protective colloid and for purposes where high water-binding capacity is required, stabilizers, etc.), ice cream, paper coatings, emulsion paints, drilling fluids, ceramics, pharmaceuticals, and cosmetics.

### 9.6.4 Cyanoethylcellulose

Cellulose reacts with α,β-unsaturated compounds containing strongly electron-attracting groups to form substituted ethyl ethers. The most common derivative of this type is cyanoethylcellulose. Cyanoethylation requires strongly basic catalysts and is usually carried out in the presence of sodium hydroxide with acrylonitrile as reagent:

$$
\text{(9-16)}
$$

The cellulose anion formed attacks the positive carbon atom in acrylonitrile to form a resonance-stabilized intermediate anion, which then adds a proton from water to form the product under simultaneous liberation of a hydroxyl ion. All the reaction steps are reversible and because of the regeneration of hydroxyl ions no alkali is consumed. However, the process is accompanied with the consumption of acrylonitrile in several side reactions, such as formation of 3,3'-oxydipropionitrile:

$$
2\ CH_2 = CHCN + H_2O \rightarrow O(CH_2CH_2CN)_2 \tag{9-17}
$$

Processes have been developed to minimize these side reactions.

Highly cyanoethylated celluloses with a DS of about 2.5 are soluble in polar organic solvents. Because of an unusually high dielectric constant and low dissipation factor they can be used as the resin matrix for the phosphorous and electroluminescent lamps. Cyanoethylated kraft wood pulp with a DS of only about 0.2 is used for making insulation paper for transformers. Cyanoethylated paper has also a good thermal and dimensional stability.

## 9.7 Cellulose Xanthate

The preparation of viscose *rayon* fibers and *cellophane* proceeds via the xanthate, which therefore is an extremely important derivative of cellulose. Treatment of alkali cellulose with carbon disulfide results in the formation of cellulose xanthate (dithiocarbonate):

$$\text{Cell} - \text{O}^{\ominus} + \overset{S}{\underset{S}{\overset{\displaystyle \parallel}{C}}} \rightleftharpoons \text{Cell} - \text{O} - \underset{\underset{S}{\parallel}}{C} - S^{\ominus} \tag{9-18}$$

In the first step cellulose is treated with 18% sodium hydroxide at 15°–30°C. After removing excess sodium hydroxide from the fibers by pressing the alkali cellulose is shredded and subjected to alkaline ripening to bring the DP down to 200–400. Xanthation is then carried out at 25°–30°C for ca. 3 hours resulting in a DS of approximately 0.5. The cellulose xanthate is dissolved in aqueous sodium hydroxide resulting in an orange-colored viscous liquid, called *viscose*. After ripening the viscose solution is filtered and forced through a spinnerette into an acid (sulfuric acid and salts) bath where the cellulose is regenerated in the form of fine filaments resulting in rayon fibers:

$$\underset{\underset{S}{\parallel}}{\text{Cell} - \text{O} - C - S^{\ominus}} \overset{H^{\oplus}}{\rightarrow} \text{Cell} - \text{OH} + CS_2 \tag{9-19}$$

Cellophane is prepared by pressing the viscose through a narrow slit into an acid bath to form thin sheets.

## 9.8 Cross-Linking of Cellulose

In order to improve the properties of cellulose fibers mainly for applications in textiles, the cross-linking of cellulose has been studied extensively in heterogeneous systems to maintain the fibrous structure. Usually agents capable of forming cross-links through ether bonds with the hydroxyl groups of cellulose are used; ester cross-links are also possible but not so useful because of their low stability against alkali. New reactive groups can also be introduced into cellulose as coupling points for cross-linking agents.

The most common type of cellulose cross-linking is accomplished by using aldehydes and dialdehydes, such as formaldehyde and glyoxal as reactants. Already at the beginning of this century it was observed that treatment of cellulose with formaldehyde in the presence of acid results in improved wet strength of the fibers. As shown later formaldehyde evidently condenses with the hydroxyl groups of cellulose producing intermolecular ether cross-links, although it is not known to what extent the adjacent secondary hydroxyl groups of a glucopyranose unit participates in this reaction:

$$2 \text{ Cell—OH} + \text{HCHO} \xrightarrow{\text{H}^{\oplus}} \text{Cell—O—CH}_2\text{—O—Cell} \qquad (9\text{-}20)$$

In the 1950s formaldehyde was commercially applied to improve the dimensional stability of rayon fabrics, but it has largely been replaced by other cross-linking agents. The most important technology for cross-linking of cellulose is based on the use of formaldehyde precondensates with amides including ureas, triazines, and carbamates. For example, urea and its derivatives condense with formaldehyde according to the following equation:

$$2 \text{ CH}_2\text{O} + \underset{\underset{R}{\overset{|}{\phantom{.}}}}{\text{HN}} - \underset{\underset{O}{\overset{\|}{\phantom{.}}}}{\text{C}} - \underset{\underset{R}{\overset{|}{\phantom{.}}}}{\text{NH}} \longrightarrow \text{HOCH}_2\underset{\underset{R}{\overset{|}{\phantom{.}}}}{\text{N}} - \underset{\underset{O}{\overset{\|}{\phantom{.}}}}{\text{C}} - \underset{\underset{R}{\overset{|}{\phantom{.}}}}{\text{N}}\text{CH}_2\text{OH} \qquad (9\text{-}21)$$

(R = H or substituent)

After impregnation of the cellulose substrate with an aqueous solution containing urea-formaldehyde precondensates and subsequent short heating at 130°–160°C, cross-linking takes place:

$$2 \text{ Cell—OH} + \text{HOCH}_2\underset{\underset{R}{\overset{|}{\phantom{.}}}}{\text{N}} - \underset{\underset{O}{\overset{\|}{\phantom{.}}}}{\text{C}} - \underset{\underset{R}{\overset{|}{\phantom{.}}}}{\text{N}}\text{CH}_2\text{OH} \longrightarrow \text{Cell—OCH}_2\underset{\underset{R}{\overset{|}{\phantom{.}}}}{\text{N}} - \underset{\underset{O}{\overset{\|}{\phantom{.}}}}{\text{C}} - \underset{\underset{R}{\overset{|}{\phantom{.}}}}{\text{N}}\text{CH}_2\text{O—Cell} \qquad (9\text{-}22)$$

Such a treatment of fabrics ("finishing") results in improved dimensional stability, crease resistance, wash-and-wear performance, and durable-press properties, all important for textiles. A variety of agents including cyclic urea derivatives has been applied to find the best effects. Another approach includes the use of activated vinyl compounds and their derivatives such as acrylamides and vinyl ketones. Furthermore, polyfunctional compounds containing oxirane and aziridinyl groups have been used for creation of cross-links into cellulose.

## 9.9   Grafting on Cellulose

In addition to being the basis of traditional ether and ester derivatives, cellulose can be modified by cross-linking or by preparing so-called graft copolymers. Cross-linking improves certain properties, such as crease and shrink resistance, but because of the three-dimensional network formed the structure of cellulose becomes more rigid. The product is then less suitable for textile purposes because it is more brittle. By grafting, however, polymer branches can be created to the cellulose backbone without destroying the desirable properties of the original cellulose fibers. Graft copolymerization has been applied to a number of cellulosic materials, including cotton, rayon, paper, cellophane, and wood. However, only minor commercial success has so far been attained despite all research.

Although cellulose can be grafted homogeneously using soluble cellulose derivatives or suitable solvents, grafting is usually performed in a hetero-geneous system and is greatly influenced by the physical structure of the cellulose. The vast majority of grafting methods involve polymerization of vinyl monomers of type $CH_2{=}CH{-}X$ here X is an inorganic moiety, such as halide, $-CN$, $-NO_2$, or an organic substituent.

The grafting methods can in principle be divided into three categories, namely, radical polymerization, ionic polymerization, and condensation or addition polymerization. Only the first case is discussed in the following since the most common grafting methods belong to this category.

A free radical-initiated grafting reaction can be generated when a vinyl monomer is polymerized in the presence of cellulose. Grafting is initiated by abstraction of a hydrogen atom from cellulose either by a growing chain radical or directly by a radical created from the catalyst. The first pathway is termed *chain transfer*. In both cases an unpaired electron is left on the cellulose chain which initiates grafting. In common with other radical reactions, grafting reactions involve stages of initiation, propagation, and termination as illustrated by the following equations:

Creation of initiator radicals from catalyst $R-R \rightarrow 2\ R\cdot$ \hfill (9-23)

Initiation of homopolymer radicals from monomer (M) $\quad R\cdot + M \rightarrow RM\cdot$ \hfill (9-24)

Homopolymer chain growth (propagation) $\quad RM\cdot + M \rightarrow RM_2\cdot$, etc. \hfill (9-25)

Chain transfer to cellulose $\quad RM_{\dot{x}} + \text{Cell}-H \rightarrow RM_xH + \text{Cell}\cdot$ \hfill (9-26)

Graft copolymer chain initiation $\quad \text{Cell}\cdot + M \rightarrow \text{Cell}-M\cdot$ \hfill (9-27)

Graft side chain growth (propagation) $\quad \text{Cell}-M\cdot + M \rightarrow \text{Cell}-M_2\cdot$, etc. \hfill (9-28)

Homopolymer termination $\quad M_x\cdot + M_{\dot{x}+n} \rightarrow M_{2x+n}$ \hfill (9-29)

Graft side chain termination $\quad \text{Cell}-M_x\cdot + M_{\dot{x}+n} \rightarrow \text{Cell}-M_{2x+n}$ \hfill (9-30)

Cross-linking $\quad \text{Cell}-M_x\cdot + \text{Cell}-M_{\dot{x}+n} \rightarrow \text{Cell}-M_{2x+n}-\text{Cell}$ \hfill (9-31)

For initiation, use is made of peroxides or azo compounds, which can be decomposed to free radicals ($R\cdot$) generating growing monomer radicals (equations 9-23, 9-24, and 9-25). Because of chain transfer to cellulose (equation 9-26) a branch is initiated (equation 9-27) and the side-chain begins to grow (equation 9-28).

Certain compounds, such as those bearing thiol groups (—SH), facilitate chain transfer. Thiol groups, leading to higher grafting yields, can be introduced into the cellulose by reaction with ethylene sulfide:

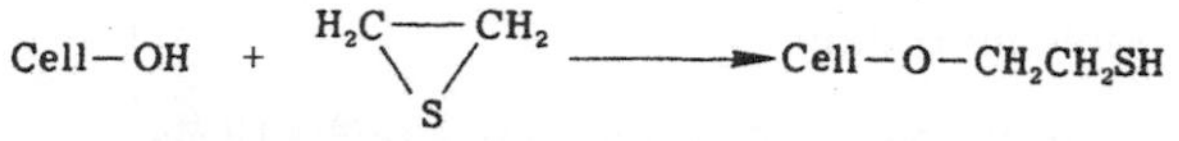

$$Cell-OH \quad + \quad \underset{S}{\overset{H_2C----CH_2}{\diagdown\diagup}} \quad \longrightarrow Cell-O-CH_2CH_2SH \qquad (9\text{-}32)$$

After abstraction of hydrogen from the thiol group by a growing chain radical, the chain is terminated under simultaneous generation of a thiol radical:

$$Cell-O-CH_2CH_2SH + R\!\!\sim\!\cdot \;\rightarrow\; R\!\!\sim\!H + Cell-O-CH_2CH_2S\cdot \qquad (9\text{-}33)$$

Formation of graft polymer is then started:

$$Cell-O-CH_2CH_2S\cdot \;+\; n(CH_2\!=\!CHX) \rightarrow$$

$$Cell-O-CH_2CH_2S-(CH_2-CHX)_n\cdot \qquad (9\text{-}34)$$

Another method to improve the yield of grafting involves generation of the initiating species in the swollen cellulose substrate itself. Because the monomer concentration in this system is low, the reacting species have a greater chance to initiate graft copolymerization (equation 9-35) than homopolymerization of the monomers (equation 9-24).

$$Cell-H + R\cdot \rightarrow Cell\cdot + RH \qquad (9\text{-}35)$$

To produce radicals the cellulose substrate can be saturated with potassium persulfate. Another method includes impregnation of cellulose with ferrous salts followed by treatment with a monomer solution in the presence of hydrogen peroxide. Hydroxyl radicals are produced in this system according to the following reaction:

$$Fe^{2\oplus} + H_2O_2 \rightarrow Fe^{3\oplus} + HO^{\ominus} + HO\cdot \qquad (9\text{-}36)$$

The hydroxyl radicals can react with cellulose, initiating graft copolymerization, or react with monomer, resulting in homopolymerization. A similar redox system is based on the use of ceric ions, which produce radicals by direct oxidation of the cellulose chains and thus initiate graft polymerization:

$$\text{Cell—H} + \text{Ce}^{4\oplus} \rightarrow \text{Cell}\cdot + \text{Ce}^{3\oplus} + \text{H}^{\oplus} \qquad (9\text{-}37)$$

$$\text{Cell}\cdot + \text{M} \rightarrow \text{graft copolymer} \qquad (9\text{-}38)$$

More specific initiators are $Mn^{3+}$ ions in aqueous solution which give efficient grafting of acrylic and methacrylic monomers onto cellulose and very little homopolymer ($\leqslant 2\%$).

In another method radicalizable groups, such as hydroperoxides, are introduced into cellulose by ozone treatment. The hydroperoxides are decomposed directly or in the presence of reducing agents to radicals which initiate the graft copolymerization:

$$\text{Cell—O}_2\text{H} + \text{Fe}^{2\oplus} \rightarrow \text{Cell—O}\cdot + \text{Fe}^{3\oplus} + \text{HO}^{\ominus} \qquad (9\text{-}39)$$

$$\text{Cell—O}_2\text{H} + \text{Fe}^{2\oplus} \rightarrow \text{Cell—O}^{\ominus} + \text{Fe}^{3\oplus} + \text{HO}\cdot \qquad (9\text{-}40)$$

$$\text{Cell—O}\cdot + \text{M} \rightarrow \text{graft copolymer} \qquad (9\text{-}41)$$

$$\text{HO}\cdot + \text{M} \rightarrow \text{homopolymer} \qquad (9\text{-}42)$$

Both ultraviolet light and high-energy radiation such as gamma rays from a $^{60}$Co source or highly accelerated electrons have been used for initiation of grafting. The radical sites created on the cellulose initiate the copolymerization in the presence of vinyl monomers. However, cellulose radicalized by irradiation is degraded because of the cleavage of glucosidic bonds.

Radicals can finally be created by mechanical means, such as milling or by using electrical discharges, but these methods have found only very limited use.

## 9.10   Cellulose Ion Exchangers

Cellulose is suitable for certain applications for which a solid and inert matrix with a large surface area is required. Introduction of acidic or basic substituents into the backbone of cellulose results in cation and anion exchangers, respectively (Table 9-8). Although the cellulose ion exchangers are chemically less stable than the synthetic ion exchange resins and their capacity is relatively low, they are useful particularly for biochemical separation problems involving large molecules such as proteins. Cellulose ion exchangers are commercially available as powder as well as in the form of fiber and paper and they can thus be used for various types of separation techniques including column, paper, and thin-layer chromatography.

*Cation Exchange Celluloses*   The cation exchange celluloses are of two types: a weakly acidic type containing carboxylic acid groups and a strongly

**TABLE 9-8.   Cellulose Ion Exchangers**

| Name | Functional group |
|---|---|
| Cation exchangers | |
|   Cellulose phosphate (P-cellulose) | $-OPO_3^{2-}$ |
|   Sulfoethylcellulose (SE-cellulose) | $-OC_2H_5SO_3^{-}$ |
|   Carboxymethylcellulose (CM-cellulose) | $-OCH_2COO^-$ |
|   Oxidized cellulose | $-COO^-$ |
| Anion exchangers | |
|   Aminoethylcellulose (AE-cellulose) | $-OC_2H_4\overset{+}{N}H_3$ |
|   Diethylaminoethylcellulose (DEAE-cellulose) | $-OC_2H_4\overset{+}{N}H(C_2H_5)_2$ |
|   Triethylaminoethylcellulose (TEAE-cellulose) | $-OC_2H_4\overset{+}{N}(C_2H_5)_3{}^a$ |
|   ECTEOLA-cellulose (condensation product of epichlorohydrin, triethanolamine, and cellulose) | |

$^a$ Only a part of the functional groups are of this type.

acidic type containing sulfonic acid or phosphoric acid groups. The most common type of a carboxylic acid ion exchanger is carboxymethylcellulose with a low DS giving a cation exchange capacity of 0.4–0.7 meq/g. Products of higher capacity cannot be applied to chromatography unless they are cross-linked to prevent extensive swelling. Other carboxylic acid cellulose cation exchangers can be obtained by oxidation, for example, with nitrogen dioxide which converts the primary hydroxyl groups into carboxyls rather selectively. Sulfoethylcellulose represents a cellulose cation exchanger of strongly acidic type and is prepared by reacting α-chloroethanesulfonic acid with cellulose in the presence of sodium hydroxide:

$$Cell-OH + ClCH_2CH_2SO_3Na + NaOH \rightarrow Cell-OCH_2CH_2SO_3Na + NaCl + H_2O \quad (9-43)$$

The capacity of the product is about 0.5 meq/g. Sulfomethylcellulose has been prepared by using monochloromethanesulfonate, made from dichloromethane and sodium sulfite. Phosphonomethylcellulose, a bifunctional cation exchanger, is prepared by reacting disodium chloromethylphosphonate and cellulose:

$$Cell-OH + ClCH_2PO_3Na_2 + NaOH \rightarrow Cell-OCH_2PO_3Na_2 + NaCl + H_2O \quad (9-44)$$

*Anion Exchange Celluloses*   Aminoethylcellulose, a weakly basic anion exchanger, is prepared in the reaction of cellulose with α-aminoethylsulfuric acid in the presence of sodium hydroxide:

$$Cell-OH + H_2NCH_2CH_2OSO_3H + NaOH \rightarrow$$
$$Cell-OCH_2CH_2NH_2 + NaHSO_4 + H_2O \quad (9-45)$$

The capacity is about 0.2 meq/g but can be increased after cross-linking up to about 0.7 meq/g. Diethylaminoethylcellulose (DEAE-cellulose) is made from cotton linters, previously cross-linked with formaldehyde or 1,3-dichloropropanol, by using 2-chlorotriethylamine as reagent. Quarternary cellulose anion exchangers of strongly basic type have also been prepared by reacting DEAE-cellulose with alkyl halides under anhydrous conditions to yield, for example, triethylaminoethylcellulose (TEAE-cellulose). After reaction of cellulose with epichlorohydrin and triethanolamine in the presence of excess alkali, another anion exchanger or so-called ECTEOLA-cellulose can be produced having an ion exchange capacity of 0.3–0.4 meq/g.

In addition to the types mentioned above, a variety of special cellulose ion exchangers have been prepared including materials containing complexing groups showing strong preference for heavy metal ions. Furthermore, specific immunological adsorbents for isolation of antibodies from blood serum have been made by attaching the corresponding antigens to ion exchange celluloses or to unmodified celluloses.

# Chapter 10

# WOOD-BASED CHEMICALS AND PULPING BY-PRODUCTS

On a worldwide basis, wood constitutes an enormous, renewable raw material resource (biomass) for production of both energy and chemicals. Traditionally and globally, wood has played an important role in the production of various commodities, but most of them have been gradually replaced by products from the petrochemical industry. Apart from the established position of petrochemicals, the resources of oil and other fossil materials from which they are produced have their limits. In addition to the increased oil prices and the shortage of fossil materials there are also environmental concerns which have focused new interest on the possibilities of converting wood to various forms of fuels and chemical products. In the long run new applications can be expected to emerge from the research which was tremendously intensified during 1970s because of the "oil crisis." However, irrespective of this, it should be kept in mind that the function of wood as a construction material and as a source for fibers presumably remains unchanged or even becomes more important in the future.

Presently, without being an inherent part of the pulping process, converting wood directly to useful "silvichemicals" is economically attractive only in exceptional cases. However, solid wood residues after harvesting constitute an enormous biomass resource remaining mainly in the forest. In fact, 40–50% of the dry wood weight remains unused in the form of stumps, roots, branches, needles, and debarking residues. Besides agricultural wastes these wood residues are important biomass resources.

## 10.1    Wood-Based Chemicals and Fuels

The wood-based chemicals can be grouped according to their origin or the technique used for their isolation and processing (Fig. 10-1). Most of the nonstructural wood constituents can be removed by extraction with suitable solvents or collected as exudates from the tree. The extractives can be used as such or processed further to a large variety of end products.

Strong mineral acids hydrolyze the wood polysaccharides into monosaccharides, which can further be converted to various end products by fermentation or other treatments.

By heat treatment at elevated temperatures the wood substance is broken down to low-molecular-weight products. The gases can be used for chemical synthesis or as fuels either directly or after conversion to a liquid form.

All kinds of wood conversion, especially thermal treatment, result in complex product mixtures from which single components can be isolated only by applying adequate separation techniques.

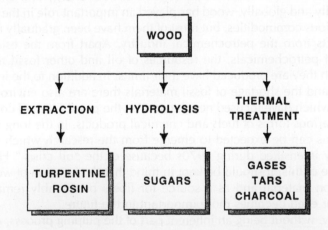

**Fig. 10-1.** Product groups obtained from wood after different treatments.

### 10.1.1 Extractives

Wood extractives have been utilized by humans since ancient times. A frequently cited example is the fossilized resin named *amber,* which originates from coniferous trees and was a product valued by the Greek and Roman ancients. Most of the extractives utilized for various purposes today are still recovered from softwoods, particularly from pines. So-called turpentining is an operation in which living pine trees are wounded and the exuded *oleoresin,* named "gum," is collected manually. The degradation products of this gum, such as tar and pitch, were originally found to be useful for protection and tightening of the hulls of wooden ships and for preservation of ropes. This type of use initiated the concept *"naval stores,"* which was later extended to cover more generally all types of extractives and resinous products based on wood. The *naval stores industry* has a long history, dating back to the early decades of the seventeenth century. For example, in the United States this industry started in southeastern longleaf and slash pine areas. Another naval stores industry was established in the Les Landes region of southwestern France utilizing maritime pine.

*Oleoresin*  (For sulfate turpentine and tall oil rosin, see Section 10.2.2.) The major naval stores products are *turpentine,* primarily composed of volatile mono- and sesquiterpenoids, and *rosin,* mainly a mixture of resin acids. These are the basic products of oleoresin, which is recoverable from resin-rich trees either by tapping ("gum oleoresin") or by extraction (mainly from virgin pine stumps). Both of these techniques still play an important role in the world's production of oleoresins, but in the North American and Scandinavian countries most of the turpentine and rosin originates from the kraft pulp industry. An interesting technique enabling a significant increase of the oleoresin production in pine trees is to induce "lightwood" (oleoresin-soaked wood) formation by injecting into the wood xylem suspensions of pitch canker fungus or dipyridyl fungicides such as "paraquat" (1,1'-dimethyl-4,4'-bipyridinium salt). However, this technique has not yet reached full readiness for commercial applications.

Traditionally, turpentine has been used as a solvent for varnishes and paints, but today it is being used mainly as a raw material for making chemicals. The raw turpentine is first purified by distillation into fractions of monoterpenes and a separate *pine oil* fraction constituting mainly of hydroxylated monoterpenes (monoterpenoids). The monoterpene fractions are either used as raw materials for preparation of various chemicals or hydrated to *synthetic pine oil* ($\alpha$-terpineol) (Fig. 10-2). Pine oil is used for a variety of purposes, such as for applications needing solvents with good emulsifying and dispersing properties and for flotation of minerals.

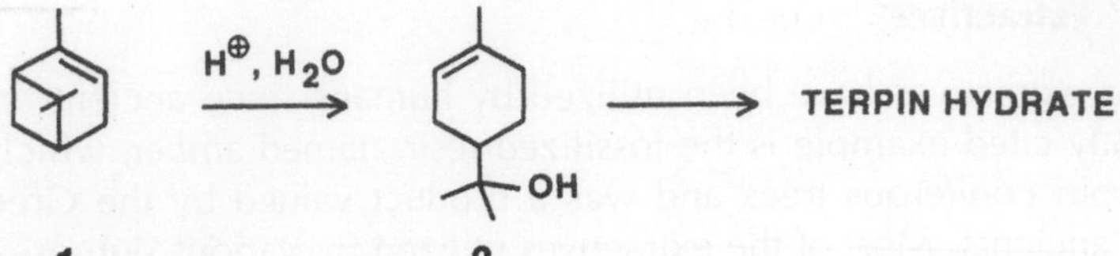

**Fig. 10-2.** Conversion of α-pinene to α-terpineol and terpin hydrate.

Another important use of turpentine and its fractions is to polymerize them into polyterpene resins. Such resins are used for various purposes such as for preparation of pressure-sensitive or hot-melt adhesives.

The needles of certain pine species contain attractive flavor constituents, which can be recovered by water-steam distillation. However, the flavor and fragrance products are predominantly prepared synthetically from monoterpenes. The most important starting material is β-pinene. Since the prominent component in most turpentines is α-pinene, it is first isomerized to β-pinene, which can be pyrolyzed to myrcene. This is the starting material for a number of products including linalool (lilac-like fragrances) and geraniol (rose-like) or their esters. However, menthol, which because of its cooling effect [especially the (−)-isomer] is used in cigarettes and cosmetic products and also for flavoring, is directly synthesized from α-pinene. Even phenolic-type chemicals, e.g., anisaldehyde, can be prepared from turpentine.

Among the various types of chemicals prepared from turpentine are insecticides. However, today most of these products have lost their importance and have been replaced by corresponding petroleum-based chemicals.

Interesting substances possessing juvenile hormone activity against hemipteran bug have been detected in some conifers. Such a substance is the acyclic sesquiterpenoid (+)-juvabione, which has been found to be present in balsam fir and in the heartwood of other *Abies* species. Juvabione is extractable from the wood and it may have potential use for insecticidal applications.

Rosin mainly consists of resin acids in addition to some neutral substances. The biggest market for rosin is as a paper sizing agent. For this purpose a number of various rosin formulations are commercially available. In addition, rosin and its modifications are used for many other purposes such as polymerization emulsifiers, adhesives, and additives in inks (see further details in Section 10.2.2).

**Waxes and Steroids**   Douglas fir is the dominant timber species in the western United States and Canada and the volume of the bark is especially high. Interest has been directed toward the use of bark wastes also because these represent a considerable environmental problem. A marketable wax product, predominantly consisting of mixed esters of aliphatic alcohols and fatty acids, has been isolated from Douglas fir bark and used for different purposes such as a binder or a substitute for carbon wax. However, in the long run, this product has not been commercially competitive. In addition, the wax fraction contains a variety of other substances, predominantly mixed esters derived from higher aliphatic alcohols and fatty acids. Suberin and steroids, particularly sitosterol, can also be recovered. The interest in sitosterol is related to its potential use in the synthesis of hormones, such as cortisone derivatives. The bark of birches contains significant amounts of betulinol, but so far no commercial uses have been found for this bark constituent.

*Phenolic Compounds*   Various monomeric as well as oligo- and polymeric phenolic compounds are widely distributed in woody plants, especially in heartwood, barks, leaves, fruits, and roots.

Commercially the most important phenolic constituents belong to the group of flavonoids. Taxifolin (dihydroquercetin) is a simple representative which can be recovered by extraction from the abundant bark supplies of Douglas fir. However, the most important constituents of this group are the condensed tannins, based on polymerized flavonoids. Their ability to form complexes with proteins makes them useful for many applications in the fields of nutrition and leather tannage. Another important area of applications includes the use of the condensed tannins as adhesives. Tannin can be extracted from many plants including hemlock (*Tsuga* spp.), oak (*Quercus* spp.), and chestnut (*Castanea* spp.). However, the most important tannin sources in the world are the South American quebracho tree (*Schinopsis* spp.) (heartwood) and the South African wattle tree (*Acasia mearnsii*) (bark). For example, in the United States most of the vegetable tannins are imported from these sources.

Other interesting phenolic constituents belong to the group of lignans, but commercial markets for them have not been established. Extensive research has been conducted to extract and recover conidendrin, which is a typical component in hemlock and some spruce species. One potential use of conidendrin is related to its antioxidant properties. Another example of lignans is plicatic acid, present in western cedar (*Thuja plicata*) heartwood, from which it can be isolated by extraction with water. Plicatic acid can be converted to derivatives with antioxidant properties having potential uses in food products, including fats and oils.

*Carbohydrates*   From most wood species only minute amounts of carbohydrates are extractable with water. However, the heartwood of *Larix* species contains high amounts of water-soluble arabinogalactan. Some arabinogalactan is produced and marketed for a few applications mainly in the pharmaceutical and food industry.

Examples of rare carbohydrate products present in some wood species are the exudate gums, such as gum arabic and tragacanth, as well as maple syrup. Limited amounts of such products are recovered and marketed for rather specific purposes.

*Rubber*    Practically all of the natural rubber (*cis*-1,4-polyisoprene) marketed today in the world is derived from *Hevea brasiliensis,* even though rubber is produced by more than 200 other plant species. Although the production of synthetic rubber (*trans*-1,4-polyisoprene) has enormously increased, about 30% of the total (natural + synthetic) rubber production in the world is still natural rubber. Most of the gutta percha and balata (*trans*-1,4-polyisoprene), however, are made synthetically.

### 10.1.2    Hydrolysis Products and Their Further Processing

*Acid Hydrolysis*    The glycosidic linkages are easily cleaved by strong mineral acids in the presence of which polysaccharides can be hydrolyzed to simple sugars. However, for a complete hydrolysis of cellulose, concentrated acid solutions must be used in order to bring about the necessary swelling and at least a partial destroying of the ordered regions (cf. Section 9.1). Much work has been done to find out realistic hydrolysis conditions so that dilute acids could be used. Process applications have been developed in which aqueous hydrochloric or sulfuric acid solutions are passed through the wood material at elevated temperatures (140–180°C) with continuous withdrawal of hydrolyzate to prevent decomposition of the liberated monosaccharides. The hydrolyzate containing usually 4–6% sugars is neutralized and concentrated by evaporation. An average yield of 50–55% sugars of the dry wood weight is typical, although it varies depending on the raw material. A considerable portion of cellulose remains unhydrolyzed when dilute acid solutions are used, while the hemicelluloses are converted almost completely to monosaccharides.

At more drastic acidic conditions and higher temperatures, pentoses are converted to furfural and hexoses to hydroxymethylfurfural, which reacts further to levulinic and formic acid (cf. Section 2.5.4). Furfural is used as an industrial solvent and as a starting material for production of various chemicals and polymers. It can be produced from hardwoods or other xylan-rich materials, such as oat hulls and corncobs, by two-stage hydrolysis via xylose as intermediate product or directly in one stage. The starting material is heated in a digester, usually in the presence of an acid catalyst. The steam introduced is continuously withdrawn from the digester (water-steam distillation). The condensed vapors consist mainly of furfural and water and they are separated by azeotropic distillation.

Hydroxymethylfurfural is not volatile by steam. It is prepared from hexoses in the presence of an acid catalyst by short heat treatment to avoid further degradation to levulinic and formic acid. After recovery by solvent extraction hydroxymethylfurfural is purified by distillation. Levulinic acid can be prepared in good yield from hexose-based polysaccharides by heating with acids. In this reaction formic acid is liberated and levulinic acid is easily cyclized to form α- and β-angelica lactones (Fig. 2-30, p. 46).

The solid wood residue remaining after extraction and hydrolysis of wood is mainly composed of lignin, which is more or less condensed depending on the conditions during the acid treatment. It can be used as fuel or possibly be subject to various treatments to yield low-molecular-weight degradation products, including phenols (cf. Section 10.2.2).

**Enzymatic Hydrolysis**   Interest has also been directed to the use of enzymes instead of acids for the hydrolysis of cellulosic wastes. Screening and genetic improvements of the cellulolytic fungi such as *Trichoderma* species have successively resulted in highly increased cellulase activities compared with those which were earlier obtained with the enzyme preparations from the same fungi. Genetic manipulation and other attempts might lead to further improvements, but today cellulose can be hydrolyzed to glucose only partially. The real obstacle is the restricted accessibility of enzymes because of morphological reasons and the ordered cellulose structure. Various types of pretreatments of wood and cellulosic materials have been made in order to increase the accessibility, but no practical applications have emerged from these experiments.

**Further Processing by Fermentation**   Monosaccharides can be fermented to alcohols, carboxylic acids, or other more complex products (Fig. 10-3). Ethanol, usually produced by yeast (*Saccharomyces cerevisiae*), is the most important fermentation product of hexoses. Much interest has been directed to the use of ethanol as a motor fuel, usually mixed with gasoline (gasohol). Ethanol can be further converted to a number of useful chemicals and polymers by various chemical treatments (Fig. 10-4). However, considering the present oil prices, most of these products cannot today compete with the corresponding petrochemicals.

Carboxylic acids belong to the second type of fermentation products and gluconic acid is an important commercial product. However, like many other carbohydrate-based chemicals, it is produced industrially from other carbohydrate sources, such as starch, and not from wood.

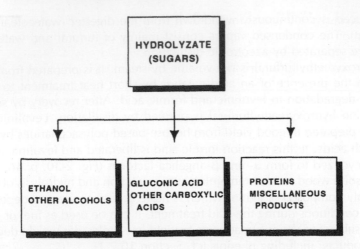

**Fig. 10-3.** Fermentation of sugar hydrolyzates to various products.

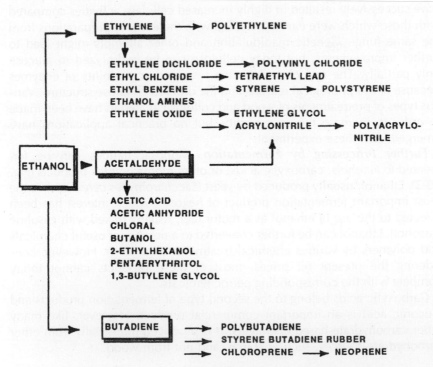

**Fig. 10-4.** Possible routes for conversion of ethanol to various chemicals and polymers (adopted from Alén, 1990).

The hydrolysis products may also find uses as fermentation substrates for production of single-cell proteins (CEL) and even other complex and high-value products including antibiotics, enzymes, and hormones.

*Chemical Modification*   The monosaccharides obtained after hydrolysis of wood can be modified by various chemical treatments even if this type of processing so far has resulted in only some industrial applications (Fig. 10-5). However, a successful industrial process for the production of xylitol was started in the 1970s in Finland. Xylose, obtained from birch wood hydrolysates or other xylose sources, such as sulfite spent liquors (cf. Section 10.2.1), is catalytically hydrogenated to xylitol. A powerful separation technique based on ion exclusion chromatography and crystallization decisively contributed to the success of this industrial process. Xylitol is a useful sweetener because its ability to prevent dental caries.

The reduction products of other monosaccharides do not possess such unique properties as xylitol, although glucitol and mannitol have also found use in the food industry.

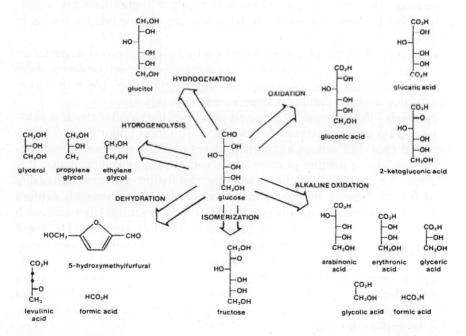

**Fig. 10-5.**  Examples of products obtained from glucose by various chemical and biochemical treatments (from Alén, 1990).

### 10.1.3    Thermal Treatment

Heating wood to temperatures slightly above 100°C already initiates some thermal decomposition. Around 270°C this thermal decomposition does not require any external heat source because the process becomes exothermic. Wood is typically decomposed in a stepwise manner: the hemicelluloses are degraded in the temperature range of 200–260°C, cellulose at 240–350°C, and lignin at 280–500°C. The two main thermal conversion methods, *pyrolysis* and *gasification*, differ in that the former is carried out in an inert atmosphere, whereas the latter refers to conditions under which controlled amounts of oxygen or other oxidizing agents are present. This strict definition is not always used, and sometimes thermal degradation even in the presence of air or other additives is named pyrolysis. However, for industrial applications, pyrolysis is done at a lower temperature range (about 500°C), whereas much higher temperatures (above 800°C) are typical of gasification processes. To maximize the formation of liquid products the pyrolysis temperature and gas residence time should be decreased, but the heating rate increased. The opposite conditions will result in higher fuel gas yields. When both the temperature and heating rate are low the process results in high char production.

Compared with the processes based on hydrolysis, thermal degradation can be performed in the presence of low amounts of solvent (water) and the reactions proceed faster. However, the degradation takes place in a very unselective manner yielding a large variety of products.

*Pyrolysis*    In conventional pyrolysis processes the wood material is slowly heated up to the maximum temperature. The most important product is the solid char, but various amounts of gases and tars are also formed. Besides pyrolysis, a number of other terms such as carbonization, wood distillation, destructive distillation, and dry distillation are interchangeably used for this type of thermal processing. During the nineteenth century "wood distillation" was locally practiced to produce various chemicals such as methanol, acetic acid, acetone, turpentine, phenols, and wood tar. These products were important commodities in many communities. Today petrochemicals have almost completely displaced them and wood pyrolysis is no longer economical. However, the techniques developed in more recent times including fast pyrolysis have created new interest in the possibilities of wood pyrolysis, and processes developed for municipal solid wastes may also be applied for wood.

In the traditional "wood distillation industry" hardwood was preferred for production of chemicals because it gave high yields of acetic acid, methanol, and acetone. In the thermal degradation of wood the volatile compo-

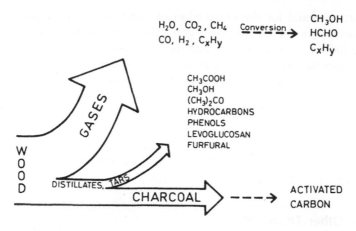

**Fig. 10-6.** Main product fractions obtained on wood pyrolysis.

nents are distillable and can be recovered as liquids after condensation (Fig. 10-6). The solid residue, *charcoal,* is mainly composed of carbon. At higher temperatures the carbon content is increased because of more complete dehydration and removal of volatile degradation products. Charcoal is mainly used as combustible material for special purposes. A number of charcoal products are known, including activated carbon for adsorption purposes.

The settled tars can be fractionated into light and heavy oil fractions. The former consists of aldehydes, ketones, acids, and esters, whereas various phenols, including a high proportion of cresols and pitch, are present in the heavy oil fraction.

The noncondensable gases formed in the distillation of hardwoods usually contain carbon dioxide, carbon monoxide, hydrogen, methane, and other hydrocarbons as the main constituents.

Destructive distillation of softwoods gives lower yields of acetic acid and methanol than hardwood distillation. Longleaf and slash pine, especially in the form of recovered stumps, were used because of their high resin content. Turpentine and pine oil were the principal products. However, like hardwood distillation, the old softwood distillation process is no longer practiced because high-quality tall oil and turpentine can be produced more economically as by-products of the kraft pulp industry (cf. Section 10.2.2).

Even if rather high yields of liquid products can be obtained under optimized conditions, pyrolysis technique seems not very attractive for production of pure chemicals because of the complexity of the reaction mixtures.

**Gasification** Gasification of wood means its thermal degradation by means of partial combustion. Although gasification has much broader application potential for the conversion of available coal supplies to more convenient fuels, the technology is generally applicable to biomass gasification, including wood waste. In steam-oxygen gasification the wood material is converted to synthesis gas, which besides water is composed mainly of hydrogen, carbon monoxide, carbon dioxide, and methane. This gas can be used either directly as fuel or as raw material for production of chemicals, especially methanol and ammonia. Another possibility is to convert this gas catalytically, for example, to aliphatic hydrocarbons according to the Fischer–Tropsch process. Gasification technology, especially gasification under pressure, is currently under extensive development to maximize the yields of both liquid and gaseous products.

### 10.1.4 Other Treatments

Organic matter is decomposed in the presence of certain microorganisms to a methane-rich gas and carbon dioxide. This type of anaerobic degradation is commercially applied for production of useful biogas fuel from agricultural, animal, and urban wastes. Wood wastes represent carbon sources which can be converted in a similar way into biogas, although a special technique involving pretreatments is necessary.

Another type of degradation of the polysaccharides takes place when wood is treated with alkali at elevated temperatures. In this case the carbohydrate material is degraded to aliphatic carboxylic acids. Although much work has been done to study the possibilities of producing useful chemicals from wood, cellulose, and other carbohydrate materials by alkali treatment, kraft pulping liquors undoubtedly represent the most potential sources from which this type of degradation products can be isolated (see further details in Section 10.2.2).

## 10.2  Chemicals from Pulping Liquors

Because of extensive energy-saving measures during the last two decades, practically all the process steam and electrical power needed for production of chemical sulfite or kraft pulp can be generated from the organic solids dissolved in pulping (spent) liquors. According to modern technology even considerable amounts of excess energy can be produced, which is useful for making bleaching chemicals or can be delivered to·an integrated paper mill or to external consumers.

The dissolved organic solids thus represent a considerable fuel value and the combustion is an integral part of the recovery process of inorganic chemicals. It can therefore be understood why only rather few industrial applications exist for producing chemicals from the pulping liquors. However, other factors contribute that can make at least a partial recovery of the organic solids attractive or even necessary. So is the situation in the sulfite pulp industry, when evaporation and combustion of the spent liquor often cause severe air pollution. Also, the situation can be changed in the long run depending on the research in this interesting area, which indeed offers a great challenge for new innovations with respect to the future.

### 10.2.1  Sulfite Spent Liquors

The majority of the organic material in sulfite spent liquors originates from lignin (lignosulfonates) and hemicelluloses (Tables 7-6 and 10-1). The value of the lignosulfonates as fuel is rather high, roughly half of that of oil, whereas the heat value of the carbohydrate fraction is much lower, roughly only one-third of that of oil.

Various useful products are made industrially both from lignosulfonates and carbohydrates (Table 10-2). Traditionally, the fermentation methods play a dominating role in the industrial processing of the carbohydrate fraction. In most cases the isolation of single components from these liquors is not

**TABLE 10-1.**  Main Substance Groups In Sulfite Spent Liquors[a]

| Component | Spruce | Birch |
|---|---|---|
| Lignosulfonates[b] | 480 (540) | 370 |
| M > 5000 | 245 (280) | 55 |
| Carbohydrates[c] | 280 | 375 |
| Arabinose | 10 | 10 |
| Xylose | 60 | 340 |
| Mannose | 120 | 10 |
| Galactose | 50 | 10 |
| Glucose | 40 | 5 |
| Aldonic acids | 50 | 95 |
| Acetic acid | 40 | 100 |
| Extractives | 40 | 40 |
| Other compounds | 40 | 60 |

[a] Approximate values given in kilograms per ton of pulp.
[b] Calculated as lignin. The estimated values for lignosulfonates are given in parentheses (calculated by assuming the degree of sulfonation to be 2.5 meq/g lignin).
[c] 80–85% of carbohydrates comprises monosaccharides, the rest is oligo- and polysaccharides.

**TABLE 10-2. Main By-Products Based on the Sulfite Spent Liquor and Condensates**

| Product | Quantity (kg/ton of pulp) | Separation method | Utilization |
|---|---|---|---|
| Lignosulfonates | 400–550 | Evaporation, precipitation, ultrafiltration, electrodialysis, ion exclusion | Additive (oil well drilling muds, Portland cement concrete), dispersing agent and binder (textiles, products of printing industry, mineral slurries), raw material for production of vanillin, etc. |
| Sugars and aldonic acids | 300–450 | Evaporation, ultrafiltration, ion exchange, electrodialysis, ion exclusion | Food and chemical industries |
| Ethanol | 40–60 | Fermentation, distillation | Food and chemical industries (solvent, raw material) |
| Protein | 90–110 | Fermentation | Food and fodder industries |
| Acetic acid | 30–80 | Extraction, distillation | Food and chemical industries (solvent, raw material) |
| Furfural | 10–15 | Adsorption, distillation | Plastic and lacquer industries |
| Butanol, acetone, lactic acid | 30–40 | Fermentation, extraction, distillation | Plastic and lacquer industries (solvent, raw material) |
| Cymene | 0.3–1 | Distillation | Plastic and lacquer industries (solvent) |

possible without using tedious and complex separation methods. Therefore, the value of the final products must be sufficiently high to compensate for the costs caused by separation.

*Cymene*   During acid sulfite pulping α-pinene and some monocyclic monoterpenes are partially converted to *p*-cymene (Fig. 7-18). Crude cymene can be separated from the digester gas relief condensates and purified by distillation. In the case of spruce and fir the distilled product is 99% *p*-cymene and only traces of other products, such as sesquiterpenoids, are present. Pine wood extractives contain 3-carene, which affords appreciable quantities of *m*-cymene besides *p*-cymene. The crude cymene can be used within the mill as a resin-cleaning solvent. The distilled product finds use in the paint and varnish industry.

*Lignosulfonates*   Lignosulfonates can be isolated from spent sulfite liquors in more or less pure forms by different methods. Simply when adding calcium hydroxide (lime) to the spent liquor, lignosulfonates are precipitated as calcium salts. Quaternary ammonium salts can also be used as precipitation agents. In more recent years membrane processes and ion exclusion technique have been applied for isolation and purification of lignosulfonates. Particularly ultrafiltration has found to be useful for further fractionation of the lignosulfonates according to their molecular weight. After isolation and purification the solution is concentrated by evaporation. Usually the lignosulfonates are marketed in a powder form after spray-drying.

Lignosulfonates are useful for a number of applications especially because of their adhesion and dispersion properties. For example, when added to concrete, lignosulfonates are adsorbed on the mineral surface and less water is needed to provide the fluidity and plasticity necessary for handling. This results in a less permeable and stronger concrete. Similarly, when lignosulfonates are added to mineral slurries and drilling muds, the viscosity is reduced. Lignosulfonates are also used in oil drilling muds and as binders in animal food pellets. Concentrated sulfite spent liquor also finds widespread use as a dust binder for gravel roads. The use of polymeric lignosulfonates includes a number of other applications as well. One interesting possibility is the preparation of ion exchangers from lignosulfonates. Despite intensive studies, however, no commercially attractive products have been developed. The resulting products are so far unsatisfactory with respect to ion exchange capacity, insolubility, etc., and are therefore not competitive with synthetic ion exchange resins.

Another category of lignin-based chemicals is obtained when lignosulfonates are degraded to low-molecular-weight products. The most important commercial product is vanillin, which is obtained by alkaline oxidation of softwood lignosulfonates. In addition to vanillin, hardwood lignin yields syringaldehyde because of its content of syringyl groups. Hardwood lignin is therefore not a useful raw material for vanillin production.

*Carbohydrates and Aliphatic Acids*   The industrial use of the carbohydrates present in sulfite spent liquors is mainly limited to fermentation processes. The most common product is ethanol fermented from hexoses by yeast (*Saccharomyces cerevisiae*) and separated afterwards by distillation. Because some contaminants, such as sulfur dioxide, inhibit the growth of the yeast, they must be removed from the liquor prior to the fermentation. Among the other possible, but not common, fermentation products are acetone, *n*-butanol, and lactic acid, which can be produced by other microorganisms. For fermentation of ethanol to other products, see Section 10.1.2.

For overall utilization of both carbohydrates and aliphatic acids in the sulfite spent liquors, aerobic cultivation using yeast (*Candida utilis*) or other fungi (e.g., *Paecilomyces varioti*) can be applied. These organisms not only consume hexoses but also pentoses, aldonic acids, and acetic acid. The value of the respective product (Torula yeast or Pekilo protein) is rather low in relation to the production costs, but these fermentation processes effectively reduce the pollution load from the mill.

The spent cooking liquors after neutral sulfite high-yield pulping of hardwood contain appreciable amounts of acetic acid. A process has been developed in which the acetic acid is extracted with an organic solvent from the spent liquor after acidification. The formic acid present in small quantities can be removed by azeotropic distillation if a pure product is needed. However, the main reason for the acetic acid recovery is to avoid pollution problems caused by the discharge of the condensates to the receiving waters.

In principle, various monosaccharides and their degradation products, such as furfural, can be isolated from the sulfite spent liquors. However, so far such processes have been of very limited practical interest because a complex and expensive separation technique is needed.

### 10.2.2   Kraft Black Liquors

After kraft pulping of softwood, turpentine is recovered from the digester relief condensates. The remaining kraft black liquor principally consists of three different fractions, namely, lignin, carbohydrate degradation products, and resin and fatty acids (tall oil) (Table 10-3). In addition to turpentine, only the tall oil fraction has considerable by-product value. The lignin and carbohydrate degradation products are today practically utilized only as fuel for production of energy for the process needs. In the future, however, a partial

TABLE 10-3. Main Substance Groups in Black Liquors[a]

| Fraction | Pine | Birch |
|---|---|---|
| Lignin | 490 | 330 |
| Carbohydrate degradation products | | |
|    Hydroxy carboxylic acids | 320 | 230 |
|    Acetic acid | 50 | 120 |
|    Formic acid | 80 | 50 |
| Turpentine[b] | 10 | — |
| Resin and/or fatty acids | 50 | 40 |
| Miscellaneous products | 60 | 80 |

[a] Approximate values calculated in kilograms per ton of produced pulp.
[b] From digester relief condensates.

utilization of these degradation products as chemical feedstocks can be attractive. Also, in the long run, gasification and liquifaction techniques are considered interesting alternatives for the processing of black liquors (cf. Section 10.1.3).

*Sulfate Turpentine* The composition (cf. Table 10-4) and yield of crude turpentine varies considerably depending on the wood species, growth conditions of the tree, and storage time of the logs or chips. Typically, however, the yield of crude turpentine after kraft pulping of Scots pine in Finland and Sweden is around 10 kg/ton of pulp. The crude turpentine is purified in the distillation process by which impurities such as methyl mercaptan and dimethyl sulfide as well as higher terpenoids are removed. At least half of the distilled sulfate turpentine from Scots pine is pinene (mostly α-pinene), and the next dominating component is 3-carene. This component is absent in

TABLE 10-4. Relative Proportions (Weight %)
of the Main Components in Commercial Sulfate Turpentine[a]

| Component | Turpentine A | Turpentine B |
|---|---|---|
| α-Pinene | 50–80 | 60–70 |
| β-Pinene | 2–7 | 25–35 |
| 3-Carene | 10–30 | — |
| Other terpenes | 5–10 | 6–12 |

[a] Average values for typical commercial products from Scandinavia (Turpentine A) and from southeastern United States (Turpentine B); the latter values are according to Zinkel (1989).

the sulfate turpentine from Southern pine; instead, high amounts of β-pinene are present. Sulfate turpentines are further processed and used as described in Section 10.1.1.

*Tall Oil*\*    The tall soap is recovered from the softwood kraft black liquor as "skimmings" and the resin and fatty acids are liberated by adding sulfuric acid to yield crude tall oil (CTO). The composition and yield of the CTO depend largely on the raw material used for pulping. In Scandinavia where Scots pine is the main softwood in the kraft pulp industry, a normal yield of CTO in northern regions is at least 50 kg/ton of pulp, but considerably lower yields are obtained in the middle or southern regions. In the United States, southern pines also give rather high CTO yields (about 50 kg/ton of pulp), whereas the corresponding yields in the Douglas fir area of the Pacific coast are only about 30 kg/ton.

The predominating resin acids in commercial rosins are of the pimarane and abietane types, but lesser amounts of the labdane type can also be present in the U.S. rosins (for structures, see Section 5.1.2 and Fig. 5-8). The composition of commercial rosins can vary considerably, depending first of all on the wood species and other factors. Some examples are given in Table 10-5. Many of the resin acids are unstable, undergoing isomerization and oxidation reactions. For example, with the exception of gum rosins, no levopimaric acid is present in commercial rosins because it is easily isomerized to other resin acids under the conditions prevailing under kraft pulping and recovery processes. The principal reactions leading to losses of resin acids during distillation are decarboxylation, dehydration, and depolymerization. In addition, artifact resin acid isomers are formed.

Vacuum distillation for purification of CTO and fractionation of it into resin and fatty acids became a successful commercial process relatively recently, about 30 years ago. The neutral components present are carried over to the "light oil fraction" whereas the material of higher molecular weight including that polymerized during distillation remains as a separate "pitch residue" (Fig. 10-7). However, the neutral components interfere, resulting in decreased yields and lower product quality.

In the case of chip mixtures containing hardwoods, the amount of neutral substances is increased (Table 10-6). To avoid resulting difficulties in fractionation, extraction methods have been developed for removing the neutral components from the black liquor or tall soap prior to distillation. The quality of the tall oil is considerably improved after this extraction. From the fraction remaining after extraction, sitosterol and other neutral substances

---

\*The name tall oil originates from the Swedish word "tall" which means pine.

**TABLE 10-5.** Relative Proportions (Weight %) of the Main Resin Acids in Scots Pine Wood and Respective Tall Oil Products[a]

| Resin acid | Wood | Crude tall oil | Rosin |
|---|---|---|---|
| Pimaric | 8–10 | 9–11 | 4–5 |
| Sandaracopimaric | 2–3 | 2 | 1–2 |
| Isopimaric | 4–9 | 5–7 | 3–5 |
| Dehydroabietic | 14–21 | 13–16 | 21–22 |
| Levopimaric | 24–30 | <0.5 | <0.5 |
| Palustric | 15–24 | 8–13 | 10–12 |
| Abietic | 16–23 | 40–49 | 42–43 |
| Neoabietic | 11–12 | 10–12 | 4–5 |
| Other resin acids | <0.5 | <0.5 | 8–12 |

[a] Data from Holmbom (1978).

can be separated for possible uses or as raw materials for further processing (cf. Section 10.1.1).

*Fatty Acids*  Various commercial fatty acid products from tall oil (TOFA) are available varying in both purity and composition. Oleic (monoenoic) and linoleic (dienoic) acids are common in most TOFA products. Scandina-

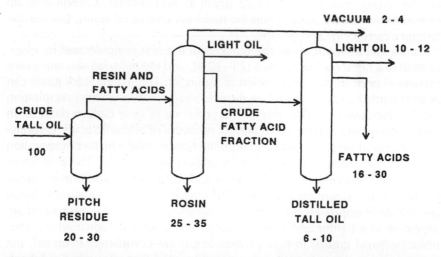

Fig. 10-7.  Principal scheme of tall oil distillation. The distribution of the product fractions indicated in the figure (% of crude tall oil) varies greatly depending on the composition of the crude tall oil. The data given for the light oil means the sum of these two fractions (data from Holmbom, personal communication). See also Zinkel (1989).

**TABLE 10-6.** Characteristics of Tall (Pine) Oil and the Corresponding Products Originating from Spruce and Birch Wood[a]

| Characteristics | Pine | Spruce | Birch |
|---|---|---|---|
| Acid number (mg KOH/g) | 160 | 140 | 100 |
| Unsaponifiables (%) | 7 | 10 | 30 |
| Resin acids (%) | 40 | 25 | 0 |
| Fatty acids (%) | 50 | 60 | 70 |

[a] From Kahila (1971).

vian TOFA products typically contain pinolenic (trienoic) acid not present in the U.S. TOFA. These products are applied for a great number of purposes as such or as intermediates. Some examples of these applications are coatings, polymers, adhesives, soaps, and lubricants.

*Kraft Lignin*    Most of the lignin degradation products or so-called "kraft lignin" can be precipitated from the black liquor by acids, but the low-molecular-weight lignin fragments remain in the solution. The precipitation yield depends on several factors, first of all on the final pH of the liquor. For technical applications it is advantageous to use carbon dioxide, but it is not possible to reach a value below pH ~ 8.5. When the liquor is acidified by adding strong mineral acid, more lignin is precipitated. Carbon dioxide liberates the weakly acidic phenolic hydroxyl groups of lignin, but not the stronger carboxylic acid groups.

For technical applications the black liquor is first concentrated by evaporation to a higher solids content (25–30%), and the tall soap skimmings are recovered prior to the precipitation of lignin (cf. Fig. 10-9). Stack gases can at least partly be used as a carbon dioxide source, although the precipitation yield is increased considerably in the presence of pure carbon dioxide. In order to precipitate more lignin and for recovery of aliphatic acids (carbohydrate degradation products) (see later), the liquor, after a further evaporation stage, can be made strongly acidic by adding sulfuric acid. The kraft lignin precipitates obtained after carbonation and sulfuric acid addition are separated from the liquor by filtration. The filterability of the gelatinous lignin precipitate is improved at elevated temperatures (60–80°C), because of aggregation to a tighter and a less hydrated form. After carbonation the phenolic hydroxyl groups in lignin (weak acids) are completely liberated, but the stronger carboxylic acid groups are liberated first after addition of sulfuric acid.

Kraft lignins find similar uses as lignosulfonates, but their recovery and purification are relatively expensive and at present only marginal quantities are produced. Kraft lignins or their modified forms can be used as dispersing agents and as additives in rubber, resins, and plastics. Condensation of kraft lignin with formaldehyde and cross-linking with phenols may yield thermosetting polymers useful as adhesives for different products such as paper laminates and plywood. The solubility of the precipitated lignin can be increased by sulfonation. The resulting products compete with lignosulfonates, but because of the high content of phenolic hydroxyl groups they are more useful in certain applications, including tannin agents. More definite and narrow fractions with respect to the molecular weight can be obtained by fractionating kraft or alkali lignins by ultrafiltration. Compared to unfractionated lignin, the fractionated product is superior in many applications, for example, for adhesives.

The lignin polymer can also be degraded into low-molecular-weight chemicals. Lignin is partly demethylated in the kraft process because of the nucleophilic action of the hydrogen sulfide ions resulting in volatile sulfur compounds, mainly methyl mercaptan and dimethyl sulfide. When concentrated kraft black liquors are heated at 200–250°C in the presence of additional sodium sulfide, a more extensive demethylation occurs, which affords considerable quantities of dimethyl sulfide. Dimethyl sulfide is produced in the United States on an industrial scale. It can be used for several purposes or oxidized by nitrogen tetroxide and oxygen to dimethyl sulfoxide, which is a useful product owing to its excellent solvent properties. On further oxidation dimethyl sulfoxide can be converted to dimethyl sulfone.

Numerous attempts have been made to produce other low-molecular-weight products directly from kraft black liquors by drastic heat treatments in the presence of sodium hydroxide and sodium sulfide. Such conditions result in the formation of low-molecular-weight phenols and aliphatic acids. Processes based on precipitation and extraction have been proposed for the recovery and fractionation of the reaction products. The phenol fraction can be converted to condensation polymers with certain reagents, such as formaldehyde. Other attempts include oxidation and hydrogenation of kraft lignins. With few exceptions, however, no commercial processes have emerged from all this research. One great difficulty is the separation of individual reaction products from the complex mixture of compounds resulting from such treatments.

*Carbohydrate Degradation Products*   Large quantities of hydroxy carboxylic acids as well as acetic and formic acids are formed from wood polysaccharides during kraft pulping (Table 10-7; see Fig. 10-8 for structures). Especially when considering that the average heat value of the fraction of aliphatic acids is only 50% or less of that of lignin, at least their partial recovery for nonfuel purposes seems attractive.

**TABLE 10-7. Main Hydroxy Carboxylic Acids in Kraft Black Liquors**[a]

| Acid[b] | Pine | Birch |
|---|---|---|
| Glycolic | 30 | 25 |
| Lactic | 50 | 25 |
| 3,4-Dideoxypentonic | 30 | 10 |
| Glucoisosaccharinic | 110 | 35 |
| 2-Hydroxybutanoic | 15 | 60 |
| 3-Deoxypentonic | 15 | 10 |
| Xyloisosaccharinic | 15 | 25 |
| Others | 55 | 40 |
| Total | 320 | 230 |

[a] Approximate values calculated in kilograms per ton of produced pulp.
[b] For structures, see Fig. 10-8.

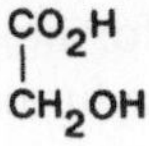

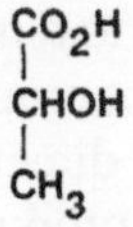

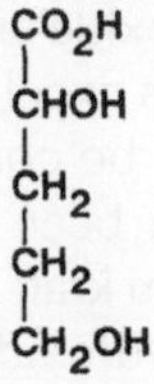

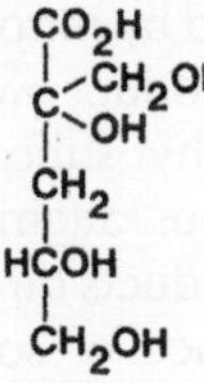

Fig. 10-8. Main hydroxy carboxylic acids in kraft black liquors: glycolic (1), lactic (2), 3,4-dideoxypentonic (3), glucoisosaccharinic (4), 2-hydroxybutanoic (5), 3-deoxypentonic (6), and xyloisosaccharinic (7) acids. Compounds 1–4 are more typical in softwood liquors, whereas compounds 5–7 are typical in hardwood liquors (predominantly degradation products of xylans).

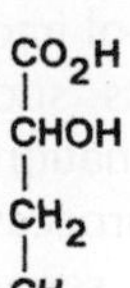

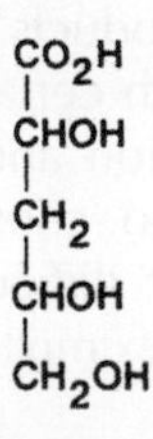

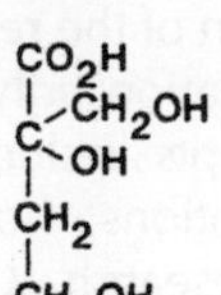

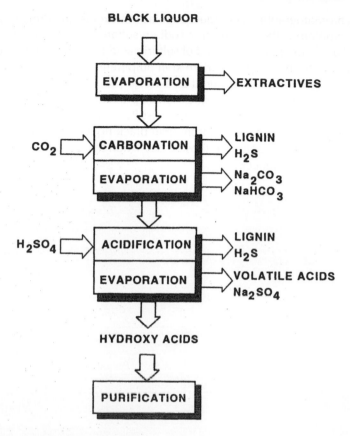

**BLACK LIQUOR**

EVAPORATION → **EXTRACTIVES**

$CO_2$ → CARBONATION → **LIGNIN** **$H_2S$**

EVAPORATION → **$Na_2CO_3$** **$NaHCO_3$**

$H_2SO_4$ → ACIDIFICATION → **LIGNIN** **$H_2S$**

EVAPORATION → **VOLATILE ACIDS** **$Na_2SO_4$**

**HYDROXY ACIDS**

**PURIFICATION**

**Fig. 10-9.** Principal scheme for recovery of lignin and carboxylic acids from kraft black liquors. From Alén and Sjöström (1991).

Based on laboratory experiments a process for the recovery of aliphatic acids from kraft black liquors has been outlined. As already described (cf. "kraft lignin" and Fig. 10-9), lignin can be precipitated from kraft black liquors with carbon dioxide and sulfuric acid. After recovery of the volatile acids (acetic and formic acid) from the evaporation condensates, a crude fraction composed mainly of hydroxy carboxylic acids is obtained. Of these, glycolic and lactic acid are commercial chemicals. Potential markets for the other hydroxy carboxylic acids could possibly be found in applications needing sequestering agents or they could be used as starting materials for production of a variety of specialty chemicals and polymers. However, the separation and purification of the crude hydroxy acid fraction is a difficult problem. Possible methods include distillation under reduced pressure or special chromatographic techniques such as ion exclusion chromatography. Another problem is the formation of sodium sulfate, even though the use of carbon dioxide decreases the need of sulfuric acid for neutralization considerably. The sulfur makeup in modern kraft pulp mills is fairly low, and excess sodium sulfate must be considered as a waste product of low value. However, a certain amount of black liquor can be processed for recovery of lignin and aliphatic acids without disturbing the sulfur and sodium balance in the mill. Recovery of these products from kraft black liquors also offers a way to increase the pulp production in those cases when the capacity of the recovery furnace is a bottleneck.

# APPENDIX

**Chemical Composition of Various Wood Species[a]**

| Species | Common name | Total extractives[b] | Lignin | Cellulose | Glucomannan[c] | Glucuronoxylan[d] | Other polysaccharides | Residual constituents |
|---|---|---|---|---|---|---|---|---|
| *Softwoods* | | | | | | | | |
| *Abies balsamea* | Balsam fir | 2.7 | 29.1 | 38.8 | 17.4 | 8.4 | 2.7 | 0.9 |
| *Pseudotsuga menziesii* | Douglas fir | 5.3 | 29.3 | 38.8 | 17.5 | 5.4 | 3.4 | 0.0 |
| *Tsuga canadensis* | Eastern hemlock | 3.4 | 30.5 | 37.7 | 18.5 | 6.5 | 2.9 | 0.5 |
| *Juniperus communis* | Common juniper | 3.2 | 32.1 | 33.0 | 16.4 | 10.7 | 3.2 | 1.4 |
| *Pinus radiata* | Monterey pine | 1.8 | 27.2 | 37.4 | 20.4 | 8.5 | 4.3 | 0.4 |
| *Pinus sylvestris* | Scots pine | 3.5 | 27.7 | 40.0 | 16.0 | 8.9 | 3.6 | 0.3 |
| *Picea abies* | Norway spruce | 1.7 | 27.4 | 41.7 | 16.3 | 8.6 | 3.4 | 0.9 |
| *Picea glauca* | White spruce | 2.1 | 27.5 | 39.5 | 17.2 | 10.4 | 3.0 | 0.3 |
| *Larix sibirica* | Siberian larch | 1.8 | 26.8 | 41.4 | 14.1 | 6.8 | 8.7 | 0.4 |
| *Hardwoods* | | | | | | | | |
| *Acer rubrum* | Red maple | 3.2 | 25.4 | 42.0 | 3.1 | 22.1 | 3.7 | 0.5 |
| *Acer saccharum* | Sugar maple | 2.5 | 25.2 | 40.7 | 3.7 | 23.6 | 3.5 | 0.8 |
| *Fagus sylvatica* | Common beech | 1.2 | 24.8 | 39.4 | 1.3 | 27.8 | 4.2 | 1.3 |
| *Betula verrucosa*[e] | Silver birch | 3.2 | 22.0 | 41.0 | 2.3 | 27.5 | 2.6 | 1.4 |
| *Betula papyrifera* | Paper birch | 2.6 | 21.4 | 39.4 | 1.4 | 29.7 | 3.4 | 2.1 |
| *Alnus incana* | Gray alder | 4.6 | 24.8 | 38.3 | 2.8 | 25.8 | 2.3 | 1.4 |
| *Eucalyptus camaldulensis* | River red gum | 2.8 | 31.3 | 45.0 | 3.1 | 14.1 | 2.0 | 1.7 |
| *Eucalyptus globulus* | Blue gum | 1.3 | 21.9 | 51.3 | 1.4 | 19.9 | 3.9 | 0.3 |
| *Gmelina arborea* | Yemane | 4.6 | 26.1 | 47.3 | 3.2 | 15.4 | 2.5 | 0.9 |
| *Acacia mollissima* | Black wattle | 1.8 | 20.8 | 42.9 | 2.6 | 28.2 | 2.8 | 0.9 |
| *Ochroma lagopus* | Balsa | 2.0 | 21.5 | 47.7 | 3.0 | 21.7 | 2.9 | 1.2 |

[a] J. Janson, P. Haglund, and E. Sjöström, unpublished data. All values are given as % of the dry wood weight.
[b] $CH_2Cl_2$ followed by $C_2H_5OH$.
[c] Including galactose and acetyl in softwood.
[d] Including arabinose in softwood and acetyl in hardwood.
[e] Also known as *Betula pendula*.

# BIBLIOGRAPHY

## General Books in Wood Chemistry and Closely Related Topics

Blažej, A., Sütý, L., Košík, M., Krkoška, P., and Golis, E. (1979). "Chemie des Holzes." VEB Fachbuchverlag, Leipzig.

Browning, B. L., ed. (1963). "The Chemistry of Wood." Wiley (Interscience), New York.

Browning, B. L., (1967). "Methods of Wood Chemistry," Vols. I and II. Wiley (Interscience), New York.

Fengel, D., and Wegener, G. (1984). "Wood—Chemistry, Ultrastructure, Reactions." Walter de Gruyter, Berlin.

Higuchi, T., ed. (1985). "Biosynthesis and Biodegradation of Wood Components." Academic Press, London.

Hon, D. N.-S., and Shiraishi, N., eds. (1991). "Wood and Cellulosic Chemistry." Marcel Dekker, New York.

Hägglund, E. (1951). "Chemistry of Wood." Academic Press, New York (Printed in Sweden, Esselte).

Lewin, M., and Goldstein, I. S., eds. (1991). "Wood Structure and Composition," Marcel Dekker, New York.

Rowell, R., ed. (1984). "The Chemistry of Solid Wood." Advances in Chemistry Series (M. J. Comstock, ed.), No. 207. Am. Chem. Soc., Washington, D.C.

Rydholm, S. A. (1965). "Pulping Processes." Wiley (Interscience), New York.

Timell, T. E. (1986). "Compression Wood in Gymnosperms," Vols. 1–3. Springer, Berlin.

Wise, L. E., and Jahn, E. C., eds. (1952). "Wood Chemistry," 2nd ed., Vols. I and II. ACS Monograph Series, No. 97, Reinhold Publishing, New York.

# Literature by Chapters

## Chapter 1.   Wood Structure

### Books

Butterfield, B. G., and Meylan, B. A. (1980). "Three-Dimensional Structure of Wood. An Ultrastructural Approach," 2nd ed. Chapman and Hall, London.

Carlquist, S. (1988). "Comparative Wood Anatomy. Systematic, Ecological, and Evolutionary Aspects of Dicotyledon Wood." Springer, Berlin.

Core, H. A., Côté, W. A., Jr., and Day, A. C. (1979). "Wood: Structure and Identification," 2nd ed. Syracuse Univ. Press, Syracuse, N.Y.

Côté, W. A., Jr. (1967). "Wood Ultrastructure." University of Washington Press, Syracuse, NY.

Koch, P. (1985). "Utilization of Hardwoods Growing on Southern Pine Sites, Vol. 1, The Raw Material." U.S. Dept. of Agriculture Forest Service, Agric. Handb. No. 605. U.S. Govt. Printing Office, Washington, D.C.

Hakkila, P. (1989). "Utilization of Residual Forest Biomass." Springer, Berlin.

Panshin, A. J., and de Zeeuw, C. (1980). "Textbook of Wood Technology, Vol. 1, Structure, Identification, Uses and Properties of the Commercial Woods of the United States and Canada," 4th ed. McGraw-Hill, New York.

Timell, T. E. (1986). "Compression Wood in Gymnosperms," Vols. 1–3. Springer, Berlin.

Tsoumis, G. (1968). "Wood as Raw Material." Pergamon, Oxford.

Wagenführ, R. (1980). "Anatomie des Holzes unter besonderer Berücksichtigung der Holztechnik," 2nd ed. VEB Fachbuchverlag, Leipzig.

### Reviews and Articles

Chattaway, M. M. (1949). The development of tyloses and secretion of gum in heartwood formation. *Aust. J. Sci. Res. Ser.* **B2**, 227–240.

Bucher, H. (1965). Das Geheinmnis des Holzes. *Hespa Mitt.* **15**(3), 1–24.

Harada, H., and Côté, W. A., Jr. (1985). Structure of wood. *In* "Biosynthesis and Biodegradation of Wood Components" (T. Higuchi, ed.), pp. 1–42. Academic Press, London.

Ilvessalo-Pfäffli, M.-S. (1967). The structure of wood. *In* "Wood Chemistry" (W. Jensen, ed.), 1st ed., pp. $B_1$1–$B_1$52. Text Handb. Finn. Paper Eng. Assoc., Helsinki. (In Finn.)

Ilvessalo-Pfäffli, M.-S. (1977). The structure of wood. *In* "Wood Chemistry" (W. Jensen, ed.), 2nd ed., pp. 7–81. Text Handb. Finn. Paper Eng. Assoc., Helsinki. (In Finn.)

Mark, H. (1940). Intermicellar hole and tube system in fiber structure. *J. Phys. Chem.* **44**, 764–787.

Meier, H. (1958). The fine structure of wood fibers. *Svensk Papperstidn.* **61**, 633–640. (in Swed.)

# Chapter 2.    Introduction to Carbohydrate Chemistry

## Books

Aspinall, G. O., ed. (1982). "The Polysaccharides," Vol. 1. Academic Press, New York.

Biermann, C. J., and McGinnis, G. D., eds. (1988). "Analysis of Carbohydrates by GLC and MS." CRC Press, Boca Raton, Fla.

Binkley, R. W. (1988). "Modern Carbohydrate Chemistry." Marcel Dekker, New York.

Dey, P. M., ed. (1990). "Methods in Plant Biochemistry" (P. M. Dey and J. B. Harborne, eds.), Vol. 2 , "Carbohydrates." Academic Press, San Diego.

El Khadem, H. S. (1988). "Carbohydrate Chemistry. Monosaccharides and Their Oligomers." Academic Press, San Diego.

Kennedy, J. F., ed. (1988). "Carbohydrate Chemistry." Clarendon Press, Oxford.

Pigman, W., and Horton, D., eds. (1972). "The Carbohydrates: Chemistry and Biochemistry," 2nd ed., Vol. IA. Academic Press, New York.

Pigman, W., and Horton, D., eds. (1980). "The Carbohydrates: Chemistry and Biochemistry," 2nd ed., Vol. IB. Academic Press, New York.

Pigman, W., and Horton, D., eds. (1970). "The Carbohydrates: Chemistry and Biochemistry," 2nd ed., Vols. IIA and IIB. Academic Press, New York.

Stoddart, J. F. (1971). "Stereochemistry of Carbohydrates." Wiley (Interscience), New York.

Whistler, R. L., BeMiller, J. N., and Wolfrom, M. L., eds. (1962–1980). "Methods in Carbohydrate Chemistry," Vols. I–VIII. Academic Press, New York.

## Reviews and Articles

Angyal, S. J. (1984). The composition of reducing sugars in solution. Adv. Carbohydr. Chem. Biochem. 42, 15–68.

Biermann, C. J. (1988). Hydrolysis and other cleavages of glycosidic linkages in polysaccharides. Adv. Carbohydr. Chem. Biochem. 46, 251–271.

Churms, S. C. (1990). Recent developments in the chromatographic analysis of carbohydrates. J. Chromatogr. 500, 555–583.

Dutton, G. G. S. (1973, 1974). Applications of gas-liquid chromatography to carbohydrates. Parts I–II. Adv. Carbohydr. Chem. Biochem. 28, 11–160; 30, 9–110.

Hicks, K. B. (1988). High-performance liquid chromatography of carbohydrates. Adv. Carbohydr. Chem. Biochem. 46, 17–72.

Isbell, H. S. (1973). Enolization and oxidation reactions of reducing sugars. Adv. Chem. Ser. 117, 70–87.

Lindberg, B., Lönngren, J., and Svensson, S. (1975). Specific degradation of polysaccharides. Adv. Carbohydr. Chem. Biochem. 31, 185–240.

Robards, K., and Whitelaw, M. (1986). Chromatography of monosaccharides and disaccharides. J. Chromatogr. 373, 81–110.

Theander, O. (1980). Acids and other oxidation products. In "The Carbohydrates. Chemistry and Biochemistry" (W. Pigman and D. Horton, eds.), Vol. IB, pp. 1013–1099. Academic Press, New York.

Vuorinen, T. (1988). Mechanisms and kinetics of isomerization, degradation, and oxidation of reducing carbohydrates; reaction paths in alkaline solutions containing oxygen and 2-anthraquinonesulfonic acid. Diss., Helsinki Univ. of Technology. Ann. Acad. Sci. Fenn. Ser. A, II, No. 220.

# Chapter 3. Wood Polysaccharides

## Books

Atalla, R. H., ed. (1987). "The Structures of Cellulose. Characterization of the Solid States." ACS Symposium Series, No. 340. Am. Chem. Soc., Washington, D.C.

Bikales, N. M., and Segal, L., eds. (1971). "Cellulose and Cellulose Derivatives," Parts IV–V. Wiley (Interscience), New York.

Karlson, P., Doenecke, D., Fuchs, G., Koolman, J., and Schäfer, G. (1988). "Kurzes Lehrbuch der Biochemie für Mediziner und Naturwissenschaftler," 13th ed., Georg Thieme Verlag, Stuttgart.

Miller, L. P., ed. (1973). "Phytochemistry—The Process and Products of Photosynthesis," Vol. I. Van Nostrand Reinhold, New York.

Nevell, T. P., and Zeronian, S. H., eds. (1985). "Cellulose Chemistry and Its Applications." Ellis Horwood, Chichester.

Rawn, J. D. (1989). "Biochemistry." Neil Patterson Publ., Burlington, North Carolina.

Timell, T. E. (1986). "Compression Wood in Gymnosperms," Vol. 1, Springer, Berlin.

Young, R. A., and Rowell R. M., eds. (1986). "Cellulose. Structure, Modification and Hydrolysis." Wiley (Interscience), New York.

## Reviews and Articles

Andersson, S-I., Samuelson, O., Ishihara, M., and Shimizu, K. (1983). Structure of the reducing end-groups in spruce xylan. Carbohydr. Res. 111, 283–288.

Aspinall, G. O. (1970). Pectins, plant gums and other plant polysaccharides. In "The Carbohydrates: Chemistry and Biochemistry" (W. Pigman and D. Horton, eds.), 2nd ed., Vol. 2B, pp. 515–536. Academic Press, New York.

Billmeyer, F. W., Jr. (1965). Characterization of molecular weight distributions in high polymers. J. Polym. Sci. C. 8, 161–178.

Blackwell, J., Kolpak, F. J., and Gardner, K. H. (1977). Structures of native and regenerated celluloses. In "Cellulose Chemistry and Technology" (J. C. Arthur, Jr., ed.), ACS Symposium Series, No. 48, pp. 42–55. Am. Chem. Soc., Washington, D.C.

Brown, W. J. (1966). The configuration of cellulose and derivatives in solution. Tappi 49(8), 367–373.

Delmer, D. P. (1987). Cellulose Biosynthesis. Annu. Rev. Plant Physiol. 38, 259–290.

Delmer, D. P. (1988). Biosynthesis of plant cell walls. In "The Biochemistry of Plants," Vol. 14, pp. 373–420. Academic Press, San Diego.

Dey, P. M., and Brinson, K. (1984). Plant cell-walls. Adv. Carbohydr. Chem. Biochem. 42, 265–382.

Gardner, K. H., and Blackwell, J. (1974). The hydrogen bonding in native cellulose. Biochim. Biophys. Acta 343, 232–237.

Goring, D. A. I. (1962). The physical chemistry of lignin. Pure Appl. Chem. 5, 233–254.

Goring, D. A. I. (1971). Polymer properties of lignin and its derivatives. In "Lignins" (K. V. Sarkanen and C. H. Ludwig, eds.), pp. 695–768. Wiley (Interscience), New York.

Hassid, W. Z. (1970). Biosynthesis of sugars and polysaccharides. In "The Carbohydrates:

Chemistry and Biochemistry" (W. Pigman and D. Horton, eds.), 2nd ed., Vol. 2A, pp. 301–373. Academic Press, New York.

Johansson, M. H., and Samuelson, O. (1977). Reducing end groups in birch xylan and their alkaline degradation. Wood Sci. Technol. 11, 251–263.

Kolpak, F. J., and Blackwell, J. (1976). Determination of the structure of cellulose II. Macromolecules 9, 273–278.

Kolpak, F. J., Weih, M., and Blackwell, J. (1978). Mercerization of cellulose: 1. Determination of the structure of mercerized cotton. Polymer 19, 123–131.

Kuo, C-M., and Timell, T. E. (1969). Isolation and characterization of a galactan from tension wood of American beech (Fagus grandifolia Ehrl.). Svensk Papperstidn. 72, 703–716.

Leloir, L. F. (1964). The biosynthesis of polysaccharides. Proc. Plenary Sess. 6th Int. Congr. Biochem., New York, pp. 15–29.

Lindberg, B., Rosell, K.-G., and Svensson, S. (1973). Positions of the O-acetyl groups in birch xylan. Svensk Papperstidn. 76, 30–32.

Lindberg, B., Rosell, K.-G., and Svensson, S. (1973). Positions of the O-acetyl groups in pine glucomannan. Svensk Papperstidn. 76, 383–384.

Meier, H. (1962). Studies on a galactan from tension wood of beech (Fagus silvatica L.). Acta Chem. Scand. 16, 2275–2283.

Meier, H. (1985). Localization of polysaccharides in wood cell walls. In "Biosynthesis and Biodegradation of Wood Polysaccharides" (T. Higuchi, ed.), pp. 43–50. Academic Press, London.

Meier, H., and Wilkie, K. C. B. (1959). The distribution of polysaccharides in the cell-wall of tracheids of pine (Pinus silvestris L.). Holzforschung 13, 177–182.

Nikaido, H., and Hassid, W. Z. (1971). Biosynthesis of saccharides from glucopyranosyl esters of nucleoside pyrophosphates ("sugar nucleotides"). Adv. Carbohydr. Chem. Biochem. 26, 351–483.

O'Neill, M., Albersheim, P., and Darvill, A. (1990). The pectic polysaccharides of primary cell walls. In "Methods in Plant Biochemistry" (R. M. Dey and J. B. Harborne, eds.), Vol. 2, Carbohydrates pp. 415–441. Academic Press London.

Overend, W. G. (1972). Glycosides. In "The Carbohydrates: Chemistry and Biochemistry" (W. Pigman and D. Horton, eds.), 2nd ed., Vol. 1A, pp. 279–353. Academic Press, New York.

Rosell, K-G., and Svensson, S. (1975). Studies on the distribution of the 4-O-methyl-D-glucuronic acid residues in birch xylan. Carbohydr. Res. 42, 297–304.

Shimizu, K., and Samuelson, O. (1973). Uronic acids in birch hemicellulose. Svensk Papperstidn. 76, 150–155.

Sumi, Y., Hale, R. D., and Rånby, B. G. (1963). The accessibility of native cellulose microfibrils. Tappi 46(2), 126–130.

Sumi, Y., Hale, R. D., Meyer, J. A., Leopold, B., and Rånby, B. G. (1964). Accessibility of wood and wood carbohydrates measured with tritiated water. Tappi 47(10), 621–624.

Timell, T. E. (1964, 1965). Wood hemicelluloses. Parts I and II. Adv. Carbohydr. Chem. Biochem. 19, 247–302; 20, 409–483.

Timell, T. E. (1965). Wood and bark polysaccharides. In "Cellular Ultrastructure of Woody Plants" (W. A. Côté, Jr., ed.), pp. 127–156. Syracuse Univ. Press, Syracuse, N.Y.

Timell, T. E. (1967). Recent progress in the chemistry of wood hemicelluloses. Wood Sci. Technol. 1, 45–70.

Westermark, U., Hardell, H.-L., and Iversen, T. (1986). The content of protein and pectin in the lignified middle lamella/primary wall of spruce fibers. Holzforschung 40, 65–68.

Whistler, R. L., and Chen, C.-C. (1991). Hemicelluloses. In "Wood Structure and Composition" (M. Lewin and I. S. Goldstein, eds.), pp. 287–319. Marcel Dekker, New York.

# Chapter 4.   Lignin

## Books

Freudenberg, K., and Neish, A. C. (1968). "Constitution and Biosynthesis of Lignin." Springer, Berlin-Heidelberg.

Glasser, W. G., and Sarkanen, S., eds. (1989). "Lignin—Properties and Materials." ACS Symposium Series, No. 397. Am. Chem. Soc., Washington, D.C.

Pearl, I. A. (1967). "The Chemistry of Lignin." Marcel Dekker, New York.

Sarkanen, K. V., and Ludwig, C. H., eds. (1971). "Lignins. Occurrence, Formation, Structure and Reactions," Wiley (Interscience), New York.

## Reviews and Articles

Adler, E. (1977). Lignin chemistry—past, present and future. *Wood Sci. Technol.* **11**, 169–218.

Brunow, G., Ede, R. M., Simola, L. K., and Lemmetyinen, J. (1990). Lignins released from *Picea abies* suspension cultures—True native lignins? *Phytochemistry* **29**(8), 2535–2538.

Chen, C.-L., and Robert, D. (1988). Characterization of lignin by $^1$H and $^{13}$C NMR spectroscopy. *Methods Enzymol.* **161**, 137–174.

Chen, C.-L. (1991). Lignins: Occurrence in woody tissues, isolation, reactions, and structure. *In* "Wood Structure and Composition" (M. Lewin and I. S. Goldstein, eds.), pp. 183–261. Marcel Dekker, New York.

Ede, R. M., and Brunow, G. (1990). Two-dimensional $^1$H-$^1$H chemical shift correlation and *J*-resolved NMR studies on isolated and synthetic lignins. *Holzforschung* **44**, 95–101.

Eriksson, I., Lidbrandt, O., and Westermark, U. (1988). Lignin distribution in birch (*Betula verrucosa*) as determined by mercurization with SEM- and TEM-EDXA. *Wood Sci. Technol.* **22**, 251–257.

Eriksson, Ö., and Lindgren, B. O. (1977). About the linkage between lignin and hemicelluloses in wood. *Svensk Papperstidn.* **80**, 59–63.

Faix, O. (1991). Classification of lignins from different botanical origins by FT-IR spectroscopy. *Holzforschung* **45**(Suppl.), 21–27.

Fergus, B. J., and Goring, D. A. I. (1970). The distribution of lignin in birch wood as determined by ultraviolet microscopy. *Holzforschung* **24**, 118–124.

Fergus, B. J., Procter, A. R., Scott, J. A. N., and Goring, D. A. I. (1969). The distribution of lignin in sprucewood as determined by ultraviolet microscopy. *Wood Sci. Technol.* **3**, 117–138.

Fukushima, K., and Terashima, N. (1990). Heterogeneity in formation of lignin. XIII. Formation of *p*-hydroxyphenyl lignin in various hardwoods visualized by microautoradiography. *J. Wood Chem. Technol.* **10**, 413–433.

Gellerstedt, G., and Northey, R. A. (1989). Analysis of birch wood lignin by oxidative degradation. *Wood Sci. Technol.* **23**, 75–83.

Goring, D. A. I. (1962). The physical chemistry of lignin. *Pure Appl. Chem.* **5**, 233–254.

Goring, D. A. I. (1971). Polymer properties of lignin and lignin derivatives. *In* "Lignins" (K. V. Sarkanen and C. H. Ludwig, eds.), pp. 695–768. Wiley (Interscience), New York.

Higuchi, T., Shimada, M., Nakatsubo, F., and Tanahashi, M. (1977). Differences in biosynthesis of guaiacyl and syringyl lignins in wood. *Wood Sci. Technol.* **11**, 153–167.

Higuchi, T. (1990). Lignin biochemistry. Biosynthesis and biodegradation. *Wood Sci. Technol.* **24**, 23–63.

Higuchi, T. (1989). Biodegradation of lignin and its potential applications. *In* "Bioprocess Engineering" (T. K. Ghose, ed.), pp. 39–58. Ellis Horwood, Chichester.

Iversen, T. (1985). Lignin-carbohydrate bonds in a lignin-carbohydrate complex isolated from spruce. *Wood Sci. Technol.* **19**, 243–251.

Kirk, T. K., and Obst, J. R. (1988). Lignin determination. *Methods Enzymol.* **161**, 87–101.

Kirk, T. K., and Shimada, M. (1985). Lignin biodegradation: The microorganisms involved and the physiology and biochemistry of degradation by white-rot fungi. *In* "Biosynthesis and Biodegradation of Wood Components" (T. Higuchi, ed.), pp. 579–605. Academic Press, London.

Koshijima, T., Watanabe, T., and Yaku, F. (1989). Structure and properties of the lignin–carbohydrate complex polymer as an amphipathic substance. *In* "Lignin: Properties and Materials" (G. Glasser and S. Sarkanen, eds.), ACS Symposium Series, No. 397, pp. 11–28. Am. Chem. Soc., Washington, D.C.

Landucci, L. (1991). Application of modern liquid-state NMR to lignin characterization. *Holzforschung* **45**, 55–60.

Lai, Y.-Z., and Guo, X.-P. (1991). Variation of the phenolic hydroxyl group content in wood lignins. *Wood Sci. Technol.* **25**, 467–472.

Lapierre, C., Pollet, B., Monties, B., and Rolando, C. (1991). Thioacidolysis of spruce lignins: GC-MS analysis of the main dimers recovered after Raney nickel desulphuration. *Holzforschung* **45**, 61–68.

Lapierre, C., Pollet, B., and Monties, B. (1991). Lignin structural fingerprint by thioacidolysis followed with Raney nickel desulphuration. *Proc. 6th Int. Symp. Wood & Pulping Chem.*, Melbourne, Vol. 1, pp. 543–550.

Lundquist, K. (1974). Low-molecular weight lignin hydrolysis products. *J. Appl. Polym. Sci. Appl. Polym. Symp.* **28**, 1393–1407.

Lundquist, K. (1991). ¹H NMR spectral studies of lignin. *Proc. 6th Int. Symp. Wood & Pulping Chem.*, Melbourne, Vol. 1, pp 65–70.

Lundquist, K., Simonson, R., and Tingsvik, K. (1983). Lignin carbohydrate linkages in milled wood lignin preparations from spruce wood. *Svensk Papperstidn.* **86**, R44–R47.

Lundquist, K., Simonson, R., and Tingsvik, K. (1990). On the composition of dioxane-water extracts of milled spruce wood: Characterization of hydrophilic constituents. *Nord. Pulp Pap. Res. J.* **5**, 107–113.

Minor, J. L. (1991). Location of lignin-bonded pectic polysaccharides. *J. Wood Chem. Technol.* **11**, 159–169.

Monties, B. (1989). Lignins. *In* "Methods in Plant Biochemistry." (R. M. Day and J. B. Harborne, eds.), Vol. 1, pp. 113–157. Academic Press, San Diego.

Monties, B. (1989). Molecular structure and biochemical properties of lignins in relation to possible self-organization of lignin networks. *Ann. Sci. For.* **46**(suppl.), 848–855.

Nimz, H. (1974). Beech lignin—Proposal of a constitutional scheme. *Angew. Chem. Int. Ed. Engl.* **13**, 313–321.

Nimz, H. H., Tschirner, U., Stähle, M., Lehmann, R., and Schlosser, M. (1984). Carbon-13 NMR spectra of lignins, 10. Comparison of structural units in spruce and beech lignin. *J. Wood Chem. Technol.* **4**, 265–284.

Obiaga, T. I. (1972). Lignin molecular weight and molecular weight distribution during alkaline pulping of wood. Diss., Univ. of Toronto, Toronto.

Obst, J. R. (1990). Lignins: Structure and distribution in wood and pulp. *In* "Materials Interactions Relevant to the Pulp, Paper, and Wood Industries" (D. F. Caulfield, J. D. Passaretti, and S. F. Sobczynski, eds.), *Mater. Res. Soc. Symp. Proc.*, Vol. 197, pp. 11–20.

Obst, J. R., and Kirk, T. K. (1988). Isolation of lignin. *Methods Enzymol.* **161**, 3–12.

Obst, J. R., and Landucci, L. L. (1986). Quantitative ¹³C NMR of lignins—Methoxyl:aryl ratio. *Holzforschung* **40**(Suppl.), 87–92.

Obst, J. R., and Ralph, J. (1983). Characterization of hardwood lignin: Investigation of syringyl/guaiacyl composition by ¹³C nuclear magnetic resonance spectrometry. *Holzforschung* **37**, 297–302.

Saka, S., and Goring, D. A. I. (1985). Localization of lignins in wood cell walls. In "Biosynthesis and Biodegradation of Wood Components" (T. Higuchi, ed.), pp. 51–62. Academic Press, London.

Sakakibara, A. (1991). Chemistry of lignin. In "Wood and Cellulosic Chemistry" (D. N.-S. Hon and N. Shiraishi, eds.), pp. 113–175. Marcel Dekker, New York.

Sederoff, R. and Chang, H.-M. (1991). Lignin biosynthesis. In "Wood Structure and Composition" (M. Lewin and I. S. Goldstein, eds.), pp. 263–285. Marcel Dekker, New York.

Sipilä, J. (1990). On the reactions of quinone methide intermediates during lignin biosynthesis. A study with model compounds. Diss., University of Helsinki, Helsinki.

Sipilä, J., and Brunow, G. (1991). On the mechanism of formation of non-cyclic benzyl ethers during lignin biosynthesis. Part 4. The reactions of a β-O-4 type quinone methide with carboxylic acids in the presence of phenols. The formation and stability of benzyl esters between lignin and carbohydrates. *Holzforschung* **45**(Suppl.), 9–14.

Sorvari, J., Sjöström, E., Klemola, A., and Laine, J. E. (1986). Chemical characterization of wood constituents, especially lignin, in fractions separated from middle lamella and secondary wall of Norway spruce (*Picea abies*). *Wood Sci. Technol.* **20**, 35–51.

Terashima, N. (1989). An improved radiotracer method for studying formation and structure of lignin. In "Plant Cell Wall Polymers: Biogenesis and Biodegradation" (N. G. Lewis and M. G. Paice, eds.). *ACS Symp. Ser.* No. 399, pp. 148–159. Am. Chem. Soc., Washington, D. C.

Terashima, N. (1990). A new mechanism for formation of a structurally ordered protolignin macromolecule in the cell wall of tree xylem. *J. Pulp Pap. Sci.* **16**, J150–J155.

Terashima, N., and Fukushima, K. (1989). Biogenesis and structure of macromolecular lignin in the cell wall of tree xylem as studied by microautoradiography. In "Plant Cell Wall Polymers: Biogenesis and Biodegradation" (N. G. Lewis and M. G. Paice, eds.). ACS Symposium Series, No. 399, pp. 160–168. Am. Chem. Soc., Washington, D.C.

Watanabe, T. (1989). Isolation and binding-site analysis of lignin-carbohydrate complexes from *Pinus densiflora* wood. Diss., Kyoto University, Kyoto.

Westermark, U. (1985). The occurrence of *p*-hydroxyphenylpropane units in the middle-lamella lignin of spruce (*Picea abies*). *Wood Sci. Technol.* **19**, 223–232.

Westermark, U., Lidbrandt, O., and Eriksson, I. (1988). Lignin distribution in spruce (*Picea abies*) determined by mercurization with SEM- and TEM-EDXA. *Wood Sci. Technol.* **22**, 243–250.

# Chapter 5.  Extractives

## Books

Ayres, D. C., and Loike, J. D. (1990). "Lignans." Cambridge Univ. Press, Cambridge.

Harborne, J. B., ed. (1989). "Methods in Plant Biochemistry" (P. M. Dey and J. B. Harborne, eds.), Vol. 1. Academic Press, San Diego.

Hillis, W. E., ed. (1962). "Wood Extractives and Their Significance to the Pulp and Paper Industries." Academic Press, New York.

Hillis, W. E. (1987). "Heartwood and Tree Exudates." Springer, Berlin.

Miller, L. P., ed. (1973). "Phytochemistry," Vol. II, Organic Metabolites. Van Nostrand-Reinhold, New York.

Ralston, A. W. (1948). "Fatty Acids and Their Derivatives." Wiley (Interscience), New York.

Rao, C. B. S., ed. (1978). "Chemistry of Lignans." Andhra Univ. Series, No. 149. Andhra Univ. Press, India.

Rowe, J. W., ed. (1989). "Natural Products of Woody Plants: Chemicals Extraneous to the Lignocellulosic Cell Wall," Vols. I and II. Springer, Berlin.

Zinkel, D. F. and Russel, J., eds. (1989). "Naval Stores. Production, Chemistry, Utilization." Pulp Chemicals Association, New York.

## Reviews and Articles

Back, E. (1969). Wood-anatomical aspects on resin problems. Svensk Pappperstidn. 72, 109–21. (In Swed.)

Clayton, R. B. (1970). The chemistry of nonhormonal interactions: Terpenoid compounds in ecology. In "Chemical Ecology" (E. Sondheimer and J. B. Simeone, eds.), pp. 235–280. Academic Press, New York.

Croteau, R., and Johnson, M. A. (1985). Biosynthesis of terpenoid wood extractives. In "Biosynthesis and Biodegradation of Wood Components," (T. Higuchi, ed.), pp. 379–439. Academic Press, London.

Dev, S. (1989). Terpenoids. In "Natural Products of Woody Plants" (J. W. Rowe, ed.), Vol. II, pp. 691–807. Springer, Berlin.

Ekman, R. (1976). Analysis of lignans in Norway spruce by combined gas chromatography-mass spectrometry. Holzforschung 30, 79–85.

Ekman, R., Peltonen, C., Hirvonen, P., Pensar, G., and von Weissenberg, K. (1979). Distribution and seasonal variation of extractives in Norway spruce. Acta Acad. Åboens. (Ser. B) 39(8), 1–26.

Ekman, R. (1980). Wood extractives of Norway spruce. A study of nonvolatile constituents and their effects on Fomes annosus. Diss., Åbo Akademi, Åbo.

Erdtman, H. (1973). Molecular taxonomy. In "Phytochemistry" (L. P. Miller, ed.), Vol. 3, pp. 327–350. Van Nostrand-Reinhold, New York.

Gottlieb, O. R. (1989). Evolution of natural products. In "Natural Products of Woody Plants" (J. W. Rowe, ed.), Vol. 1, pp. 125–153. Springer, Berlin.

Hafizoğlu, H. (1983). Wood extractives of Pinus sylvestris L., Pinus nigra Arn. and Pinus brutia Ten. with special reference to nonpolar components. Holzforschung 37, 321–326.

Hillis, W. E. (1972). Formation and properties of some wood extractives. Phytochemistry 11, 1207–1218.

Holmbom, B. (1978). Constituents of tall oil. A study of tall oil processes and products. Diss., Åbo Akademi, Åbo.

Kimland, B., and Norin, T. (1972). Wood extractives of common spruce, Picea abies (L.) Karst. Svensk Papperstidn. 75, 403–409.

Landucci, L. L., and Zinkel, D. F. (1991). The $^1$H and $^{13}$C NMR spectra of the abietadienoic resin acids. Holzforschung 45, 341–346.

Lindgren, B., and Norin, T. (1969). The chemistry of extractives. Svensk Pappperstidn. 72, 143–153. (In Swed.)

Nes, W. R. (1989). Steroids. In "Natural Products of Woody Plants" (J. W. Rowe, ed.), Vol. II, pp. 808–842. Springer, Berlin.

Odashi, H., and Imai, T. (1990). Characterization of physiological functions of sapwood: Synthesis and accumulation of heartwood extractives in the withering process of immature Japanese cedar trunk. *Holzforschung* **44**, 317–323.

Pardos, J. A., Lange, W., and Weissmann, G. (1990). Morphological and chemical aspects of *Pinus sylvestris* L. from Spain. *Holzforschung* **44**, 143–146.

Piretti, M. V., and Doghieri, P. (1990). Separation of peracetylated flavanoid and flavonoid polyphenols by normal-phase high-performance liquid chromatography on a cyano-silica column and their determination. *J. Chromatogr.* **514**, 334–342.

Saranpää, P., and Höll, W. (1989). Soluble carbohydrates of *Pinus sylvestris* L. sapwood and heartwood. *Trees* **3**, 138–143.

Saranpää, P., and Nyberg, H. (1987). Lipids and sterols of *Pinus sylvestris* L. sapwood and heartwood. *Trees* **1**, 82–87.

Suckling, I. D., Gallagher, S. S., and Ede, R. M. (1990). A new method for softwood extractives analysis using high performance liquid chromatography. *Holzforschung* **44**, 339–345.

Zavarin, E., and Cool, L. (1991). Extraneous materials from wood. *In* "Wood Structure and Composition" (M. Lewin and I. S. Goldstein, eds.), pp. 321–407. Marcel Dekker, New York.

Zavarin, E., and Snajberk, K. (1980). Oleoresins of pinyons. *J. Agric. Food Chem.* **28**, 829–834.

# Chapter 6.   Bark

## Reviews and Articles

Cole, B. J. W., Bentley, M. D., and Hua, Y. (1991). Triterpenoid extractives in the outer bark of *Betula lenta* (black birch). *Holzforschung* **45**, 265–268.

Ekman, R. (1983). The suberin monomers and triterpenoids from the outer bark of *Betula verrucosa* Ehrh. *Holzforschung* **37**, 205–211.

Fechtal, M., and Riedl, B. (1991). Analyse des extraits tannants des écorces des Eucalyptus après hydrolyse acide par la chromatographie en phase gazeuse couplée avec la spectrométrie de masse (GC-MS). *Holzforschung* **45**, 269–273.

Hemingway, R. W. (1981). Bark: Its chemistry and prospects for chemical utilization. *In* "Organic Chemicals from Biomass" (I. S. Goldstein, ed.), pp. 189–248. CRC Press, Boca Raton, Fla.

Kolattukudy, P. E., and Espelie, R. E. (1985). Biosynthesis of cutin, suberin, and associated waxes. *In* "Biosynthesis and Biodegradation of Wood Components" (T. Higuchi., ed.), pp. 161–207. Academic Press, London.

Laks, P. E. (1991). Chemistry of Bark. *In* "Wood and Cellulosic Chemistry" (D. N.-S. Hon and N. Shiraishi, eds.), pp. 257–330. Marcel Dekker, New York.

Laver, M. L. (1991). Bark. *In* "Wood Structure and Composition" (M. Lewin and I. S. Goldstein, eds.), pp. 409–434. Marcel Dekker, New York.

Lorbeer, E., and Zelman, N. (1988). Investigation of the distribution of the non-volatile lipophilic part of rosin in spruce (*Picea abies*). Part 1. Chemical composition of the rosin using samples of needles, twigs and bark. *Holzforschung* **42**, 241–246.

Nurmesniemi, H. (1983). Isolation and characterization of the main constituents of birch (*Betula verrucosa*) inner bark. Diss., University of Oulu, Oulu.

Yazaki, Y., and Aung, T. (1988). Alkaline extraction of *Pinus radiata* bark and isolation of aliphatic dicarboxylic acids. *Holzforschung* **42**, 357–360.

Ånäs, E., Ekman, R., and Holmbom, B. (1983). Composition of nonpolar extractives in bark of Norway spruce and Scots pine. *J. Wood Chem. Technol.* **3**, 119–130.

# Chapter 7.  Wood Pulping

## Books

Casey, J. P., ed. (1980). "Pulp and Paper Chemistry and Chemical Technology," 3rd ed., Vols 1–3, Wiley (Interscience), New York.

Grace, T. M., Leopold, B., and Malcolm, E. W., eds. (1989). "Alkaline Pulping," Vol. 5 in Series "Pulp and Paper Manufacture," 3rd ed. (M. J. Kocurek, ed.). TAPPI/CPPA, Atlanta/Montreal.

Ingruber, O. V., Kocurek, M. J., and Wong, A., eds. (1985). "Sulfite Science & Technology," Vol. 4 in Series "Pulp and Paper Manufacture," 3rd ed. (M. J. Kocurek, ed.). TAPPI/CPPA, Atlanta/Montreal.

Kocurek, M. J., and Stevens, C. F. B., eds. (1983). "Properties of Fibrous Raw Materials and their Preparation for Pulping," Vol. 1 in Series "Pulp and Paper Manufacture," 3rd ed. (M. J. Kocurek, ed.). TAPPI/CPPA, Atlanta/Montreal.

Leask, R. A., ed. (1987). "Mechanical Pulping," Vol. 2 in Series "Pulp and Paper Manufacture," 3rd ed. (M. J. Kocurek, ed.) TAPPI/CPPA, Atlanta/Montreal.

Mimms, A., Kocurek, M. J., Pyatte, J. A., and Wright, E. E., eds. (1989). "Kraft Pulping." TAPPI, Atlanta.

Rydholm, S. A. (1965). "Pulping Processes." Wiley (Interscience), New York.

## Reviews and Articles

Alén, R., Hentunen, P., Sjöström, F., Paavilainen, L., and Sundström, O. (1991). A new approach for process control of kraft pulping. *J. Pulp Pap. Sci.* **17**(1), J6–J9.

Alén, R., Lahtela, M., Niemelä, K., and Sjöström, E. (1985). Formation of hydroxy carboxylic acids from softwood polysaccharides during alkaline pulping. *Holzforschung* **39**, 235–238.

Alén, R., and Vikkula, A. (1989). Formation of lignin monomers during delignification of softwood. *Holzforschung* **43**, 397–400.

Alén, R., and Vikkula, A. (1989). Formation of lignin monomers during kraft pulping of birch wood. *Cellulose Chem. Technol.* **23**, 579–583.

Allen, L. H., Sitholé, B. B., MacLeod, J. M., Lapointe, C. L., and McPhee, F. J. (1991). The importance of seasoning and barking in the kraft pulping of aspen. *J. Pulp Pap. Sci.* **17**(3), J85–J91.

Aminoff, H., Brunow, G., Miksche, G. E., and Poppius, K. (1979). A Mechanism for the delignifying effect of anthraquinone in soda pulping. *Pap. Puu* **61**, 441–442.

Anonymous. (1984). Organosolv pulping processes—Boon or boondoggle? *Pulp Pap. Can.* **85**(7), 15–17.

Arbin, F. L. A., Schroeder, L. R., Thompson, N. S., and Malcolm, E. W. (1980). Anthraquinone-induced scission of polysaccharide chains. *Tappi* **63**(4), 152–153.

Arbin, F. L. A., Schroeder, L. R., Thompson, N. S., and Malcolm, E. W. (1981). The effects of oxygen and anthraquinone on the alkaline depolymerization of amylose. *Cellulose Chem. Technol.* **15**, 523–534.

Aziz, S., and Sarkanen, K. (1989). Organosolv pulping—A review. *Tappi J.* **72**(3), 169–175.

Bihani, B., and Samuelson, O. (1980). Carbohydrate stabilization in wood by quinones and oxygen. *Polym. Bull.* **3,** 425–430.

Blackwell, B. R., MacKay, W. B., Murray, F. E., and Oldham, W. K. (1979). Review of kraft foul condensates. Sources, quantities, chemical composition, and environmental effects. *Tappi* **62**(10), 33–37.

Brunow, G., and Poppius, K. (1982). Cleavage of β-aryl ether bonds in phenolic lignin model compounds with anthrahydroquinone and anthrone. *Acta Chem. Scand. Ser. B* **36,** 377–379.

Chiang, V. L., and Funaoka, M. (1990). The difference between guiacyl and guiacyl-syringyl lignins in their responses to kraft delignification. *Holzforschung* **44,** 309–313.

Chiang, V. L., and Stokke, D. D. (1989). Lignin fragmentation and condensation reactions in middle lamella and secondary wall regions during kraft pulping of Douglas fir. *J. Wood Chem. Technol.* **9,** 61–83.

Dimmel, D. R. (1985). Electron transfer reactions in pulping systems (I): Theory and applicability to anthraquinone pulping. *J. Wood Chem. Technol.* **5,** 1–14.

Dimmel, D. R., Perry, L. F., Palasz, P. D., and Chum, H. L. (1985). Electron transfer reactions in pulping systems (II): Electrochemistry of anthraquinone/lignin model quinonemethides. *J. Wood Chem. Technol.* **5,** 15–36.

Edel, E. (1989). Das Organocell-Verfahren—Bericht über den Betrieb einer Demonstrationsanlage. *Das Papier* 43(10A), V116–V123.

Ekman, R., Eckerman, C., and Holmbom, B. (1990). Studies on the behavior of extractives in mechanical pulp suspensions. *Nord. Pulp Pap. Res. J.* **5,** 96–103.

Ekman, R., and Holmbom, B. (1989). Analysis by gas chromatography of the wood extractives in pulp and water samples from mechanical pulping of spruce. *Nord. Pulp Pap. Res. J.* **4,** 16–24.

Enkvist, T., Alfredson, B., and Martelin, J.-E. (1957). Determinations of the consumption of alkali and sulfur at various stages of sulfate, soda, and alkaline and neutral sulfite digestion of spruce wood. *Svensk Papperstidn.* **60,** 616–620.

Evans, R., Wallis, A. F. A., and Wearne, R. H. (1987). Influence of additives on the alkaline degradation of cellulose. *In* "Wood and Cellulosics. Industrial Utilization, Biotechnology, Structure and Properties," (J. F. Kennedy, G. O. Philips, and P. A. Williams, eds.) pp. 165–172. Ellis Horwood, Chichester.

Fullerton, T. J. (1987). The condensation reactions of lignin model compounds in alkaline pulping liquors. *J. Wood Chem. Technol.* **7,** 441–462.

Fullerton, T. J., Watson, P. A., and Wright, L. J. (1990). Catalysts for alkaline pulping—beyond AQ. *Appita* **43**(1), 23–28.

Gellerstedt, G. (1976). The reactions of lignin during sulfite pulping. *Svensk Papperstidn.* **79,** 537–543.

Gellerstedt, G., and Gierer, J. (1971). The reactions of lignin during acidic sulphite pulping. *Svensk Papperstidn.* **74,** 117–127.

Gellerstedt, G., Gustafsson, K., and Northey, R. A. (1988). Structural changes in lignin during kraft cooking. Part 8. Birch lignins. *Nord. Pulp Pap. Res. J.* **3,** 87–94.

Gellerstedt, G., and Robert D. (1987). Quantitative $^{13}C$ NMR analysis of kraft lignins. *Acta Chem. Scand.,* Ser. B **41,** 541–546.

Ghosh, K. L., Venkatesh, V., Chin, W. J., and Gratzl, J. S. (1977). Quinone additives in soda pulping of hardwoods. *Tappi* **60**(11), 127–131.

Gierer, J. (1970). The reactions of lignin during pulping. A description and comparison of conventional pulping processes. *Svensk Papperstidn.* **73,** 571–596.

Gierer, J. (1982). The chemistry of delignification. A general concept. Parts I and II. *Holzforschung* **36**, 43–51; 55–64.

Gierer, J., Lindeberg, O., and Norén, I. (1979). Alkaline delignification in the presence of anthraquinone/anthrahydroquinone. *Holzforschung* **33**, 213–214.

Gierer, J., and Wännström, S. (1984). Formation of alkali-stable C-C-bonds between lignin and carbohydrate fragments during kraft pulping. *Holzforschung* **38**, 181–184.

Gierer, J., and Wännström, S. (1986). Formation of ether bonds between lignins and carbohydrates during alkaline pulping processes. *Holzforschung* **40**, 347–352.

Goliath, M., and Lindgren, B. O. (1961). Reactions of thiosulphate during sulphite cooking. Part 2. Mechanism of thiosulphate sulphidation of vanillyl alcohol. *Svensk Papperstidn.* **64**, 469–471.

Gustafsson, L., and Teder, A. (1969). Alkalinity in alkaline pulping. *Svensk Papperstidn.* **72**, 795–801.

Hansson, J.-Å. (1970). Sorption of hemicelluloses on cellulose fibers. Part 3. The temperature dependence on sorption of birch xylan and pine glucomannan at kraft pulping conditions. *Svensk Papperstidn.* **73**, 49–53.

Heikkilä, H., and Sjöström, E. (1975). Introduction of aldonic acid end-groups into cellulose by various oxidants. *Cellulose Chem. Technol.* **9**, 3–11.

Holmbom, B., and Eckerman, C. (1983). Tall oil constituents in kraft pulping—Effect of pulping temperature. *Tappi J.* **66**(5), 108–109.

Holton, H. H., and Chapman, F. L. (1977). Kraft pulping with anthraquinone. Laboratory and full-scale trials. *Tappi* **60**(11), 121–125.

Hon, D. N.-S., and Glasser, W. (1979). On possible chromophoric structures in wood and pulps—A survey of the present state of knowledge. *Polym. Plast. Technol. Eng.* **12**, 159–179.

Ingruber, O. V. (1958). The influence of the pH factor in sulphite pulping. *Tappi* **41**(12), 764–772.

Iversen, T., and Wännström, S. (1986). Lignin-carbohydrate bonds in a residual lignin isolated from pine kraft pulp. *Holzforschung* **40**, 19–22.

Janson, J. (1980). Pulping processes based on autocausticizable borate. *Svensk Papperstidn.* **83**, 392–395.

Janson, J. (1988). Influence of carbohydrates on the alkaline degradation of lignin. *Holzforschung* **42**, 105–109.

Janson, J., and Fullerton, T. (1987). Influence of carbohydrates and related compounds on the alkaline cleavage of the β-aryl ether linkage in a phenolic lignin model compound. *Holzforschung* **41**, 359–362.

Janson, J., and Sjöström, E. (1964). Behaviour of xylan during sulphite cooking of birchwood. *Svensk Papperstidn.* **67**, 764–771.

Johansson, A., Aaltonen, O., and Ylinen, P. (1987). Organosolv pulping—methods and pulp properties. *Biomass* **13**, 45–65.

Johansson, M. H., and Samuelson, O. (1974). The formation of end groups in cellulose during alkali cooking. *Carbohydr. Res.* **34**, 33–43.

Johansson, M. H., and Samuelson, O. (1977). Alkaline destruction of birch xylan in the light of recent investigations of its structure. *Svensk Papperstidn.* **80**, 519–524.

Kaufmann, Z. (1951). Über die chemischen Vorgänge beim Aufschluss von Holz nach dem Sulfiprozess. Diss., Eidg. Tech. Hochsch. Zürich, Zürich.

Kiiskilä, E. (1980) Recovery of sodium hydroxide from alkaline pulping liquors by causticizing molten sodium carbonate with amphoteric oxides. *Pap. Puu* **62**, 339–350.

Kleppe, P. J. (1970). Kraft pulping. *Tappi* **53**(1), 35–47.

Kubes, G. J., Fleming, B. I., McLeod, J. M., and Bolker, H. I. (1980). Alkaline pulping with additives. A review. *Wood Sci. Techn.* **14**, 207–228.

Landucci, L. L. (1980). Quinones in alkaline pulping. Characterization of an anthrahydro-quinone-quinone methide intermediate. *Tappi* **63**(7), 95–99.

Lee, H.-B., and Peart, T. E. (1990). Gas chromatographic and mass spectrometric determination of some resin and fatty acids in pulpmill effluents as their pentafluorobenzyl ester derivatives. *J. Chromatogr.* **498**, 367–379.

Lindenfors, S. (1980). Additives in alkaline pulping—What reduces what? *Svensk Papperstidn.* **83**, 165–173.

Lindgren, B. O. (1952). The sulphonatable groups of lignin. Svensk Papperstidn. **55**, 78–89.

Lindström, M., Ödberg, L., and Stenius, P. (1988). Resin and fatty acids in kraft pulp washing. Physical state, colloid stability and washability. *Nord. Pulp Pap. Res. J.* **3**, 101–106.

Löwendahl, L., and Samuelson, O. (1977). Carbohydrate stabilization during kraft cooking with addition of anthraquinone. *Svensk Papperstidn.* **80**, 549–551.

Maddern, K. N. (1985). Mill-scale development of the DARS direct caustization process. *TAPPI Proc. Int. Chem. Recovery Conf.*, New Orleans, La., Book 2, pp. 227–234.

Maddern, K. N. (1991). Bleached market pulp: an assessment of alternatives to the kraft process. *Proc. Int. Conf. Bleached Kraft Pulp Mills*, CSRIO, Melbourne, pp. 81–94.

Malinen, R., and Sjöström, E. (1975). The formation of carboxylic acids from wood polysaccharides during kraft pulping. *Pap. Puu* **57**, 728–736.

Marton, R. (1971). Reactions in alkaline pulping. *In* "Lignins" (K. V. Sarkanen and C. H. Ludwig, eds.), pp. 639–694. Wiley (Interscience), New York.

Mortimer, R. D. (1982). The formation of coniferyl alcohol during alkaline delignification with anthraquinone. *J. Wood Chem. Technol.* **2**, 383–415.

Mörck, R., and Kringstad, K. P. (1985). [13]C-NMR spectra of kraft lignins. II. Kraft lignin acetates. *Holzforschung* **39**, 109–119.

Mörck, R., Yoshida, H., Kringstad, K. P., and Hatakeyama, H. (1986). Fractionation of kraft lignin by successive extraction with organic solvents. I. Functional groups, [13]C- NMR spectra and molecular weight distributions. *Holzforschung* **40** Suppl., 51–60.

Niemelä, K. (1990). Low-molecular-weight organic compounds in birch black liquor. Diss., Helsinki Univ. of Technology, *Ann. Acad. Scient. Fenn. Ser. A* **II**, 229.

Niemelä, K. (1991). Organic compounds in birch kraft black liquor. *Proc. 6th Int. Symp. Wood & Pulping Chem.*, Melbourne, Vol. 1, pp. 313–317.

Niemelä, K., Alén, R., and Sjöström, E. (1985). The formation of carboxylic acids during kraft and kraft-anthraquinone pulping of birch wood. *Holzforschung* **39**, 167–172.

Nimz, H. H., Berg, A., Granzow, C., Casten, R., and Muladi, S. (1989). Zellstoffgewinnung und- Bleiche nach dem Acetosolv-Verfahren. *Das Papier* **43**, V102–V108.

Obst, J. R. (1983). Kinetics of alkaline cleavage of β-aryl ether bonds in lignin models: Significance to delignification. *Holzforschung* **37**, 23–28.

Paszner, L. (1989). Topochemistry of softwood delignification by alkali earth metal salt catalyzed organosolv pulping. *Holzforschung* **43**, 159–168.

Patt, R., Knoblauch, J., Faix, O., Kordsachia, O., and Puls, J., (1991). Lignin and carbohydrate reactions in alkaline sulfite, anthraquinone, methanol (ASAM) pulping. *Das Papier* **45**, 389–396.

Patt, R., Kordachia, O., and Kopfmann, K. (1989). Wirtschaftliche, technologische und ökologische Aspekte der Zellstoffherstellung nach dem ASAM-Verfahren. *Das Papier* **43**(10A), V108–V115.

Pekkala, O., and Palenius, I. (1973). Hydrogen sulphide pretreatment in alkaline pulping. *Pap. Puu* **55**, 659–668.

Pfister, K., and Sjöström, E. (1977). The formation of monosaccharides and aldonic and uronic acids during sulphite pulping. *Pap. Puu*, 711–720.

Poppius-Levlin, K., Mustonen, R., Huovila, T., and Sundquist, J. (1991). MILOX pulping with acetic acid/peroxyacetic acid. *Pap. Puu* **73**, 154–158.

Samuelson, O., and Sjöberg, L.-A. (1972). Oxygen-alkali cooking of wood meal. *Svensk Papperstidn.* **75**, 583–588.

Sarkanen, K. V., Hrutfiord, B. F., Johanson, L. N., and Gardner, H. S. (1970). Kraft odor. *Tappi* **53**, 766–783.

Saukkonen, M., and Palenius, I. (1975). Soda-oxygen pulping of pinewood for different end products. *Tappi* **58**(7), 117–120.

Schöön, N.-H. (1962). Kinetics of the formation of thiosulphate, polythionates and sulphate by the thermal decomposition of sulphite cooking liquors. *Svensk Papperstidn.* **65**, 729–754.

Simonson, R. (1963). The hemicellulose in the sulfate pulping process. Part 1. The isolation of hemicellulose fractions from pine sulfate cooking liquors. *Svensk Papperstidn.* **66**, 839–845.

Simonson, R. (1965). The hemicellulose in the sulfate pulping process, Part 3. The isolation of hemicellulose fractions from birch sulfate cooking liquors. *Svensk Papperstidn.* **68**, 275–280.

Sjöström, E. (1964). Chemical aspects of high-yield pulping processes. *Norsk Skogind.* **18**, 212–223. (In Swed.)

Sjöström, E. (1977). The behavior of wood polysaccharides during alkaline pulping processes. (1977) *Tappi* **60**(9), 151–154.

Sjöström, E. (1989). The origin of charge on cellulosic fibers. *Nord. Pulp Pap. Res. J.* **4**, 90–93.

Sjöström, E., and Enström, B. (1967). Characterization of acidic polysaccharides isolated from different pulps. *Tappi* **50**(1), 32–36.

Sjöström, E., and Haglund, P. (1963). Dissolution of carbohydrates during beating of chemical pulps. *Svensk Papperstidn.* **66**, 718–720.

Sjöström, E., Haglund, P., and Janson, J. (1962). Changes in cooking liquor composition during sulphite pulping. *Svensk Papperstidn.* **65**, 855–869.

Sjöström, E., Janson, J., Haglund, P., and Enström, B. (1965). The acidic groups in wood and pulp as measured by ion exchange. *J. Polym. Sci. C* **11**, 221–241.

Sjöström, E., Sorvari, J., Klemola, A., and Laine, J. E. (1987). Delignification studies on spruce wood tissue fractions isolated from outer and inner cell wall regions. *Nord. Pulp Pap. Res. J.* **2**, 92–96.

Sjöström, J. (1990). Fractionation and characterization of organic substances dissolved in water during groundwood pulping of spruce. (1990). *Nord. Pulp Pap. Res. J.* **5**, 9–15.

Stenius, P., Palonen, H., Ström, G., and Ödberg, L. (1984). Micelle formation and phase equilibria of surface active components in wood. In "Surfactants in Solution" (B. Lindman and K. L. Mittal, eds.), Vol. 1, pp. 153–174. Plenum Press, New York.

Stone, J. E. (1957). The effective capillary cross-sectional area of wood as a function of pH. *Tappi* **40**(7), 539–541.

Teder, A. (1969). Some aspects of the chemistry of polysulfide pulping. *Svensk Papperstidn.* **72**, 294–303.

Teder, A., and Tormund, D. (1973). The equilibrium between hydrogen sulfide and sulfide ions in kraft pulping. *Svensk Papperstidn.* **76**, 607–609.

Tormund, D., and Teder, A. (1989). New findings on sulfide chemistry in kraft pulping liquors. *Tappi J.* **72**(5), 205–210.

Virkola, N.-E., Pusa, R., and Kettunen, J. (1981). Neutral sulfite AQ pulping as an alternative to kraft pulping. *Tappi* **64**(5), 103–107.

Vroom, J. R. (1957). The "H" factor: A means of expressing cooking times and temperatures as a single variable. *Pulp Pap. Mag. Can.* **58**(3), 228–231.

Vuorinen, T. (1988). Mechanisms and kinetics of isomerization, degradation, and oxidation of reducing carbohydrates; Reaction paths in alkaline solutions containing oxygen and 2-anthraquinonesulfonic acid. Diss., Helsinki Univ. of Technol., *Ann. Acad. Scient. Fenn. Ser. A* **II**, No. 220.

Wallis, A. F. A., and Wearne, R. H. (1987). Oxidation of monohydric alcohols with anthraquinone and its derivatives under soda pulping conditions. *J. Wood Chem. Technol.* **7**, 513–526.

Westermark, U., and Samuelsson, B. (1986). The reactivity of lignin in highly lignified parts of spruce during kraft pulping. *Holzforschung* **40**(Suppl.), 139–146.

Westermark, U., Samuelsson, B., Simonson, R., and Pihl, R. (1987). Investigation of a selective sulfonation of wood chips. Part 5. Thermomechanical pulping with low addition of sulfite. *Nord. Pulp. Pap. Res. J.* **2**, 146–151.

Wood, J. R., and Goring, D. A. I. (1973). The distribution of lignin in fibres produced by kraft and acid sulphite pulping from spruce wood. *Pulp Pap. Mag. Can.* **74**(9), T309–T313.

Yllner, S., and Enström, B. (1956). Studies on the adsorption of xylan on cellulose fibres during the sulphate cook, Part 1. *Svensk Papperstidn.* **59**, 229–232.

Young, R. A. (1989). Ester pulping: a status report. *Tappi J.* **72**(4), 195–200.

Zarubin, M. Ya., Dejneko, I. P., Evtuguine, D. V., and Robert, A. (1989). Delignification by oxygen in acetone–water media. *Tappi J.* **72**(11), 163–168.

Zimmermann, M., Patt, R., Kordsachia, O., and Hunter, W. D. (1991). ASAM pulping of Douglas-fir followed by a chlorine-free bleaching sequence. *Tappi J.* **74**(11), 129–134.

# Chapter 8.    Pulp Bleaching

## Books

Gratzl, J. S., Nakano, J., and Singh, R. P., eds. (1980). "Chemistry of Delignification with Oxygen, Ozone, and Peroxides" (Symposium, Raleigh, N.C., 1975). UNI Publ. Co., Tokyo.

Rodgers, M. A. J., and Powers, E. L., eds. (1981). "Oxygen and Oxy-Radicals in Chemistry and Biology." Academic Press, New York.

Rydholm, S. (1965). "Pulping Processes." Wiley (Interscience).

Rånby, B., and Rabek, J. F., eds. (1978). "Singlet Oxygen. Reactions with Organic Compounds and Polymers." Wiley (Interscience), Chichester.

Singh, R. P., ed. (1979). "The Bleaching of Pulp," 3rd ed., Tappi Press, Atlanta.

## Reviews and Articles

Abrahamsson, K., Löwendahl, L., and Samuelson, O. (1981). Pretreatment of kraft pulp with nitrogen dioxide before oxygen bleaching. *Svensk Papperstidn.* **84**, R152–R158.

Agnemo, R., Francis, R. C., Alexander, T. C., and Dence, C. W. (1991). Studies on the photoyellowing of bleached mechanical and chemimechanical pulps. III. The role of hydroxyl radicals. *Holzforschung* **45**(Suppl.), 101–108.

Alén, R., and Sjöström, E. (1991). Formation of low-molecular-mass compounds during the oxygen delignification of pine kraft pulps. *Holzforschung* **45** (Suppl.), 83–86.

Axegård, P. (1988). Improvement of bleach plant effluent by cutting back on $Cl_2$. *TAPPI Proc. Int. Pulp Bleaching Conf.*, Orlando, Fla., pp. 69–76.

Berry, R. M., Luthe, C. E., Voss, R. H., Wrist, P. E., Axegård, P., Gellerstedt, G., Lindblad, P-O., and Pöpke, I. (1991). The effects of recent changes in bleached softwood kraft mill technology on organochlorine emissions: An international perspective. *Pulp Pap. Can.* **92**(6), T155–T165.

Bowen, I. J., and Hsu, J. C. L. (1990). Overview of emerging technologies in pulping and bleaching. *Tappi J.* **73**(9), 205–217.

Brage, C., Eriksson, T., and Gierer, J. (1991). Reactions of chlorine dioxide with lignins in unbleached pulps. Parts I and II. *Holzforschung* **45,** 25–30, 147–152.

Brännland, R., Nordén, S., and Lindström, L.-Å. (1990). Implementation in full scale—The next step for Prenox[R]. *Tappi J.* **73**(5), 231–237.

Carlberg, G. E., Johnsen, S., Landmark, L. H., Bengtsson, B.-E, Bergström, B., Skramstad, J., and Storflor, H. (1988). Investigations of chlorinated thiophenes: A group of bioaccumulable compounds identified in the effluents from kraft bleaching. *Water Sci. Techn.* **20**(2), 37–48.

Chang, H.-M., and Allan, G. G. (1971). Oxidation. In "Lignins" (K. V. Sarkanen and C. H. Ludwig, eds.), pp. 433–485. Wiley (Interscience), New York.

Daneault, C., Robert, S., and Lévesque, M. (1991). The prevention of light-induced yellowing of paper: The inhibition of reversion by mercaptans of TMP and CTMP pulp from balsam fir (*Abies balsamea*) and black spruce (*Picea mariana*). *J. Pulp Pap. Sci.* **17,** J187–J193.

Ek, M., Lennholm, H., Lindblad, G., and Iversen, T. (1991). The light-induced color reversion of groundwood pulps. A study on the mechanism of chromophore formation. *Proc. 6th Int. Symp. Wood & Pulping Chem.*, Melbourne, Vol. 1, pp. 439–442.

Eriksson, K.-E. L. (1990). Biotechnology in the pulp and paper industry. *Wood Sci. Technol.* **24,** 79–101.

Eriksson, T., and Gierer, J. (1985). Studies on the ozonization of structural elements in residual kraft lignins. *J. Wood Chem. Technol.* **5,** 53–84.

Eriksson, T., Gierer, J., and Brage, C. (1991). Reactions of chlorine dioxide with stilbenes and styrenes. *Proc. 6th Int. Symp. Wood & Pulping Chemistry*, Melbourne, Vol. 1, pp. 337–339.

Fischer, K., Schmidt, I., and Koch, H. (1991). The role of oxygen species at light-induced yellowing and possibilities to reduce their action. *Proc. 6th Int. Symp. Wood & Pulping Chem.*, Melbourne, Vol. 1, pp. 431–449.

Fujita, K., Kondo, R., and Sakai, K. (1991). Biobleaching of kraft pulp by the hyperligninolytic fungus IZU-154. *Proc. 6th Int. Symp. Wood & Pulping Chem.*, Melbourne, Vol. 1, pp. 475–480.

Gellerstedt, G. (1983). Light-induced and heat-induced yellowing of mechanical pulps. *Svensk Papperstidn.* **86,** R157–R163.

Gellerstedt, G. (1992). Chemical aspects on kraft pulp bleaching. In "Lignocellulosics Science, Technology, Development and Use" (J. F. Kennedy, G. O. Phillips, and P. A. Williams, eds.), pp. 291–304. Ellis Horwood, Chichester.

Gellerstedt, G., Gustafsson, K., and Lindfors, E.-L. (1986). Structural changes of lignin during oxygen bleaching. *Nord. Pulp Pap. Res. J.* **1,** 14–17.

Gellerstedt, G., and Lindfors, E.-L. (1987). Hydrophilic groups in lignin after oxygen bleaching. *Tappi J.* **70**(6), 119–122.

Gellerstedt, G., and Lindfors, E.-L. (1991). On the structure and reactivity of residual lignin in kraft pulp fibers. *Proc. Int. Pulp Bleaching Conf.*, Stockholm, Vol. 1, pp. 73–88.

Gellerstedt, G., Lindfors, E.-L., Pettersson, M., Sjöholm, E., and Robert, D. (1991). Chemical aspects on chlorine dioxide as a bleaching agent for chemical pulps. *Proc. 6th Int. Symp. Wood & Pulping Chemistry,* Melbourne, Vol. 1, pp. 331–336.

Gellerstedt, G., and Pettersson, I. (1982). Chemical aspects of hydrogen peroxide bleaching. Part II. The bleaching of kraft pulps. *J. Wood Chem. Technol.* **2,** 231–250.

Gellerstedt, G., Pettersson, I., and Sundin, S. (1981). Chemical aspects of hydrogen peroxide bleaching. *Proc. Int. Symp. Wood & Pulping Chem., The Ekman-days,* Stockholm, Vol. 2, pp. 120–124.

Gellerstedt, G., Pettersson, I., and Sundin, S. (1983). Sodium sulfite and DTPA. A brightening mixture for high yield pulps. *Proc. Int. Symp. Wood & Pulping Chem.,* Tsukuba Science City, Vol. 2., pp. 90–96.

Germgård, U. (1989). Chlorate discharges from bleach plants—How to handle a potential environmental problem. *Pap. Puu* **71,** 255–260.

Gierer, J. (1990). Basic principles of bleaching. Part 1: Cationic and radical processes. Part 2: Anionic processes. *Holzforschung* **44,** 387–394, 395–400.

Gierer, J., and Imsgard, F. (1977). The reactions of lignins with oxygen and hydrogen peroxide in alkaline media. *Svensk Papperstidn.* **80,** 510–518.

Gierer, J., Jansbo, K., Yang, E., Yoon, B.-H., and Reitberger, T. (1991). On the participation of hydroxyl radicals in oxygen and peroxide bleaching processes. *Proc. 6th Int. Symp. Wood & Pulping Chem.,* Melbourne, Vol. 1, pp. 93–97.

Gierer, J., and Lin, S. Y. (1972). Photodegradation of lignin. A contribution to the mechanism of chromophore formation. *Svensk Papperstidn.* **75,** 233–239.

Godsay, M. P., and Pearce, E. M. (1985). Mechanism of inhibition & retardation of ozone-lignocellulose reactions. *TAPPI Proc. Pulping Conf.,* Hollywood, Fla., Book 2, pp. 245–262.

Gratzl, J. S. (1985). Lichtinduzierte Vergilbung—Ursachen und Verhütung. *Das Papier* **39** (10A), V14–V23.

Gratzl, J. S. (1987). Abbaureaktionen von Kohlenhydraten und Lignin durch chlorfrei Bleichmittel—Mechanismen sowie Möglichkeiten der Stabilisierung. *Das Papier* **41,** 120–130.

Grundelius, R. (1991). Oxidation equivalents, OXE—An alternative to active chlorine. *Proc. Int. Pulp Bleaching Conf.,* Stockholm, Vol. 1, pp. 49–58.

Haglind, I. A. K., Hultman, B. G., Landner, L., Lövblad, R. B., and Strömberg, L. M. (1991). Environmental impact from modern Swedish bleached kraft pulp mills. *Proc. Int. Pulp Bleaching Conf.,* Stockholm, Vol. 1, pp. 59–71.

Hall, E. R., Fraser, J., Garden, S., and Cornacchio, L.-A. (1989). Organo-chlorine discharges in wastewaters from kraft mill bleach plants. *Pulp Pap. Can.* **90**(11), T421–T425.

Heitner, C., and Schmidt, J. A. (1991). Light-induced yellowing of wood-containing-papers. A review of fifty years research. *Proc. 6th Int. Symp. Wood & Pulping Chem.,* Melbourne, Vol. 1, pp. 131–149.

Hise, R. G., Streisel, R. C., and Bills, A. M. (1992). The effect of brownstock washing, split addition of chlorine, and pH control in the C stage on formation of AOX and chlorophenols during bleaching. *Tappi J.* **75**(2), 57–62.

Holmbom, B. (1990). Mutagenic compounds in chlorinated pulp bleaching waters and drinking waters. *In* "Complex Mixtures and Cancer Risk," (H. Vainio, M. Sorsa, and A. J. McMichael, eds.), pp. 333–339. International Agency for Research on Cancer, Lyon.

Holmbom, B., Ekman, R., Sjöholm, R., Eckerman, C., and Thornton, J. (1991). Chemical changes in peroxide bleaching of mechanical pulps. *Das Papier* **45**(10A), V16–V22.

Holmbom, B., Sjöholm, R., and Åkerback, N. (1991). Analysis of chemical changes in lignin on irradiation of papers by light. *Proc. 6th Int. Symp. Wood & Pulping Chem.,* Melbourne, Vol. 1, pp. 443–449.

Holmbom, B., Voss, R. H., Mortimer, R. D., and Wong, A. (1984). Fractionation, isolation, and characterization of Ames mutagenic compounds in kraft chlorination effluents. *Environ. Sci. Technol.* **18**, 333–337.

Hong, Q., Shin, N. H., and Chang, H.-M. (1989). Effects of oxygen extraction on organic chlorine contents in bleach plant effluents. *Tappi J.* **72**(6), 157–162.

Hrutfiord, B. F., and Negri, A. R. (1990). Chemistry of chloroform formation in pulp bleaching: a review. *Tappi J.* **73**(6), 219–225.

Hrutfiord, B. F., and Negri, A. R. (1991). Chlorinated dibenzofurans and dibenzodioxins from lignin models. *Proc. 6th Int. Symp. Wood & Pulping Chem.*, Melbourne, Vol. 1. pp. 551–556.

Janson, J., and Forsskåhl (1989). Farbveränderungen der Holzstoffe und Erhöhung infolge von Lichteinwirkung. *Zellst. Pap. (Leipzig)* **38**, 47–50.

Janson, J., and Forsskåhl, I. (1989). Color changes in lignin-rich pulps on irradiation by light. *Nord. Pulp Pap. Res. J.* **4**, 197–205.

Kantelinen, A., Sundquist, J., Linko, M., and Viikari, L. (1991). The role of reprecipitated xylan in the enzymatic bleaching of kraft pulp. *Proc. 6th Int. Symp. Wood & Pulping Chem.*, Melbourne, Vol. 1, pp. 493–500.

Kolar, J. J., Lindgren, B. O., and Pettersson, B. (1983). Chemical reactions in chlorine dioxide stages of pulp bleaching. *Wood Sci. Technol.* **17**, 117–128.

Kratzl, K., Claus, P., and Reichel, G. (1976). Reactions of lignin model compounds with ozone. *Tappi* **59**(11), 86–87.

Kringstad, K. P. (1989). Environmental aspects on the future developments of pulp bleaching. *In* "Wood Processing and Utilization" (J. F. Kennedy, G. O. Phillips, and P. A. Williams, eds.), pp. 31–42. Ellis Horwood, Ltd., Chichester.

Kringstad, K., Johansson, L. Kolar, M-C., and de Sousa, F. (1989). The influence of chlorine ratio and oxygen bleaching on the formation of PCDFs and PCDDs in pulp bleaching. Part 2: A full mill study. *Tappi J.* **72**(6), 163–170.

Kringstad, K. P., and Lin, S. Y. (1970). Mechanism of the yellowing of high-yield pulps by light. Structure and reactivity of free radical intermediates in the photodegradation of lignin. *Tappi* **53**(12), 2296–2301.

Kringstad, K. P., and Lindström, K. (1984). Spent liquors from pulp bleaching. *Environ. Sci. Technol.* **18**, 236A–248A.

Lachenal, D., and Muguet, M. (1991). Degradation of residual lignin in kraft pulp with ozone. Application to bleaching. *Proc. 6th Int. Symp. Wood & Pulping Chem.*, Melbourne, Vol. 1, pp. 107–112.

Leary, G. J. (1968). The yellowing of wood by light. Part II. *Tappi* **51**(6), 257–267.

Liebergott, N., van Lieropp, B., and Skothos, A. (1992). A survey of the use of ozone in bleaching pulps. Parts 1 and 2. *Tappi J.* **75**(1,2), 145–152, 117–124.

Lin, S. Y., and Kringstad, K. P. (1970). Photosensitive groups in lignin and lignin model compounds. *Tappi* **53**(4), 658–663.

Lindholm, C-A. (1991). Some effects of treatment consistency in ozone bleaching. *Proc. Int. Pulp Bleaching Conf.*, Stockholm, Vol. 2, pp. 1–17.

Lindström, K., and Österberg, F. (1986). Chlorinated carboxylic acids in softwood kraft pulp spent bleach liquors. *Environ. Sci. Technol.* **20**, 133–138.

Luo, Ch., and Göttsching, L. (1991). Vergilbungsfunktion in Z-Richtung von Papier bei lichtinduzierter Vergilbung. Teil 1: Empirische und statistische Erfassung der Vergilbungsfunktion. *Das Papier* **45**(10), 601–609.

Malinen, R. (1975). Behaviour of wood polysaccharides during oxygen-alkali delignification. Diss., Helsinki University of Technology, Espoo, Finland. *Pap. Puu* **57**, 193–204 (summary).

Mattinen, H., and Wartiovaara, I. (1981). The pollution load of a closed D/CEDED bleachery. *Pap. Puu* **63**, 688–706.

McDonough, T. J., LaFleur, L. E., Brunck, R., and Malcolm, E. W. (1991). Factors affecting formation of PCDD/F in kraft pulp bleaching. *Proc. Int. Pulp Bleaching Conf.*, Stockholm, Vol. 2, pp. 171–193.

McKague, A. B., and Reeve, D. W. (1991). Identification of chlorinated compounds in bleached pulp extracts. *Nord. Pulp Pap. Res. J.* **6**, 35–39.

Myers, M., Edwards, L., and Haynes, J. (1989). Oxygen delignification systems: Synthesizing the optimum design. *Tappi J.* **72**(4), 131–135.

Omori, S., Francis, R. S., and Dence, C. W. (1991). Studies on the mechanism of the photoyellowing of bleached mechanical and chemimechanical pulps. II. The role of photosensitizers. *Holzforshung* **45**(Suppl.), 93–100.

Pan, X., Lachenal, D., Lapierre, C., and Monties, B. (1991). Analysis of spruce photodegraded lignins by thioacidolysis. *Proc. 6th Int. Symp. Wood & Pulping Chem.*, Melbourne, Vol. 1, pp. 451–455.

Patt, R., Hammann, M., and Kordsachia, O. (1991). The role of ozone in chemical pulp bleaching. *Holzforschung* **45**(Suppl.), 87–92.

Perrolaz, J. J., Davis, S., Gysin, B., Zimmerman, W., Casimir, J., and Fiechter, A. (1991). Elemental chlorine free bleaching with a thermostable xylanase: A new alternative to elemental chlorine for bleaching. *Proc. 6th Int. Symp. Wood & Pulping Chem.*, Melbourne, Vol. 1, pp. 485–489.

Pettersson. B., Yang, J.-L., and Eriksson, K.-E. (1988). Biotechnical approaches to pulp bleaching. *Nord. Pulp Pap. Res. J.* **3**, 198–202.

Pfister, K., and Sjöström, E. (1979). Characterization of spent bleaching liquors. Part 6. Composition of material dissolved during chlorination and alkali extraction (OCE sequence). *Pap. Puu* **61**, 619–622.

Poppius-Levlin, K., Toikkanen, L., Tuominen, I., and Sundquist, J. (1991). Chlorine-free bleaching of kraft pulps. *Proc. Int. Pulp Bleaching Conf.*, Stockholm, Vol. 3, pp. 103–120.

Rappe, C., Swanson, S., Glas, B., Kringstad, K. P., de Sousa, F., Johansson, L., and Abe, Z. (1989). On the formation of PCDDs and PCDFs in the bleaching of pulp. *Pulp Pap. Can.* **90**(8), T273–T278.

Reeve, D. W. (1992). Organochlorine in paper products. *Tappi J.* **75**(2), 63–69.

Sameshima, K., Simson, B., and Dence, C. W. (1979). The fractionation of toxic materials in kraft bleaching liquors. *Svensk Papperstidn.* **82**, 162–170.

Samuelson, O. (1991). Pretreatment of kraft pulp with nitrogen oxides without interfering attack on the carbohydrates. *Proc. Int. Pulp Bleaching Conf.*, Stockholm, Vol. 1, pp. 17–32.

Shen, X., and van Heiningen, A. (1991). Delignification mechanism during chlorine bleaching of kraft pulp; The formation of chlorinated phenolic compounds from the cleavage of alkyl aryl ether linkages. *Proc. 6th Int. Symp. Wood & Pulping Chem.*, Melbourne, Vol. 1, pp. 557–561.

Sixta, H., Götzinger, G., Schrittwieser, A., and Hendel, P. (1991). Medium consistency ozone bleaching: Laboratory and mill experience. *Das Papier* **45**, 610–625.

Sjöström, E. (1981). The chemistry of oxygen delignification. *Pap. Puu* **63**, 438–442.

Sjöström, E., and Enström, B. (1966). Spectrophotometric determination of the residual lignin in pulp after dissolution in cadoxen. *Svensk Papperstidn.* **69**, 469–476.

Sjöström, E., and Välttilä, O. (1972, 1978). Inhibition of carbóhydrate degradation during oxygen bleaching. Part I. Comparison of various additives. Part II. The catalytic activity of transition metals and the effect of magnesium and triethanolamine. *Pap. Puu* **54**, 695–705; **60**, 37–43.

Smeds, A., Holmbom, B., and Tikkanen, L. (1990). Formation and degradation of mutagens in kraft pulp mill water systems. *Nord. Pulp Pap. Res. J.* **5**, 142–147.

Strömberg, L. M., deSousa, F., Ljungquist, P., McKague, B., and Kringstad, K. P. (1987). An abundant chlorinated furanone in the spent chlorination liquor from pulp bleaching. *Environ. Sci. Technol.* **21**, 754–756.

Stuthridge, T. R., Wilkins, A. L., Langdom, A. G., Mackie, K. L., and McFarlane, P. N. (1990). Identification of novel chlorinated monoterpenes formed during kraft pulping of *Pinus radiata*. *Environ. Sci. Technol.* **24**, 903–908.

Suntio, L. R., Shin, W. Y., and Macay, D. (1988). A review of the nature and properties of chemicals present in pulp mill effluents. *Chemosphere* **17**, 1249–1290.

Swanson, S. E., Rappe, C., Malmström, J., and Kringstad, K. (1988). Emissions of PCDDs and PCDFs from the pulp industry. *Chemosphere* **17**, 681–691.

Süss, H. U., and Nimmerfroh, N. (1991). Zur Bildung halogenierter Verbindungen bei der Zellstoffbleiche mit Hypochlorit und Chlordioxid. *Das Papier* **45**, 52–62.

Talka, E. (1986). Fractionation and identification of some biologically active compounds in bleached kraft mill effluents. Part 1. Volatile compounds. *Pap. Puu* **68**, 670–673.

Talka, E., and Priha, M. (1987). Fractionation and identification of some biologically active compounds in bleached kraft mill effluents. Part 2. Neutral compounds. *Pap. Puu* **69**, 221–228.

Thakakore, A. N., and Oehlschlager, A. C. (1977). Structures of toxic constituents in kraft mill caustic extraction effluents from $^{13}C$ and $^1H$ nuclear magnetic resonance. *Can. J. Chem.* **55**, 3298–3303.

Van Lierop, B., Liebergott, N., Theoderescu, G., and Kubes, G. J. (1986). Oxygen in bleaching sequences—an overview. *Pulp Pap. Can.* **87**(5), T193–T197.

Viikari, L., Ranua, M., Kantelinen, A., Sundquist, J., and Linko, M. (1986). Bleaching with enzymes. *Proc. 3rd Int. Conf. on Biotechnology in the Pulp and Paper Industry,* Stockholm, pp. 67–69.

Österberg, F., and Lindstöm, K. (1985). Characterization of the high molecular mass chlorinated matter in spent bleach liquors (SBL). 3. Mass spectrometric interpretation of aromatic degradation products in SBL. *Org. Mass Spectrom.* **20**, 515–524.

# Chapter 9.   Cellulose Derivatives

## Books

Bikales, N. M., and Segal, L., eds. (1971). "Cellulose and Cellulose Derivatives," Parts IV and V. Wiley (Interscience), New York.

Hebeish, A., and Guthrie, J. T. (1981). "The Chemistry and Technology of Cellulosic Copolymers." Springer, Berlin.

Inagaki, H., and Phillips, G. O., eds. (1989). "Cellulosics Utilization. Research and Rewards in Cellulosics." Proceedings of the Nisshinbo International Conference on Cellulosics Utilization in the Near Future, 1988, Tokyo. Elsevier, London.

Kennedy, J. F., and White, C. A., eds. (1983). "Bioactive Carbohydrates: In Chemistry, Biochemistry and Biology." Ellis Horwood, Chichester.

Kennedy, J. F., Phillips, G. O., and Williams, P. A., eds. (1985). "Cellulose and its Derivatives: Chemistry, Biochemistry and Applications." (Cellucon 84 Conference). Ellis Horwood, Chichester.

Kennedy, J. F., Phillips, G. O., and Williams, P. A., eds. (1987). "Wood and Cellulosics. Industrial Utilisation, Biotechnology, Structure and Properties." (Cellucon 86 Conference). Ellis Horwood, Chichester.

Kennedy, J. F., Phillips, G. O., and Williams, P. A., eds. (1989). "Cellulose. Structural and Functional Aspects" (Cellucon 88 Conference). Ellis Horwood, Chichester.

Nevell, T. P., and Zeronian, S. H., eds. (1985). "Cellulose Chemistry and Its Applications." Ellis Horwood, Chichester.

Turbak, A. F., ed. (1975). "Cellulose Technology Research," ACS Symposium Series, No. 10. Am. Chem. Soc., Washington, D.C.

Turbak, A. F., ed. (1977). "Solvent Spun Rayon, Modified Cellulose Fibres and Derivatives," ACS Symposium Series, No. 58. Am. Chem. Soc., Washington, D.C.

Ward, K., Jr. (1973). "Chemical Modification of Papermaking Fibers." Marcel Dekker, New York.

## Reviews and Articles

Bikales, N. M. (1971). Ethers from $\alpha,\beta$-unsaturated compounds. In "Cellulose and Cellulose Derivatives" (N. M. Bikales and L. Segal, eds.), Part V, pp. 811–833. Wiley (Interscience), New York.

Buytenhuys, F. A., and Bonn, R. (1977). Distribution of substituents in CMC. Papier (Darmstadt) 31, 525–527.

Cassidy, H. G., and Kun, K. A. (1965). "Oxidation-Reduction Polymers," pp. 41–52. Wiley (Interscience), New York.

Demint, R. J., and Hoffpauir, C. L. (1957). Influence of pretreatment on the reactivity of cotton as measured by acetylation. Text. Res. J. 27, 290–294.

Gal'braikh, L. S., and Rogovin, Z. A. (1971). Derivatives with unusual functional groups. In "Cellulose and Cellulose Derivatives" (N. M. Bikales and L. Segal, eds.), Part V, pp. 877–905. Wiley (Interscience), New York.

Goldman, R., Goldstein, L., and Katchalski, E. (1971). Water-insoluble enzyme derivatives and artificial enzyme membranes. In "Biochemical Aspects of Reactions on Solid Supports" (G. R. Stark, ed.), pp. 1–72. Academic Press, New York.

Haines, A. H. (1976). Relative reactivities of hydroxyl groups in carbohydrates. Adv. Carbohydr. Chem. Biochem. 33, 11–109.

Hiatt, G. D., and Rebel, W. J. (1971). Esters. In "Cellulose and Cellulose Derivatives" (N. M. Bikales and L. Segal, eds.), Part V, pp. 741–784. Wiley (Interscience), New York.

Malm, C. J. (1961). Pulp for acetylation. Svensk Papperstidn. 64, 740–743.

Rånby, B. (1952). Fine structure and reactions of native cellulose. Diss., Univ. of Uppsala, Uppsala.

Rånby, B., and Rydholm, S. (1956). Cellulose and cellulose derivatives. In "Polymer Processes" (C. Schildknecht, ed.), pp. 351–428. Wiley (Interscience), New York.

Rowland, S. P. (1978). Hydroxyl reactivity and availability in cellulose. In "Modified Cellulosics" (R. M. Rowell and R. A. Young, eds.), pp. 147–167. Academic Press, New York.

Rydholm, S. A. (1965). "Pulping Processes," pp. 100–156. Wiley (Interscience), New York.

Savage, A. B. (1971). Ethers. In "Cellulose and Cellulose Derivatives" (N. M. Bikales and L. Segal, eds.), Part V, pp. 785–809. Wiley (Interscience), New York.

Segal, L. (1971). Effect of morphology on reactivity. In "Cellulose and Cellulose Derivatives" (N. M. Bikales and L. Segal, eds.), Part V, pp. 719–739. Wiley (Interscience), New York.

Sjöström, E. (1990). Characterization of carboxymethylcellulose by gas-liquid chromatography.

*In* "Cellulose, Structural and Functional Aspects" (J. F. Kennedy, G. O. Phillips, and P. A. Williams, eds.), pp. 239–249. Ellis Horwood, Chichester.

Stannett, V. T., and Hopfenberg, H. B. (1971). Graft copolymers. *In* "Cellulose and Cellulose Derivatives" (N. M. Bikales and L. Segal, eds.), Part V, pp. 907–936. Wiley (Interscience), New York.

Tesoro, G. C., and Willard, J. J. (1971). Crosslinked cellulose. *In* "Cellulose and Cellulose Derivatives" (N. M. Bikales and L. Segal, eds.), Part V, pp. 835–875. Wiley (Interscience), New York.

Timell, T. (1950). Studies on cellulose reactions. Diss., Univ. of Stockholm, Stockholm.

Tripp, V. W. (1971). Measurements of crystallinity. *In* "Cellulose and Cellulose Derivatives" (N. M. Bikales and L. Segal, eds.), Part IV, pp. 305–323. Wiley (Interscience), New York.

Wadsworth, L. C., and Cuculo, J. A. (1978). Determination of accessibility and crystallinity of cellulose. *In* "Modified Cellulosics" (R. M. Rowell and R. A. Young, eds.), pp. 117–146. Academic Press, New York.

# Chapter 10.   Wood-Based Chemicals and Pulping By-Products

## Books

Bridgwater, A. V., and Kuester, J. L., eds. (1988). "Research in Thermochemical Biomass Conversion." Elsevier, London.

Brown, R. D., Jr., and Jurasek, L., eds. (1979). "Hydrolysis of Cellulose: Mechanisms of Enzymatic and Acid Catalysis," Advances in Chemistry Series, No. 181. Am. Chem. Soc., Washington, D.C.

Campos-López, E., ed. (1980). "Renewable Resources. A Systematic Approach." Academic Press, New York.

Cheremisinoff, N. P., Cheremisinoff, P. N., and Ellerbusch, F. (1980). "Biomass. Applications, Technology, and Production." Marcel Dekker, New York.

Chum, H. L., and Baizer, M. M. (1985). "The Electrochemistry of Biomass and Derived Materials." ACS Monograph 183. Am. Chem. Soc., Washington, D.C.

Glasser, W. G., and Sarkanen, S., eds. (1989). "Lignin. Properties and Materials." ACS Symposium Series, No. 397, Washington, D.C.

Goldstein, I. S., ed. (1981). "Organic Chemicals from Biomass." CRC Press, Boca Raton, Fla.

Hakkila, P. (1989). "Utilization of Residual Forest Biomass." Springer, Berlin.

Kennedy, J. F., Phillips, G. O., and Williams, P. A., eds. (1989). "Wood Processing and Utilization" (Cellucon 88 Conference). Ellis Horwood, Chichester.

Kennedy, J. F., Phillips, G. O., and Williams, P. A., eds. (1990). "Cellulose Sources and Exploitation" (Cellucon 89 Conference). Ellis Horwood, Chichester.

Kennedy, J. F., Phillips, G. O., and Williams, P. A., eds. (1992). "Ligno-cellulosics. Science, Technology, Development, and Use" (Cellucon 90 Conference). Ellis Harwood, Chichester.

Klass, D. L., ed. (1988). "Energy from Biomass and Wastes XI." Institute of Gas Technology, Chicago.

Kirk, T. K., and Chang H.-M., eds. (1990). "Biotechnology in Pulp and Paper Manufacture. Applications and Fundamental Investigations" (Proc. Fourth Int. Conf. on Biotechnology in the Pulp & Paper Industry), Butterworth-Heineman, Stoneham, MA.

Koch, P. (1985). "Utilization of Hardwoods Growing on Southern Pine Sites, Vol. 3, Products and Prospective," pp. 2543–3710. U.S. Dept. of Agriculture Forest Service, Agric. Handb. No. 605. U.S. Govt. Printing Office, Washington, D.C.

Leatham, G. F., and Himmel, M. E. (1991). "Enzymes in Biomass Conversion." ACS Symposium Series, No. 460. Am. Chem. Soc., Washington, D.C.

Lichtenthaler, F. W., ed. (1990). "Carbohydrates as Organic Raw Materials." VCH Verlagsgesellschaft, Weinheim.

Maloney, G. T. (1978). "Chemicals from Pulp and Wood Waste. Production and Applications." Noyes Data Corp., Park Ridge, N.J.

Overend, R. P., Milne, T. A., and Mudge, L. K., eds. (1985). "Fundamentals of Thermochemical Biomass Conversion." Elsevier, London.

Rowell, R. M., Schultz, T. P., and Narayan, R., eds. (1990). "Emerging Technologies for Materials and Chemicals from Biomass," ACS Symposium Series, No. 476. Am. Chem. Soc., Washington, D.C.

Sarkanen, K. V., and Tillman, D. A., eds. (1979). "Progress in Biomass Conversion," Vol. I. Academic Press, New York.

Shafizadeh, F., Sarkanen, K. V., and Tillman, D. A., eds. (1976). "Thermal Uses and Properties of Carbohydrates and Lignins." Academic Press, New York.

Soltes, E. J., ed. (1983). "Wood and Agricultural Residues. Research on Use for Feed, Fuels, and Chemicals." Academic Press, New York.

Soltes, E. J., and Milne, T. A., eds. (1988). "Pyrolysis Oils from Biomass. Producing, Analyzing, and Uppgrading." ACS Symposium Series, No. 376. Am. Chem. Soc., Washington, D.C.

Tillman, D. A. (1978). "Wood as an Energy Resource." Academic Press, New York.

Wise, D. L., ed. (1983). "Organic Chemicals from Biomass." Benjamin/Cummings, Menlo Park, Calif.

## Reviews and Articles

Alén, R. (1990). Conversion of cellulose-containing materials into useful products. In "Cellulose Sources and Exploitation. Industrial Utilization, Biotechnology, and Physico-Chemical Properties" (J. F. Kennedy, G. O. Phillips, and P. A. Williams, eds.), pp. 453–464. Ellis Horwood, Chichester.

Alén, R., Moilanen, V.-P., and Sjöström, E. (1986). Potential recovery of aliphatic acids from kraft pulping liquors. Tappi J. 69(2), 76–78.

Alén, R., Patja, P., and Sjöström, E. (1979). Carbon dioxide precipitation of lignin from pine kraft black liquors. Tappi 62(11), 108–110.

Alén, R., and Sjöström, E. (1985). Degradative conversion of cellulose-containing materials into useful products. In "Cellulose Chemistry and its Applications" (T. P. Nevell and S. H. Zeronian, eds.), pp. 531–544. Ellis Horwood, Chichester.

Alén, R., and Sjöström, E. (1991). Utilization of black liquor organics as chemical feedstocks. Proc. 6th Int. Symp. Wood & Pulping Chem., Melbourne, Vol. 2, pp. 357–360.

Alén, R., Sjöström, E., and Suominen, S. (1990). Application of ion-exclusion chromatography to alkaline pulping liquors; Separation of hydroxy carboxylic acids from inorganic salts. J. Chem. Tech. Biotechnol. 51, 225–233.

Alén, R., Sjöström, E., and Vaskikari, P. (1986). Ultrafiltration studies on alkaline pulping liquors. Cellulose Chem. Technol. 20, 417–420.

Ander, P., and Eriksson, K.-E. (1978). Lignin degradation and utilization by micro-organisms. In "Progress in Ind. Microbiology" (M. J. Bull, ed.) Vol. 14, pp. 1–58, Elsevier, Amsterdam.

Andersen, R. F. (1979). Production of food yeast from spent sulphite liquor. *Pulp Pap. Can.* **80**(4), 43–45.

Bansal, I. K., and Wiley, A. J. (1975). Membrane processes for fractionation and concentration of spent sulfite liquors. *Tappi* **58**(1), 125–130.

Bridgwater, A. V., and Double, J. M. (1991). Production of liquid fuels from biomass. *Fuel* **70**, 1209–1224.

Brink, D. (1981). Gasification. In "Organic Chemicals from Biomass" (I. S. Goldstein, ed.), pp. 45–62. CRC Press, Boca Raton, Fla.

Collins, J. W., Boggs, L. A., Webb, A. A., and Wiley, A. A. (1973). Spent sulfite liquor reducing sugar purification with dynamic membranes. *Tappi* **56**(6), 121–124.

Faix, O., Meier, D., and Fortmann, I. (1990). Thermal degradation of wood. Gas chromatographic separation and mass spectrometric characterization of monomeric lignin products. *Holz Roh-und Werk.* **48**, 281–285.

Forss, K., and Fuhrmann, A. (1976). KARATEX—The lignin-based adhesive for plywood, particle board and fibre board. *Pap. Puu* **58**, 817–824.

Forss, K., and Passinen, K. (1976). Utilization of the spent sulphite liquor components in the Pekilo protein process and the influence of the process upon the environmental problems of a sulphite mill. *Pap. Puu* **58**, 608–618.

Hassi, H., Tikka, P., and Sjöström, E. (1981). Separation of sulfite spent liquor components by ion exclusion chromatography. *Proc. Int. Symp. Wood & Pulping Chem., The Ekman-days*, Stockholm, Vol. 5, pp. 65–69.

Herric, F. W., Casebier, R. L., Hamilton, J. K., and Wilson, J. D. (1975). Mannose chemicals. *Appl. Polym. Symp.* **28**, 93–108.

Herrick, F. W., and Hergert, H. L. (1977). Utilization of chemicals from wood: retrospect and prospect. In "The Structure, Biosynthesis, and Degradation of Wood" (F. A. Loewus and V. C. Runeckles, eds.), pp. 443–515. Academic Press, New York.

Holmbom, B. (1978). Constituents of tall oil. A study of tall oil processes and products. Diss., Åbo Akademi, Åbo.

Holmbom, B., and Eckerman, C. (1983). Tall oil constituents in kraft pulping—Effect of pulping temperature. *Tappi J.* **66**(5), 108–109.

Kahila, S. K. (1971). Yield quality and composition of crude tall oil. Kem. Teol. **28**, 745–756 (In Finn.).

Ketcham, M. (1990). Black liquor separation and acidulation: Changing some old habits. Tappi **73**(2), 107–111.

Kieboom, A. P. G., and van Bekkum, H. (1984). Aspects of the chemical conversion of glucose. *Recl. Trav. Chim. Pays-Bas* **103**, 1–12.

Lin, S. Y. (1983). Lignin utilization: Potential and challenge. In "Progress in Biomass Conversion" (D. A. Tillman and E. C. Jahn, eds., Vol. 4), pp. 31–77. Academic Press, New York.

Meshitsuka, G. (1991). Utilization of wood and cellulose for chemicals and energy. In "Wood and Cellulosic Chemistry" (D. N.-S. Hon and N. Shiraishi, eds.), pp. 977–1013. Marcel Dekker, New York.

Milsom, P. E., and Meers, J. L., eds. (1985). Gluconic and itaconic acids. In "Comprehensive Biotechnology," Vol. 3 (H. W. Blanch, S. Drew, and D. I. C. Wang, eds.), pp. 681–700. Pergamon Press, Oxford.

Molton, P. M., and Demmit, T. F. (1978). A literature survey of intermediate products formed during the thermal aqueous degradation of cellulose. *Polym.-Plast. Technol. Eng.* **11**(2), 127–157.

Niemelä, K., and Sjöström, E. (1986). The conversion of cellulose into carboxylic acids by a drastic alkali treatment. *Biomass* **11**, 215–221.

Rollings, J. (1985). Enzymatic depolymerization of polysaccharides. *Carbohydr. Polym.* **5**, 37–82.

Sharma, R. K., and Bakhshi, N. N. (1990). Efficient utilization of waste tall oil from the kraft pulp industry. *Tappi J.* **73**(9), 175–180.

Sjöström, E. (1983). Alternatives for balanced production of fibers, chemicals, and energy from wood. *J. Appl. Polym. Sci.: Appl. Polymer Symp.* **37**, 577–592.

Sjöström, E. (1991). Carbohydrate degradation products from alkaline treatment of biomass. *Biomass Bioenergy* **1**, 61–64.

Soltes, E. J., and Elder, T. J. (1981). Pyrolysis. *In* "Organic Chemicals from Biomass" (I. S. Goldstein, ed.), pp. 63–99. CRC Press, Boca Raton, Fla.

Theander, O., and Nelson, D. A. (1988). Aqueous, high temperature transformation of carbohydrates relative to utilization of biomass. *Adv. Carbohydr. Chem. Biochem.* **46**, 273–326.

Uloth, V. C., and Wearing, J. T. (1989). Kraft lignin recovery: Acid precipitation versus ultrafiltration: Part II: Technology and economics. *Pulp Pap. Can.* **90** (10), T357–T360.

Viikari, L., Kantelinen, A., Rättö, M., and Sundquist, J. (1991). Enzymes in pulp and paper processing. *In* "Enzymes in Biomass Conversion," (G. F. Leatham and M. E. Himmel, eds.), ACS Symposium Series, No. 460, pp. 12–21. Am. Chem. Soc., Washington, D.C.

Vuorinen, T., Hyppänen, T., and Sjöström, E. (1991). Oxidation of D-glucose with oxygen in alkaline methanol-water mixtures: A convenient method of producing crystalline sodium D-arabinonate. *Starch* **43**, 194–198.

Zinkel, D. F. (1989). Naval stores. *In* "Natural Products of Woody Plants. Chemicals Extraneous to the Lignocellulosic Cell Wall" (J. W. Rowe, ed.), Vol. II, pp. 953–978. Springer, Berlin.

# INDEX

## A

Abienol, 98, 99
Abietane-type resin acids, 99–100
Abietic acid, 99, 100
Absolute configuration, 25–26
Acetaldehyde, from ethanol, 232
Acetals, 34–35
Acetate rayon, 211–212
Acetic acid
  from black liquor, 245, 247
  in black liquor, 158, 240, 241
  from ethanol, 231, 232
  fermentation, 240
  from hardwood neutral sulfite spent
    liquors, 240
  from polysaccharides during kraft pulping,
    150
  from sulfite spent liquors, 237, 238, 240
  in sulfite spent liquors, 137–138, 237
  in wood pyrolysis, 234, 235
Acetone
  in wood pyrolysis, 234, 235
  as a fermentation product, 240
Acetovanillone, in black liquor, 159
Acetyl groups
  in galactoglucomannans, 63, 64, 65
  in glucuronoxylan, 67
  hydrolysis, 66, 68

in acid sulfite pulping, 132
  effect on carbohydrate yield, 133–135
in kraft pulping, 142, 150
in neutral sulfite pulping, 161–162
in two-stage sulfite pulping, 133–135
Acidic groups, in wood and pulps, 162–164
Acrylamides, cross-linking of cellulose with, 219
Activated carbon, see Charcoal
Activation energy, of delignification in kraft
  pulping, 144–145
Active alkali, in kraft cooking liquor, 140
Acute toxicity, 195
Aglycon, 33
Aldaric acids, 40
Alditols, 22
  preparation, 42, 233
Aldonic acids, 39–40
  epimerization, 46
  fermentation, 240
  from monosaccharides during sulfite pulp-
    ing, 133
  in sulfite spent liquors, 137, 237
Aldonic acid end groups, introduction to
    polysaccharides
  during alkali pulping, 152–153
  during anthraquinone–alkali pulping,
    156–157

Aldonic acid end groups (*cont.*)
 during bleaching, 191–192
 during kraft pulping, 152–153
 during oxygen–àlkali bleaching, 192
Aldoses, 22
Aldosuloses, 40–42
Aliphatic acids
 from black liquors, 245, 247–248
 in black liquors, 158, 246
 for protein production, 240
Alkali cellulose, 208–209
 reactivity, 208–209
Alkali lignin, 72, *see also* Kraft lignin
Alkaline extraction stage, *see* Bleaching
Alkali pulping, *see also* Kraft pulping
 effect of anthraquinone, 118, 156–157
 history, 116
 lignin reactions
  oxirane intermediates, 145–147
  stilbene structures, 147
  styryl aryl ether structures, 145–147
β-Alkoxy elimination, 47
Alkylidene derivatives, 34–35
Allose, 26
Allulose, 47
Altrose, 26
Aminoethylcellulose, as anion exchanger,
 223–224
Amylopectin, 54
Amylose, 54
Angelica lactones, from levulinic acid, 46
Anhydro sugars, 36–37
Annual ring, 2, 4–5
Anomeric effect, 31, 33
Anomers, 28
Anthraquinone–alkali pulping, 156–157
Antidaniellic acid, 100, 101
AOX, 195
Arabinan, in bark, 113
Arabinogalactan, 66
Arabinoglucuronoxylan, 65, 66, *see also*
 Glucuronoxylan; Xylan
 acid hydrolysis, 66, 132
 alkaline degradation, 66, 154
 content in softwood, 64, 249
Arabinonic acid end groups, introduction to
 polysaccharides during bleaching, 191,
 192

Arabinose, 22, 23, 26, 27
 in sulfite spent liquors, 137, 237
Arachidic acid, 104
Arachinol, 104
Arogenic acid, 75
Asymmetric carbon atom, 24
Asymmetric center, *see* Asymmetric carbon
 atom
Asymmetric induction, 44
Azo compounds, as initiators in cellulose
 grafting, 221

**B**

Balata, 92, 102, 230
Bark
 anatomy, 109–111
 chemical composition, 111–113
 hydrophilic fraction, 112, 113
 inorganic constituents, 113
 insoluble constituents, 113
 lipophilic fraction, 112
Bast fibers, 111
Behenic acid, 104
Behenol, 104
Benzilic acid rearrangement, 47–49
Betulaprenols, 92, 102, 103
Betulin, *see* Betulinol
Betulinol, 101, 102, 229
 in bark, 112
Biopulping, 118
Björkman lignin, *see* Milled wood lignin
Black liquor, composition, 158–160, 241,
 246
Bleach liquor
 chemical characteristics, environmental
  aspects, 194–198
 comparison of pollution loads, 197
 toxic (chlorinated) compounds, 196
Bleachability, factors affecting on, 177–179
Bleaching, *see also* Bleach liquor; Lignin-
  preserving bleaching; Oxygen–alkali
  bleaching
 alkaline extraction, 182
 background and history, 166–170
 carbohydrate reactions, 188–193
 concepts, 165–166
 conditions and performance, 168, 169

general aspects, 177–179
inorganic chemistry, 170–177
lignin reactions, 180–188
pollution load, see Bleach liquor
reactions of extractives, 193–194
sequences, 168, 169
spent liquor, see Bleach liquor
Boat conformation, 30
BOD, 195
Borate complex, of glucomannans, 69
Borneol, 97
Bornyl acetate, 97
Borohydride, redox potential of, 199
Brightness
of fully bleached pulps, 185
of prebleached pulps, 182
stability, of high-yield pulps, 201
Butanol, as a fermentation product, 232, 240

**C**

Cadalanes, 98
δ-Cadinene, 98
α-Cadinol, 98
Cadmium ethylenediamine, 207, 208
Cadoxen, see Cadmium ethylenediamine
Caffeic acid, 76
Cahn–Ingold–Prelog convention, 25
Calcium oxalate, in bark, 113
Callose, in bark, 113
Cambium, 2, 3, 4
Campesterol, 101, 102
Camphene, 97
Carbamates, cross-linking of cellulose with, 219
Carbohydrates
addition of cyanide, 43, 44
analysis by enzymic methods, 50
biosynthesis, 51–54
chromatographic methods, 50
determination of structure, 50
esters of, see also specific name, 37, 38
ethers of, 35
gas–liquid chromatography-mass spectrometry, 50
isolation from sulfite spent liquors, 240
methylation, 35

NMR spectroscopy, 50
nomenclature, 21–22
oxidation
with bromine, 39–40
with lead tetraacetate, 42
with nitric acid, 40, 41
with periodic acid, 41, 42
reduction, 42
trimethylsilylation, 35
triphenylmethylation, 35
Carbon monoxide, in wood gasification, 236
Carbonyl groups
addition reactions, 42–44
condensation reactions, 42–44
introduction to polysaccharides during bleaching, 188–191
reduction, 42
Carboxyl groups, introduction to polysaccharides during bleaching, 188–189, 191–192
Carboxylic acid end groups, introduction to cellulose
during anthraquinone–alkali pulping, 156–157
during kraft pulping, 151–153
during oxygen–alkali bleaching, 192
during polysulfide pulping, 155
Carboxymethylcellulose
as cation exchanger, 223
preparation, 216–217
solubility, 217
uses, 217
3-Carene, 97
conversion to m-cymene, 239
in sulfate turpentine, 241–242
Carotenoids, 94, 96
Caoutchouc, see Rubber
Catechin, 105, 106
Causticizing efficiency, 140
CED, see Cupriethylenediamine
α-Cedrene, 98
Cell
development, 4
division, 4
types, 5–11
Cell wall, layers, 13–17
Cellobiose, 38, 39

Cellophane, from cellulose xanthate, 218
Cellylolytic enzyme lignin, 72
Cellulose
  accessibility, 204–205
    effect on yield in kraft pulping, 150
  acetylation
    catalysts, 212–213
    by fibrous process, 213
    mechanism, 213
    morphological factors, 212
    by solution process, 212
  acid hydrolysis, 44–45
    effect of accessibility, 132
    glucose production, 230
  acidity of hydroxyl groups, 204
  activation, 209
  addition compounds, 208
  alkaline degradation, 46–49, 150–152
  alkylation, 214
  alkyl ethers, 214–215
  in bark, 113
  biosynthesis, 51–53
  carboxymethylation, 216–217
  chain conformation, see polymer properties
  content in wood, 54, 249
  cross-linking, 218–219
  crystalline structure, 54–57
  crystallinity, 205
  cyanoethylation, 217
  degree of polymerization, see molecular weight
  depolymerization during oxygen–alkali bleaching, 190–191
  derivatives
    degree of substitution, 206
    substituent distribution, 206
  diffusion rate, 59, 60, 61
  dissolution, mechanism, 207–208
  enzymatic hydrolysis, 231
  esterification, 209–213
    reactivity of hydroxyl groups, 204
  etherification, 214–217
    reactivity of hydroxyl groups, 204
  grafting, 219–222
  hydrogen bonds, 54–57
  hydroxyalkyl ethers, 215–216
  hydroxyethylation, 215–216
  immunological adsorbents from, 224

  inclusion compounds, 209
  inorganic esters, 209–211
  methylation, substituent distribution, 206
  mixed organic esters, 213
  molecular structure, 54–57
  molecular weight, 58–59, 60
  nitration, 210
  organic esters, 211–213
    with basic groups, 213
  polydispersity, 59, 60
  polymer properties, 58–62
  reactivity, 204–206
    influence of hydrogen bonds, 205
    morphological factors, 204–205
  sedimentation rate, 59, 60, 61
  swelling, 206–207
    alkali, 208, 209
    electrolytes, 207
    organic bases, 207, 208
    X-ray measurements, 209
  viscosity, 59, 60, 61
  water adsorption, 207
  yield in kraft pulping, 133, 150
  yield in sulfite pulping, 132, 133
Cellulose I, 55, 56, 57
Cellulose II, 55, 56, 57, 58
  from Cellulose I, 56, 57, 209
Cellulose III, 57
Cellulose IV, 57
Cellulose acetate
  preparation, 211–213
  raw materials for, 212
  solubility and uses, 212
Cellulose esters, 209–213
Cellulose ethers, 214–217
  commercial types, 214
  solubility, 214
Cellulose ion exchangers, 222–224
  immunological adsorbents from, 224
Cellulose nitrate
  preparation, 210
  solubility and uses, 210
  sulfate groups in, 210
Cellulose phosphate, 211
Cellulose solvents, 207, 208
Cellulose sulfate, 210–211
Cellulose xanthate, 218
Cembrene, 98, 99

Ceric ions, as initiators in cellulose grafting, 221, 222
Chain transfer, in cellulose grafting, 220, 221
Chanootin, 98
Chair conformation, 30–32
Charcoal, in wood pyrolysis, 235
Chemimechanical pulp, 114, 115, 117
Chemithermomechanical pulp, 115, 117
Chemotaxonomy, classification based on extractives, 105
Chiral center, see Asymmetric carbon atom
Chlorinated phenolics, 196
Chlorine
    equilibria in aqueous solutions, 170, 171
    in radical chain reactions, 171, 172
Chlorine bleaching
    carbohydrate reactions, 188–191
    conditions, 169
    effect of chlorine dioxide, 171, 172, 185
    kinetics, 171, 180
    lignin reactions, 180–182
Chlorine dioxide
    oxidation equivalents, 172
    oxidation pathways, 172, 173
    as radical scavanger, 171, 172
Chlorine dioxide bleaching
    conditions, 169
    lignin reactions, 185
    selectivity, 188
Chlorine radicals
    formation and action, 171, 172
    radical scavangers of, 171, 172
    reactions with carbohydrates, 188–191
Chlorolignin, 197
Chromophores
    formation during kraft pulping, 149
    in prebleached pulps, 182
    in pulps, 149, 198, 199, 201–203
Chorismic acid, 75
Chrysin, 105, 106
Cinnamate pathway, 75, 76
Cinnamic acid, 76
Cinnamyl alcohol, 73
Citrostadienol, 101, 102
CMC, see Carboxymethylcellulose
COD, 195
Combined sulfur dioxide, 120
Communic acid, 100, 101

Compression wood, 19–20
Conidendrin, 105, 106
    isolation from wood and properties, 229
Coniferin, 80
    in bark, 112
Coniferyl alcohol, 73, 76, 77
Coniferyl aldehyde, 77
Cork cells, 111
Cork layer, see Periderm
Cortex, 109, 110
Cotton cellulose
    crystallinity, 205
    degree of polymerization, 59
    as raw material for cellulose derivatives, 212
p-Coumaric acid, 76
p-Coumaryl alcohol, 73, 76
Cresols, in wood tar, 235
Crystallites, in cellulose, 12, 13
Cupriethylenediamine, 207, 208
Cutin, in bark, 113
Cyanoethylcellulose
    preparation, 217
    solubility and uses, 217
Cyclitols, 22
Cycloartenol, 101, 102
p-Cymene
    from α-pinene during sulfite pulping, 135, 136
    recovery and uses, 239

**D**

Dehydroabietic acid, 99, 100
Dehydrodiveratric acid, 74
Dehydrogenase polymer lignin, 72, 88, 89
3-Dehydroquinic acid, 73, 75
3-Dehydroshikimic acid, 75
Delignifying bleaching, see Bleaching
2-Deoxyglyceric acid, see 3-Hydroxypropanoic acid
3-Deoxypentonic acid, 158, 246
Dialdoses, 40, 41
Diastereoisomers, 24
Dichlorine monoxide, 170
3,4-Dideoxypentonic acid
    in black liquors, 158, 246
    from cellulose in alkali, 150, 151
    from glucomannan in alkali, 150, 151

Diethylaminoethylcellulose, as anion exchanger, 223, 224
Dihydroagathic acid, 100, 101
Dihydroconiferyl alcohol, in black liquor, 159
Dihydroquercetin, *see* Taxifolin
3,4-Dihydroxybutanoic acid, from polysaccharides by oxygen in alkali, 192, 193
2,5-dihydroxypentanoic acid, *see* 3,4-Dideoxypentonic acid
Dimethyl disulfide, 147, 148
Dimethyl sulfide, 147, 148
from kraft lignin, 245
Dimethyl sulfone, from dimethyl sulfoxide, 245
Dimethyl sulfoxide
as cellulose solvent, 207
from dimethyl sulfide, 245
Dioxane lignin, 72
Dioxins, *see* Bleach liquor, toxic (chlorinated) compounds
Dioxin value, 196
Dipentene, 97
Disaccharides, 38, 39
Displacement bleaching, 167
Dissolving pulp, 116, 134
Diterpenes, *see* Diterpenoids
Diterpenoids, 98–100
Dithionite
redox potential, 199
stability, 199, 200

**E**

Earlywood, 4, 5
tracheids in, 7–8
ECTEOLA-cellulose, as anion exchanger, 223, 224
Effective alkali, 140
Effective capillary cross sectional area, 142
Eicosatrienoic acid, 104
Einstein's equation, 62
Elementary fibril, 12
Ellagic acid, 105, 106
in bark, 112
Elliotinic acid, 100, 101
Enantiomers, 24, 25
Enzymes, application for bleaching, 169
Epidermis, 109, 110

β-Epimanool, 98, 99
Epimers, 28
Epithelial cells, 10, 92, 93
Epoxides, 36, 37
Erythronic acid end groups, formation during bleaching, 191, 192
Erythrose, 25, 26, 27
D-Erythrose 4-phosphate, 23, 73, 75
Ethanol
from sulfite spent liquors by fermentation, 237, 238, 240
from wood hydrolysates, 231, 232
further processing to chemicals, 231, 232
Ethanolysis, 74, 80
Ethyl alcohol, *see* Ethanol
Ethylcellulose, 214, 215
Ethylene, from ethanol, 231, 232
Ethylene oxide, from ethanol, 231, 232
Extractives, *see also* Fats; Phenolic extractives; Pitch; Rosin; Steroids; Terpenoids; Turpentine; Waxes; specific name
analysis, 91
in bark, 112–113
biosynthesis, 92, 94, 95, 96
from black liquors, 240–244
changes during wood storage, 107–108
chlorination products, 193, 194, 196
content in wood, 90, 91, 249
fungicidal properties, 105
inorganic components, 107
occurrence
as oleoresin, 92
as parenchyma resin, 92
reactions
during bleaching, 193–194
during kraft pulping, 157
during sulfite pulping, 135–136
use as chemicals, 227–229, 239, 240–244
Exudate gums, 63

**F**

Farnesyl pyrophosphate, 94, 95, 96
Fats, *see also* Extractives
in bark, 112
in wood, 103–104
Fatty acids

as components in fats and waxes, 104
  from black liquor, 243–244
Ferulic acid, 76
  in suberin, 113
Fiber, see Cell
Fiber fractionation, for removal of resin, 103
Fischer formulas, 24, 25
Flavonoids, 105, 106
  in bark, 112
  uses, 229
Flory's equation, 61
Formaldehyde
  condensation with lignin during kraft
    pulping, 148, 149
  cross-linking of cellulose with, 218–219
  formation from lignin during kraft pulp-
    ing, 146, 147
Formic acid
  in black liquors, 158, 240, 241
  from black liquors, 245, 247
  from hardwood neutral sulfite spent li-
    quors, 240
  from hexoses, 46
  from polysaccharides during kraft pulping,
    150, 151
  from polysaccharides by oxygen in alkali,
    192, 193
Free sulfur dioxide, 120
Fringe micellar model, 12, 13
Fructose, 23, 26, 27, 28, 32
Furanosides, 34
Furfural
  from pentoses, 45, 46, 230–231
  raw materials for, 230
  from sulfite spent liquors, 238
  in sulfite spent liquors, 138
  in wood pyrolysis, 235
  uses, 230

## G

Galactan, in compression wood, 66–67
Galactoglucomannans, 63, 64, 65
  acid hydrolysis, 66, 132
  changes during kraft pulping, 150–153
  content in softwood, 64, 249
  linkages to lignin, 85–86
Galactose, 22, 23, 26, 27
  in sulfite spent liquors, 137, 237

Galacturonic acid, in glucuronoxylan, 68
Gallic acid, 105, 106
  in bark, 112
Gasification, 235–236
Gas-phase bleaching, 167
GDP-D-glucose, 51, 52, 53
Gelatinous layer, in tension wood, 20
Genistein, 105, 106
Geranyl pyrophosphate, 94, 95, 96
Geranylgeranyl pyrophosphate, 94, 96
Geranyl-linalool, 98, 99
Glucaric acid, 40
Glucitol, 22, 42, 233
Glucoisosaccharinic acid
  from black liquors, 245–248
  in black liquors, 158, 246
  from cellulose in alkali, 46–47, 150–152
  from glucomannan in alkali, 150–152
Glucomannan, 68, see also Galac-
    toglucomannans
  acid hydrolysis, 132
  content in hardwood, 64, 249
  yield in kraft pulping, 133, 150, 152
  yield in sulfite pulping, 133
  yield in two-stage sulfite pulping, 134,
    135
Gluometasaccharinic acid end groups, intro-
    duction to cellulose in alkali, 47, 48,
    49, 151, 152
Gluconic acid, 39, 40, 231, 232, 233
Glucopyranosides
  acid hydrolysis, 44–45
  alkaline hydrolysis, 48, 49
Glucosaccharinic acid, see 2-C-Methyl-
    ribonic acid end groups
Glucose, 21, 22, 23, 26, 27
  from cellulose, 230, 231
  chair conformations, 30, 31, 32
  chemicals from, 232, 233
  GDP-D-glucose, 51, 52, 53
  mutarotation, 29–30
  oxidation, 39, 40, 41, 233
  reduction, 42, 233
  in sulfite spent liquors, 137, 237
  UDP-D-glucose, 51–53
β-D-Glucosidase, 73
Glucuronic acid, from glucose, 40, 41
Glucuronoxylan, 67, see also Ara-
    binoglucuronoxylan; Xylan

Glucuronoxylan (*cont.*)
  acid hydrolysis, 67–68
  content in hardwood, 64, 249
  reactions in alkali, 150–155
Glyceraldehyde, 23, 24
  from polysaccharides in alkali, 150, 151
  stereoisomers of, 24
Glyceric acid, from polysaccharides by oxygen in alkali, 192, 193
Glycitols, 22
Glycodiuloses, 40, 41
Glycolaldehyde, from polysaccharides in alkali, 151, 152
Glycolic acid
  in black liquor, 158, 246
  from black liquor, 245–248
  from polysaccharides by oxygen in alkali, 192, 193
Glycols, in hydrogenolysis of carbohydrates, 233
Glycosans, 36
Glycosides, 33–34
Glycosidic bonds
  acid hydrolysis, 44–45, 132
  alkaline hydrolysis, 48, 49, 150
Glycuronide bonds, acid hydrolysis, 45, 132
Graft copolymers, with cellulose, 219–222
Green liquor, causticizing, 160, 161
Guaiacol
  in black liquor, 147, 159
Guanosine, 51, 52
Gulose, 26
Gum arabic, 63

**H**

H Factor, 144, 145
Half-chair conformation, 30, 31
Hardwood
  cells, 10–11
  chemical composition, 249
Haworth formulas, 27, 28
Head-to-tail coupling, 94, 95
Heartwood, 11
  typical extractives in, 90, 91, 105
Hemicelluloses, 63–70, *see also* specific name
  acid hydrolysis, sulfite pulping, 132, 133

  analytical methods, 50, 69
  in bark, 113
  biosynthesis, 52, 53, 54
  chemical bonds to lignin, 84–86
  chemical composition, 63–68
  in compression wood, 66, 67
  content in wood, 64, 249
  degree of polymerization, 63, 64
  distribution in wood, 70
  extraction, see isolation
  in hardwood, 67–68
  isolation, 69
  monomeric units of, 63, 64
  precipitation and purification, 69
  in softwood, 63–67
  solubility, 64
  summarizing table, 64
  yield after pulping, 133, 135
Hexoses, fermentation, 231, 232
Hibbert ketones, 74
High-yield pulping
  acidic groups, 162–164
  definitions, 114, 115, 116–117
  material losses, 161–162
Hydrocarbons
  in wood gasification, 236
  *in wood pyrolysis*, 235
Hydrogen chloride, formation during chlorine bleaching, 180
Hydrogen peroxide
  dissociation, 174, 200
  stabilization, 201
Hydrogen peroxide-ferrous salts, as initiator in cellulose grafting, 221
Hydrogen sulfide, dissociation, 140–141
Hydroperoxides
  decomposition, 175, 176, 186, 187
  formation, 174, 175
  from lignin
    by hydrogen peroxide, 186
    by oxygen, 186
Hydroquinones, *see* Quinones
Hydroxy acids
  from black liquors, 245–248
  in black liquors, 158, 241, 246
  from polysaccharides
    during kraft pulping, 150–152
    during oxygen–alkali treatment, 192, 193

*p*-Hydroxybenzaldehyde, from lignin, 74
*p*-Hydroxybenzoic acid, 83
2-Hydroxybutanoic acid
   in black liquors, 158, 246
   from black liquors, 245–248
   from xylan in alkali, 150–152
Hydroxy carboxylic acids, see Hydroxy
   acids
*p*-Hydroxycinnamic acid, 76, 83
β-Hydroxy elimination, 47–49
Hydroxyethylcellulose
   molar substitution, 214, 216
   preparation, 215–216
   solubility, 214, 216
   uses, 216
Hydroxyethylethylcellulose, 215
Hydroxyethylmethylcellulose, 215
Hydroxyl radicals, as initiators in cellulose
   grafting, 221
Hydroxymatairesinol, 105, 106
Hydroxymethylfurfural, from hexoses, 46,
   231
*p*-Hydroxyphenylpyruvic acid, 75
2-Hydroxypropanoic acid, see Lactic acid
3-Hydroxypropanoic acid, from xylan by
   oxygen in alkali, 192, 193
Hydroxypropylcellulose, 215–216
Hydroxypropylmethylcellulose, 215
α-Hydroxysulfonic acids, 44
   in sulfite spent liquors, 138
Hypochlorite
   decomposition, 172
   oxidation equivalents, 172
   preparation, 172
Hypochlorite bleaching
   carbohydrate reactions, 188–191
   conditions, 168, 169
   lignin reactions, 182, 184, 185
Hypochlorous acid, dissociation constants,
   170, 171

I

Idose, 26
Inorganic components
   in bark, 113
   in wood, 107
Inositols, 22
Intercrystalline swelling, 207

Interfibrillar swelling, 207
Intracrystalline swelling, 207
Intrafibrillar swelling, 207
Inulin, 23
Ion exchange properties, of wood and
   pulps, 162–164
Ion exchangers
   from cellulose, 222–224
   from lignosulfonates, 239
Iron tartrate cellulose complex, 208
Isohemipinic acid, 74
Isopimaric acid, 99, 100
Isoprene rule, 94
Isoprene units, 92, 94
Isoprenoids, 92
Isosaccharinic acid, see Glucoisosaccharinic
   acid; Xyloisosaccarinic acid

J

Juniperol, 98
Juvabione, isolation from wood, 228

K

Ketoses, 22, 23
Kiliani reaction, 43, 44
Klason lignin, 72
Kraft black liquor, see Black liquor
Kraft lignin, 72
   from black liquor, 244, 247
   degradation products, 245
   demethylation, 245
   functional groups, 159
   molecular weight, 159
   precipitation, 244, 247
   sulfonation, 245
   sulfur content, 159
   ultrafiltration, 245
   uses, 244, 245
Kraft pulping
   background, 116
   by-products, 240–248
   carbohydrate losses, reasons, 150–155
   carbohydrate yield, 133
   consumption of alkali, 142
   cooking chemicals, 140–141
      diffusion into chips, 142
      recovery of, 160–161

Kraft pulping (cont.)
  cooking liquor
    equilibria in, 140–141
  delignification
    general aspects, 142–145
    kinetics, 143–145
    morphological factors, 122–123, 142
    selectivity, compared with sulfite pulping, 124
  extractives, reactions, 157
  history, 116
  hydrogen sulfide pretreatment, 155, 156
  impregnation, 142
  lignin reactions
    cleavage of β-aryl ether bonds, 145, 146
    cleavage of α-ether bonds, 146, 147
    condensation, 148, 149
    formation of chromophores, 149
    formation of conjugated structures, 146, 147
    liberation of phenolic hydroxyl groups, 145
    oxirane intermediates, 146
    quinone methide intermediates, 146, 147
    thiirane intermediates, 146
  polysulfide modification, 155, 156
Kraft spent liquor, see Black liquor

**L**

Labdane-type resin acids, 100, 101
Lactic acid
  from black liquors, 245–248
  in black liquors, 158, 246
  from polysaccharides in alkali, 150–152
  from sulfite spent liquors by fermentation, 240
Lactones, 40
Lambertianic acid, 100, 101
Laricinan, in compression wood, 67
Latewood, 5
  tracheids in, 7–11
Leaving group, 36
Leucochromophores, see Chromophores
Levoglucosan
  from hexoses by acid, 36

from β-D-glucopyranosides by alkali, 48, 49
  in wood pyrolysis, 235
Levopimaric acid, 99, 100
Levulinic acid, from hexoses, 46
Libriform cells, 11
Lignans, 105, 106
  extraction from wood, 229
Lignin
  acid-soluble, 72
  acidolysis, 80
  β-aryl ether bonds
    cleavage by alkali, 145, 146
    cleavage by anthrahydroquinone, 156, 157
    cleavage by hydrogen sulfide ions, 145, 146
    cleavage by neutral and alkaline sulfite, 129, 130
  in bark, 113
  benzyl alcohol groups, see α-hydroxyl groups
  biosynthesis, 73–80
  carbonyl groups, 82, 83, 159
  chemical structure, 80–86
  chlorination, 180–182
  classification, 86–88
  condensation reactions during kraft pulping, 148, 149
  condensation reactions during sulfite pulping, 128–129
  content in wood, 72, 249
  dehydrogenative formation, see biosynthesis
  distribution
    in wood fibers and cell walls, 86–89
    in softwood pulp fibers, 122, 123
  electrophilic side chain displacement by chlorine, 180, 181
  ethanolysis, 74, 80
  α-ether bonds
    cleavage by acid, 127, 128
    cleavage by alkali, 146, 147
  formula, 83, 84
  functional groups, 82, 83, 159
  gravimetric determination, 72
  guaiacyl-type, 86–88
  hydrogenolysis, 74

hydroxyl groups, 83, 159
α-hydroxyl groups, 83
  cleavage by acid, 127
  cleavage by alkali, 146
p-hydroxyphenyl-type, 87
isolation, 71–72
linkage types, 80–82
  biosynthetic formation, 77–80
  proportions, 82
linkages to carbohydrates, 84–86
methoxyl groups, 83
  cleavage by hydrogen sulfide ions, 147, 148
  cleavage by sulfite ions, 131
  cleavage during bleaching, 180, 181
molecular weight, 88, 89, 159
oxidation
  with chlorine, 180–182, 184, 185
  with chlorine dioxide, 185
  with hydrogen peroxide, 186, 187
  with hypochlorite, 182, 184, 185
  with nitrobenzene, 74
  with oxygen, 185–187
  with ozone, 187, 188
  with permanganate, 74
phenolic hydroxyls, 82, 83
phenylpropane unit, 73
polydispersity, 88, 89, 159
polymer properties, 88–89
precursors
  biosynthesis, 73–77
  polymerization, 77–80
structural methods, 80, 81
structural units and linkage types, 80–82
sulfonation, pH-dependence, 124, 125, 126
  in acidic conditions, 128–129
  in neutral and alkaline conditions, 129–131
synthetic preparations, 71–72
syringyl-type, 87
thioacetolysis, 80, 81
thioacidolysis, 80, 81
UV-spectrophotometry, 72
viscosity, 88–89
Lignin–carbohydrate complex, 85
Lignin precursors, 73, 76
  biosynthesis, 73–77

bulk polymerization, 78–80
end-wise polymerization, 77–80
Lignin-preserving bleaching, 198–201
  oxidative-type, 200–201
  reducing-type, 199–200
Lignin-removing bleaching, see Bleaching
Lignoceric acid, 104
Lignocerol, 104
Lignols, 78, 80
Lignosulfonates
  as cation exchangers, 239
  isolation from sulfite spent liquors, 238, 239, 240
  solubilizing effect on resin, 135
Limonene, 95, 97
Linoleic acid, 104
Linolenic acid, 104
Lobry de Bruyn-Alberda van Ekenstein transformation, 46, 47
Longifolene, 98
Lyxose, 26

M

Magnesium-salt, inhibitor in oxygen–alkali bleaching, 176, 190
Maltose, 38, 39
Manganic ions, as initiators in cellulose grafting, 222
Mannitol, 42, 233
Mannonic acid end groups, introduction to polysaccharides during oxygen–alkali bleaching, 192
Mannose, 21, 22, 23, 26, 27
  in sulfite spent liquors, 137, 237
Manoyloxide, 98, 99
Margo, 6, 18
Mark-Houwink equation, 62
Matairesinol, see Hydroxymatairesinol
Mechanical pulp types, 115, 117
Mercerization, 208, 209
Mercerized cotton, crystallinity, 205
Mercusic acid, 100, 101
Metasaccharinic acid end groups, introduction to polysaccharides in alkali, 150, 151, 152
Methane, in wood gasification, 236
Methanesulfonic acid, 131

Methanol
  formation during bleaching, 180, 181
  in wood pyrolysis, 235
Methoxyl groups, 83, see also Lignin
  in xylan, 65, 66, 67
    cleavage during kraft pulping, 154
Methylcellulose
  preparation, 214–215
  solubility and properties, 214, 215
4-O-Methylglucuronic acid groups
  in arabinoglucuronoxylan, 65, 66
  in glucuronoxylan, 67, 68
2-C-Methylglyceric acid end groups, intro-
    duction to polysaccharides in alkali,
    150, 151, 152
Methylglyoxal
  from polysaccharides in alkali, 150, 151
Methyl mercaptan, 147, 148, 245
2-C-Methylribonic acid end groups, intro-
    duction to polysaccharides in alkali,
    150, 151, 152
Microfibril, 13, 14, 54
Middle lamella, 13, 14, 15
Milled wood lignin, 72
Monosaccharides, see also Carbohydrates;
    specific name
  addition of hydrogen cyanide, 43, 44
  addition of hydrogen sulfite, 44
  configuration, 23–26
  conformation, 30–33
    nomenclature system, 31, 32, 33
  degradation by acid, 45, 46
  derivatives, 33–38
  DL-convention, 25
  fermentation products, 231–233, 240
  Fischer projection formula, 24
  furanoid structures, 27, 28
  glycosidic hydroxyl, 28
  Haworth projection formula, 27, 28
  isomerization in alkali, 46, 47
  Mills formula, 28
  mutarotation, 29–30
  oxidation to aldonic acids during sulfite
    pulping, 133
  pyranoid structures, 27, 28
  reduction, 42
  ring structures, 26–28
  in sulfite spent liquors, 137, 237
  from sulfite spent liquors, 240

Monoterpenes, see Monoterpenoids
Monoterpenoids, 94–97
  in sulfate turpentine, 241–242
Muconic acid structures, formation from lig-
    nin during bleaching, 180, 181, 185,
    186, 187, 188
α-Muurolene, 98
β-Myrcene, 95, 97
  preparation and uses, 228

N

Naval stores industry, 227
Neoabietic acid, 99, 100
Nonreducing disaccharides, 38
Nootkatin, 98
Nucleosides, 51, 52

O

Oleic acid, 104
Oleoresin, 92, 93
  in bark, 112
  recovery from trees and uses, 227–228
Oligosaccharides, 22
  occurrence, 38–39
Organic excess sulfur, 136, 137
Organosolv pulping, 118
Oxiranes, 36, 37
Oxygen
  energy states, 173, 174
  in radical chain reactions, 174–176
Oxygen–alkali bleaching, see also Bleach-
    ing
  carbohydrate reactions, 190–193
  chemicals recovery, 194
  conditions, 169
  effect on pollution load, 194
  inorganic chemistry, 173–176
  lignin reactions, 185–187
  selectivity, 190, 191, 192
Oxygen–alkali pulping, 118, 156
Ozone bleaching, 177, 187, 188

P

Palmitic acid, 104
Palustric acid, 99, 100
Paraformaldehyde, as cellulose solvent, 207

Paraquat, 227
Parenchyma cells, 5, 6
  in bark, 112
  delignification, 136
  in hardwood, 10–11
  in softwood, 7–10
Parenchyma resin, 92, 103
Pectic polysaccharides, 66, 67, 132, 133
Peeling reaction, of polysaccharides, 46–48,
  150–152
Pekilo protein, 240
Pentoses, fermentation, 240
Periderm, 110
Periodic acid oxidation, 41, 42
Peroxides, as initiators in cellulose grafting,
  221
Peroxide bleaching, lignin-preserving, 200–
  201
β-Phellandrene, 97
Phelloderm, 111
Phellogen, 110
Phenolic extractives, 104–106
  in bark, 112
  classification of wood species according
    to, 105
  condensation with lignin, 129
  isolation from wood and bark, 229
Phenols, in wood pyrolysis, 235
Phenoxy radicals
  in biosynthesis of lignin, 80
  from phenolates by oxygen, 175
Phenylalanine, 75, 76
Phenylosazone, 42, 43
Phenylpyruvic acid, 75
Phloem, 2, 109
Phosphochorismic acid, 75
Phosphonomethylcellulose, as cation ex-
  changer, 223
Phytanes, 98
Phytoene, 94
Phytol, 98
Pimaral, 98, 99
Pimarane-type resin acids, 99, 100
Pimaric acid, 99, 100
Pimarol, 98, 99
Pine oil
  synthetic, from monoterpenes, 227, 228
  in wood pyrolysis, 235
α-Pinene, 97

dehydrogenation during sulfite pulping,
  135, 136
  isomerization, 228
  in sulfate turpentine, 241–242
β-Pinene, 97
  pyrolysis to myrcene, 228
  in sulfate turpentine, 241–242
Pinolenic acid, 104
Pinoresinol, 105, 106
Pinosylvin, 105, 106
  condensation with lignin, 129
Pitch
  in tall oil distillation, 242, 243
  in wood tar, 234, 235
Pitch problems, 103, 107, 136, 157, 193–
  194
Pith, 2
Pit membrane, 6
Pit pair, 6
Pits, 6–7, 17–18
  in hardwood cells, 10–11
  in softwood cells, 7–10
  ultrastructure, 17–18
Plicatic acid, 105, 106
  extraction from wood, 229
Polychlorinated phenolics, 196
Polyestolides, see Suberin
Polyhydric alcohols, 22
Polymers
  intrinsic viscosity
    relation to hydrodynamic volume, 62
    relation to molecular weight, 62
    relation to polymer conformation, 59,
      60
  molecular weights, 58–59
  molecular weight distribution, 59
  stiffness, 60, 61, 62
  theta conditions, 61
Polyphenols, 105
  uses, 229
Polyprenols, 102, 103
Polysaccharides, see also Carbohydrates;
    Cellulose; Hemicelluloses; specific
    name
  in bark, 113
  content in wood, 64, 249
  degradation
    by acid, 44–46
    by alkali, 46–49

Polysaccharides (*cont.*)
  by bleaching agents, 188–193
  during wood storage, 107
  polydispersity, 59, 60
  reactivity, accessibility factors, 143
  stabilization against alkaline degradation,
    155, 156
Polysulfide pulping, 155, 156
Polysulfides
  preparation, 155
  stabilization of polysaccharides, 155, 156
Polyterpenes, *see* Polyterpenoids
Polyterpenoids, 102–103
Polythionates, 136
Polyuronides, 22
Pores, *see* Pits
Prenylation, 94
Prephenic acid, 11
Primary wall, 13–14
Prosenchyma cells, 5, 6
Proteins
  from sulfite spent liquors by fermentation,
    240
  from wood hydrolyzates, 231, 232
Pulp, mechanical treatment, material losses,
    162
Pulp types, 115
Pulping methods, *see* specific name
  background and definitions, 114–118
Pyranosides, 34
Pyrolysis, 234–235
Pyruvic acid, 73, 75

**Q**

Quercetin
  in bark, 112
  from taxifolin during sulfite pulping, 135,
    136
Quinones, redox-potentials of, 199

**R**

Radiation sources, as initiators for cellulose
    grafting, 222
Raffinose, in bark, 113
Ray parenchyma cells, 8, 9
Ray tracheids, 8, 9

Rayon fibers, from cellulose xanthate, 218
Reaction wood, 18–20
Reducing disaccharides, 38
Regenerated cellulose
  from cellulose xanthate, 218
  crystalline structure, 55, 56, 58
  crystallinity, 205
Resin, *see* Extractives
Resin acids, 99–100, 101
  recovery from black liquor, *see* Tall oil
  recovery from trees, *see* Rosin
Resin canals, 10, 92, 93
Rhamnose, 23
  in glucuronoxylan, 68
Rhytidome, 109, 110, 111
Ribose, 23, 26
Rosanoff's convention, 25
Rosin, *see also* Tall oil
  recovery from trees, 227
  uses, 228
Rubber, 102, 103, 230

**S**

Salicin, in bark, 112
Sandaracopimaric acid, 99, 100
Sapwood, 11
Schweizer's solution, 208
Sclereids, 111
Sclerenchymatous cells, 111
Secodehydroabietic acid, 100, 101
Secondary wall, 13–17
Serratanes, 101
Serratenediol, 101, 102
Sesquiterpenes, *see* Sesquiterpenoids
Sesquiterpenoids, 97–98
Shikimate pathway, 73, 75
Shikimic acid, 75
Sieve cells, in bark, 110
Sieve tubes, in bark, 110
Sinapic acid, 76
  in suberin, 113
Sinapyl alcohol, 76
Singlet oxygen, 173, 174
Sitostanol, 101, 102
Sitosterol, 101, 102
  recovery and uses, 229, 242–243
Skew-boat conformation, 30, 31

Soda-oxygen pulping, *see* Oxygen–alkali pulping
Soda pulping, *see* Alkali pulping
Sodium sulfide, equilibria in aqueous solutions, 140–141
Softwood, cells, 7–10
Sorbitol, *see* Glucitol
Spent bleach liquors, *see* Bleach liquor
Stachyose, in bark, 113
Starch, 21, 54, 66, 113, 132
Steam explosion pulping, 118
Stearic acid, 104
Stereoisomers, 24
Steroids, 100–102
    recovery and uses, 228–229, 242–243
Sterols, *see* Steroids
Stilbenes, 105, 106
Stone groundwood pulp, 117
Suberin
    in bark, 112, 113
    extraction from bark, 229
Sucrose, 38, 39
Sugar alcohols, *see* Alditols
Sugar nucleotides, 51, 52
Sugar sulfonic acids, 133
Sugars, *see* Carbohydrates
Sulfate black liquor, *see* Black liquor
Sulfate lignin, *see* Kraft lignin
Sulfate pulping, *see* Kraft pulping
Sulfate spent liquor, *see* Black liquor
Sulfate turpentine
    composition, 241
    recovery and uses, 241–242
Sulfidity, 140
Sulfite, redox potential, 199
Sulfite pulping
    active base, 120
    background and definitions, 114–116
    by-products, 237–240
    carbohydrate reactions, 132–135
    carbohydrate yield, 133, 135
    cooking chemicals, 119–121
        recovery, 138–140
    cooking liquor
        acidity, 119–121
        combined sulfur dioxide, 120
        decomposition, 136–137
        diffusion of chemicals, 122

        equilibria in, 119–120
        free sulfur dioxide, 120
        loosely combined sulfur dioxide, 138
        penetration, 122
    delignification
        general aspects, 124–126
        kinetics, 124–125
        morphological factors, 122–123
        pH-dependence, 125–126
        selectivity, 124
    high-yield pulps, special features, 161–164
    history, 114–116
    impregnation, 122
    lignin reactions
        acid sulfite, 128–129
        alkaline and neutral sulfite, 129–131
        cleavage of β-aryl ether bonds, 129, 130
        cleavage of α-ether bonds, 127, 128
        cleavage of methoxyl groups, 131
        condensation, 128–129
        degree of sulfonation, 125, 126
        formation of conjugated structures, 130, 131
        influence of carbonyl groups, 131
        quinone methide intermediates, 129, 130
    two-stage method, 133–135
Sulfite spent liquor
    composition, 137–138, 237
    fermentation, 240
Sulfoethylcellulose, as cation exchanger, 223
Sulfomethylcellulose, as cation exchanger, 223
Sulfur, formation during kraft pulping, 142, 143, 147
Sulfur dioxide
    combined, 120
    decomposition, 136–137
    equilibria in aqueous solutions, 119–120
    free, 120
    loosely combined, 138
Suspension culture lignin, 72
Syringaldehyde
    from hardwood lignosulfonates, 239–250
    from lignin, 74

**T**

Tail-to-end cyclication, 98
Tail-to-tail coupling, 94
Tall oil, see also Rosin
  composition, 242–244
  distillation, 242–243
Tall soap, see Tall oil
Talose, 26
Tannins
  from bark and wood, 229
  in bark, 112
  condensed, 105
  hydrolyzable, 105
Taxifolin, 105, 106
  in bark, 112
  dehydrogenation during sulfite pulping,
    136
TCDD-equivalents, 196, 197
Tension wood, 20
  formation, 19
Terpenes, see Terpenoids
Terpenoids, 92–103, see also Turpentine;
    Sulfate turpentine
  in bark, 112
  biosynthesis, 92, 94, 95, 96
  classification, 94
Thermomechanical pulp, 115, 117
Thioacetolysis, 80, 81
Thioacidolysis, 80, 81
Thioglycolic acid lignin, 72
Thiol groups, in cellulose grafting, 221
Thiosulfate
  formation during sulfite pulping, 136
  reactions with lignin, 129
Threose, 26
β-Thujaplicin, 97
Thunbergene, 98, 99
TOCl, 195
Torula yeast, 240
Torus, 6, 17, 18
Tracheids, in softwood, 7–9
Transition metal ions
  in cellulose grafting, 221, 222
  in oxygen-alkali bleaching, 175–176
Tree, growth, 2–4
Triazine, cross-linking of cellulose with, 219
Triethylaminoethylcellulose, as anion ex-
    changer, 223, 224

Trimethylsilylethers, of carbohydrates, 35
Triplet oxygen, 173
Triterpenes, see Triterpenoids
Triterpenoids, 100–102
Trityl ethers, of carbohydrates, 35
Tropolones, 97, 98
  complexes with heavy metal ions, 97
TSS, 195
Turpentine, see also Sulfate turpentine
  recovery from trees, 227
  uses, 227–228
  in wood pyrolysis, 234
Two-stage sulfite pulping, 133–135
Tyloses, 11–12
Tyrosine, 75, 76

**U**

UDP-D-glucose, 51–53
Uloses, 40, 41
Urea, cross-linking of cellulose with, 219
Uridine, 51, 52
Uronic acids, 40, 41

**V**

Vanillic acid, in black liquor, 159
Vanillin
  in black liquor, 159
  from lignin, 74
  production from lignosulfonates, 238, 239
Veratric acid, 74
Vessels, 10, 11
Viscose, see Cellulose xanthate

**W**

Warty layer, 14, 17
Waxes, see also Extractives
  in bark, 112
  extraction from bark, 228–229
  in wood, 103–104
White liquor, 140
Wood
  building elements, 12–13
  carbonization, see pyrolysis
  chemical composition, 249
  chemicals from, 225–236
  distillation, see pyrolysis

gasification, 235–236
hydrolysis products, 230–233
macroscopic structure, 1–2
pyrolysis, 234–235
ultrastructure, 12–18
Wood pulping, background and definitions,
114–118
Wood storage, effect on extractives, 107–
108
Wood tar, 234, 235

### X

Xylan, *see also* specific name
behavior during kraft pulping, 150–155
influence of arabinose units, 153
influence of galacturonic acid groups,
153
influence of glucuronic acid groups,
153–154
readsorption, 154
yield, 133

behavior during sulfite pulping, 132, 134,
135
yield, 133, 135
content in wood, 64, 249
linkages to lignin, 84–86
losses from pulp during mechanical treatment, 162
Xylem, 2
Xylitol, production of, 233
Xyloisosaccharinic acid
from black liquors, 245–248
in black liquors, 158, 246
from xylan in alkali, 150–152
Xylose, 22, 23, 26, 27
from hardwood hydrolysates, 233
in sulfite spent liquors, 137, 237

### Y

Yellowing, of high-yield pulps, 201–203

guncotton, 235–236
pyrolysis products, 230–233
mechanical properties, 1–2
pyrolysis, 234–235
softening temperature, 1–2
Wood, ageing, background and conditions,
114–115
Wood storage, effects on structures, 110–
108
Wood tar, 234, 236

X

Xylan, see also pyrolytic wax
hydrous decomposition, 150–151
influence of atmosphere, 152
influence on polyuronic acid groups,
157
influence on polyuronic acid groups,
153–154
endocellulose, 156
yield, 153

tetrahydrofurfuryl softening point, 133, 134
135
yield, 133, 135
reaction of wood, 62, 234
linkage to lignin, 84–86
xylan, from wood during mechanism of treatment, 145
Xylan, 145
Xylitol, production of, 153
Xylenes, see also acid
from black liquors, 245, 246
m-black liquors, 155, 246
from xylan in steam, 150–152
Xylose, 22, 23, 31, 39, 43
from hardwood hydrolysis, 237
in sulfite spent liquors, 147, 243

Following of high yield pulp, 301–302